Fiber and Whisker Reinforced Ceramics for Structural Applications

MATERIALS ENGINEERING

1. Modern Ceramic Engineering: Properties, Processing, and Use in Design. Second Edition, Revised and Expanded, *David W. Richerson*
2. Introduction to Engineering Materials: Behavior, Properties, and Selection, *G. T. Murray*
3. Rapidly Solidified Alloys: Processes · Structures · Applications, *edited by Howard H. Liebermann*
4. Fiber and Whisker Reinforced Ceramics for Structural Applications, *David Belitskus*

Additional Volumes in Preparation

Thermal Analysis of Materials, *Robert F. Speyer*

Friction and Wear of Ceramics, *edited by Said Jahanmir*

Mechanical Properties of Metallic Composites, *edited by Shojiro Ochiai*

Fiber
and Whisker
Reinforced
Ceramics
for
Structural
Applications

David Belitskus

Alcoa Technical Center
Alcoa Center, Pennsylvania

CRC Press
Taylor & Francis Group
Boca Raton London New York

CRC Press is an imprint of the
Taylor & Francis Group, an **informa** business

CRC Press
Taylor & Francis Group
6000 Broken Sound Parkway NW, Suite 300
Boca Raton, FL 33487-2742

© 1993 by Taylor & Francis Group, LLC
CRC Press is an imprint of Taylor & Francis Group, an Informa business

No claim to original U.S. Government works

ISBN-13: 978-0-8247-9111-7 (hbk)

Visit the Taylor & Francis Web site at
http://www.taylorandfrancis.com

and the CRC Press Web site at
http://www.crcpress.com

Preface

Ceramics reinforced with fibers or whiskers are an important new class of structural materials. A properly designed and fabricated composite material of this type can retain attractive ceramic characteristics such as high strength and stiffness at very high temperatures, but without the brittleness and lack of reliability inherent with unreinforced ceramics. Ceramic composites can result in dramatically improved performance for existing applications, such as cutting tools. In advanced aerospace applications, ceramic matrix composites might be the only materials capable of enabling performance targets to be attained.

Although research on ceramic reinforcement has been carried out for many years, the greatest strides have been made in the last 10 to 15 years. The literature in this area is becoming relatively extensive, but single sources of information that cover all aspects of fiber and whisker reinforced ceramics in detail have been lacking. Although edited multiauthor books are becoming available, such books tend to emphasize areas of individual authors' interest, resulting in a somewhat biased picture of the importance of various aspects of composite technology as well as gaps in the coverage. Hence, this book is intended to fill the need for a single-author work that covers all important aspects of fiber and whisker reinforced ceramics in a logical progression and with balanced coverage of various aspects.

The book provides an introduction to fiber and whisker reinforced ceramics for those not directly involved in this field and having minimal technical background in ceramics, but is also adequately detailed to provide useful information to those entering this field or those whose technology areas will be impacted by these materials. It could also serve as a supplementary textbook for college-level courses in ceramic matrix composites.

Rather than providing an exhaustive reference list on ceramic matrix composites, the book gives key references that permit the reader requiring more information on a particular topic to acquire the necessary information without undue effort. Also, extensive mathematical developments that are associated with some of the topic areas are not covered, but references cited can provide details for those requiring the mathematical treatments.

An unavoidable limitation in the reference sources cited warrants mention. The properties of ceramic matrix composites, particularly with regard to continuous fiber reinforcement, have important implications for the military establishment, largely with regard to aerospace applications. For this reason, a great deal of information generated in the United States is restricted in its dissemination by International Traffic in Arms Regulations (ITAR) of the Office of Defense Trade Controls, U.S. Department of State. Similar restrictions are applicable in a number of other countries as well. Thus, while I am aware of developments beyond those reported in this book, they were not mentioned if not documented in publications, reports, or patents that were approved for unlimited distribution. It is my experience that for all but the most sensitive information, general dissemination of new developments lags restricted disclosure by approximately two years.

Some background on my qualifications for writing this book and on events leading to its publication may be of interest. In the mid-1980s, I became involved in formulating programs and conducting R&D on continuous fiber reinforced ceramics at the Aluminum Company of America (Alcoa). For this purpose, a variety of background information on all aspects of ceramic matrix composite technology was collected, organized, studied, and assessed.

Over a period of several years, I had the opportunity to organize and conduct lectures on the technology of ceramic matrix composites. Included were lectures at two offerings, in 1987 and 1989, of a ceramic matrix composite short course organized by Professor Pradeep Rohatgi at the University of Wisconsin–Milwaukee. Professor Rohatgi, as well as several of the course attendees, first suggested that my lecture material could form the basis for a book. The information was expanded and refined for a lecture at the Materials Technology Center of the University of Southern Illinois at Carbondale, a short course conducted at a conference, Advanced Ceramics '90, sponsored by the Society of Manufacturing Engineers, and at a workshop, sponsored by Professor Shaik Jeelani, at Tuskegee University.

The detailed writing of the text was done outside of my normal work hours, but Alcoa Laboratories management graciously provided ancillary support. This included supplying photographic reproductions and the redrafting of a number of figures. Special thanks are due to Margo Xidis of Alcoa Laboratories Information Department for obtaining the permissions required to use the many copyrighted figures and tables utilized in the book.

David Belitskus

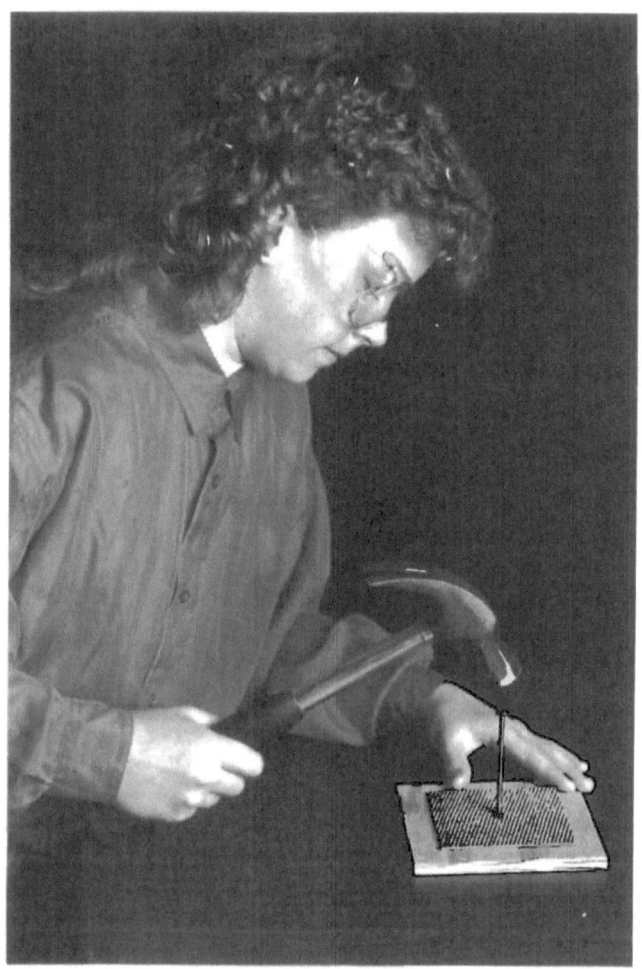

The author's daughter illustrates the remarkable toughness of a continuous fiber reinforced ceramic material. A nail driven through the material produced only a localized effect; an ordinary ceramic material would shatter under the same conditions! (Photo by Hugh Fox.)

Contents

Fiber and Whisker Reinforced Ceramics for Structural Applications

1

Introduction

I. BACKGROUND

Ceramics are inorganic, nonmetallic materials that have been subjected to high temperatures (above red heat, ~500°C) during manufacture and, often, during use. Although typical ceramics are metal oxides, borides, carbides, nitrides, or combinations of these, most ceramic products developed throughout history have been oxide materials, largely based on clay (which consists mostly of aluminum oxide and silicon dioxide). There is evidence that clay-derived pottery was used as long ago as 10,000 B.C., and that use in bricks and tile was practiced by 2000 B.C.[1]

The term "traditional ceramics" refers to ceramics made solely from unrefined clay or from clay in combination with other powdered or nonplastic materials. Traditional ceramic materials have found wide application in kitchenware such as plates, mugs, and cookware, in decorative articles such as vases, in so-called sanitary ware such as lavatories and toilet bowls, in decorative tile, and in "technical ceramics" such as electrical insulators for power lines and spark plugs. Properties such as high hardness, low thermal conductivity, low thermal expansion, and ability to be brightly colored by coating with other ceramics (in the form of "glazes") have led to development of uses of these types. However, an undesirable characteristic of any of these materials is "brittleness," or susceptibility to catastrophic cracking. Another undesirable feature is the considerable variablity in properties among apparently identical items.

Few, if any, of us have not learned about the brittleness of a ceramic in dramatic fashion relatively early in life when we accidentally dropped, and shattered, a dinner plate, drinking mug, or mother's favorite vase. We might also have acquired some insight into the variability of ceramic products if we on other occasions accidentally dropped outwardly identical ceramic items onto the same floor from the same height with the good fortune of no breakage! In contrast, probably all of us learned that dropping an item of silverware or a polymeric ("plastic") plate or mug under the same conditions invariably caused no apparent damage. Implicitly, at least, we learned that despite many useful attributes, traditional ceramics have major shortcomings relative to other materials.

Most of the applications mentioned above do not require particularly strong materials. Strengths of traditional ceramics are quite satisfactory for such structural products such as bricks, drain tile, and sewer pipe. Another widely used ceramic structural material, concrete, has a relatively low strength also. However, in more recent years, much time and money has been expended on developing stronger ceramic materials and to otherwise improve properties to compete with metal alloys for more demanding structural applications. These materials include more highly refined oxide materials as well as non-oxides such as those mentioned above. These new materials are commonly termed "advanced ceramics" or, usually in Japan, "fine ceramics." Uses and envisioned uses for advanced ceramic materials include cutting tools, wear-resistant parts, and high-temperature engine parts.

Research and development work on advanced ceramics has resulted in dramatic increases in strengths relative to traditional ceramics; other properties important for specific applications have been improved also. More quantitatively, strengths ranging from 10 to 100 times those of traditional ceramics have been attained. In many cases, strengths exceed those of the best structural metal alloys. On average, the ceramic materials are lower in density than the metal alloys, an important feature for structural applications, particularly those involving moving parts. Typically, the ceramic materials have higher melting or decomposition points than metal alloys, further increasing their attractiveness for many uses.

Nonetheless, even though some improvements have been attained, the major shortcomings of traditional ceramic materials, catastrophic fracture and considerable variability in properties among similar items, have largely remained. Thus, despite considerable effort on development of improved ceramic materials, uses have been limited by these undesirable characteristics.

For some applications, such as a cutting tool bit made from a strong, high temperature resistant advanced ceramic material, occasional catastrophic failure might be merely a tolerable nuisance with only a modest economic penalty. In other applications, such as an automobile engine part, such failure might produce serious economic consequences, if pieces of the failed part are propelled against and damage other engine parts. In a still more demanding application such as an airplane engine part or exterior structural part, catastrophic failure might well lead to loss of flying ability of the plane, with a subsequent crash and loss of life. Thus, the potential application determines just how serious catastrophic, poorly predictable failure of a ceramic material would be.

II. SOME GENERAL COMMENTS ON REINFORCED CERAMICS

Fortunately, there is available a means for retaining most of the attractive properties of advanced ceramic materials, while eliminating or at least greatly reducing the probability of catastrophic fracture, as well as increasing consistency among outwardly identical items. The method used for achieving these improvements is reinforcement with other materials. Reinforcement of ceramics, in general terms, is by no means a new concept. As an example, early bricks containing straw were composites. More familiarly, most structural concrete is reinforced with steel rods or wire to reduce the possibility of catastrophic failure.

However, the subject of this book deals with a relatively new and exciting form of a ceramic composite, consisting of a ceramic matrix reinforced with fine ceramic (or carbon) fibers or whiskers. Only this type of composite has been shown to retain many of the attractive properties of an unreinforced advanced ceramic while appreciably reducing the two major shortcomings described above.

Throughout the remainder of this book, unreinforced ceramics will generally be referred to as "monolithic ceramics." A number of terms will be used to refer to ceramics reinforced with fibers or whiskers. The terms "continuous fiber reinforced ceramics" or "continuous fiber reinforced ceramic composites" will denote ceramics in which the fibers continuously span the material in one or more directions. "Whisker reinforced ceramics" or "whisker reinforced ceramic composites" have the obvious meaning. Work on ceramics containing discontinuous fibers, as opposed to true whiskers, is limited. Although composites containing discontinuous fibers will at times be referred to as "discontinuous fiber reinforced ceramics" or "discontinuous fiber reinforced ceramic composites," more often these will be considered similar to whisker reinforced ceramics and will not be specifically mentioned.

Ceramic materials combined with other ceramic materials not in the form of fibers or whiskers ("particulate composites") are also an important class of materials that can have improved properties over monolithic ceramics. However, in general, the dramatic reduction in the brittleness and/or variability of a ceramic possible with fiber or whisker reinforcement is not attained by particulate reinforcement. One exception is a "transformation toughened ceramic," which provides appreciable improvement at ordinary temperatures but not at the elevated temperatures required for many new ceramic applications. Transformation toughened ceramics and other particulate ceramic composites will not be discussed in this book, except as useful for comparison with fiber or whisker reinforced ceramics.

In the following chapters of this book, key properties of monolithic ceramics relative to other structural materials will be quantified and described in more detail than above. Improvements in ceramic properties possible by fiber and whisker reinforcement will be indicated. Emphasis will be given to properties at high temperatures, since ceramics become more attractive relative to other materials with increasing application temperatures. The mechanisms responsible for improvement in important properties of ceramics by fiber or whisker reinforcement will then be described.

Subsequent chapters will deal with the compositions, structures, production, and properties of representative reinforcing fibers and whiskers, with emphasis on properties at high temperatures. Possibilities for arranging reinforcing fibers in a composite ("fiber architecture") will be discussed. Considerations in the selection of compatible fiber and matrix materials will be indicated, and the importance of the fiber/matrix interfaces and how these interfaces can be influenced to give good composite properties will be covered.

Other chapters will describe all of the significantly investigated ceramic matrix composite fabrication techniques and properties of composites fabricated using these techniques. Joining, machining, structural designing, and nondestructive evaluation will be covered. The book will conclude with a chapter on applications and potential applications.

REFERENCE

1. Girard, W. P., and Wachtman, J. B., Jr., Ceramics, general survey. In *Ullmann's Encyclopedia of Industrial Chemistry*, Fifth Edition, vol. A6, F. T. Campbell, R. Pfefferkorn, and J. F. Rounsaville (Ed.), VCH Publishers, New York, 1986, 1-42.

2

Properties of Structural Materials

I. BACKGROUND

In this chapter, selected properties of ceramics are compared with those of other materials. Comparison with metals is of particular significance, since for many of the potential applications to be discussed high temperature requirements rule out synthetic polymers or other non-metallic structural materials.

One type of property important in structural uses of materials is strength. Strength is expressed as the stress (or force) per cross-sectional area required to break a material. Most notable examples are stress per unit area during pulling (tensile strength), pushing (compressive strength), or bending (flexural strength). The importance of strength for engineering applications of a material is obvious and will not be discussed here in more detail.

Another important property is stiffness, expressed as the Young's modulus, and having the same units as strength. Although the importance of a high Young's modulus might not be as intuitively obvious as strength, a structure swaying appreciably in a high wind or a ladder shaking when climbed are common examples of less than desirable stiffness.

Other material properties can also be of importance for structural uses. Two other properties, fracture toughness and reliability, that were implicitly mentioned in the discussion above on shortcomings of monolithic ceramics, are treated in more detail below.

II.   FRACTURE TOUGHNESS AND RELIABILITY

Fracture toughness is a measure of the susceptibility of a material to fail due to a flaw. Several types of flaws can be present in the interior of a material, ordinarily a result of the production process. Examples are voids and inclusions of foreign matter. In addition, surface flaws can be produced either during the production process or during operations such as cutting or surface finishing.

Measurement of fracture toughness includes the introduction of a controlled flaw prior to application of mechanical stress. A simple fracture toughness test is illustrated schematically in Figure 1.[1] In this illustration, the flaw is in the form of a simple saw cut. (In practice, more complicated machining of samples to more closely direct the crack propagation is frequently used.) Using the length and cross-sectional dimensions of the specimen, the length of the cut, and the force required to fracture the specimen by propagating the crack originating at the saw cut, the fracture toughness, K_{Ic}, is calculated. The subscript "I" (Roman numeral "one") refers to "Mode I cracking," or cracking in a direction normal to the application of force. The K_{Ic} value is the most common type of fracture toughness measurement reported, although other types can be determined also.

It can be pointed out that the quality, "brittleness," attributed earlier to a ceramic material can now be more rigorously described as low fracture toughness.

The term reliability, in the present context, concerns differences in properties measured on nominally identical test specimens. The concept of reliability

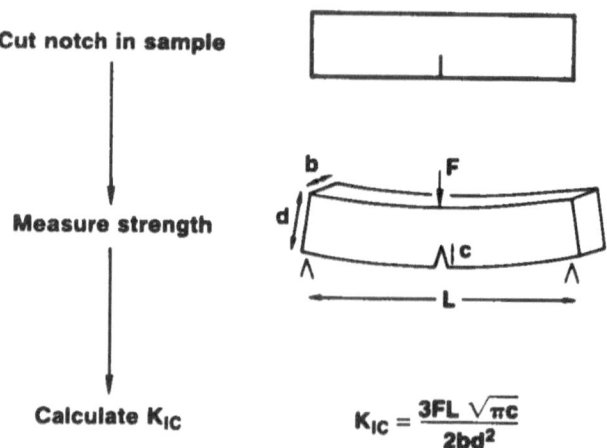

$$K_{IC} = \frac{3FL \sqrt{\pi c}}{2bd^2}$$

FIG. 1   Schematic representation of a test for fracture toughness, K_{Ic}. (From Ref. 1.)

can be quantitatively expressed by the use of the Weibull modulus. In determining the Weibull modulus, a mechanical property, such as flexural strength, is measured on a number of replicate specimens. The data are treated as indicated in Figure 2.[1] A material exhibiting a small range in values generates a steep slope and, hence, has a high Weibull modulus. A material exhibiting a large range in strength values obviously has a flatter slope and, hence, a low Weibull modulus.

A high Weibull modulus for a mechanical property such as strength enhances the utility of a material for structural applications. Consider, for example, that testing of strength for large numbers of specimens of two different materials established that both have the same average strength value of 500 MPa, but that the range of values with one material was from 490-510 MPa while that of the other was from 250-750 MPa. (For the purpose of illustration, assume that the number of specimens tested was large enough to assure a near zero probability of failure below the minimum value found in each case.) In the design of a structure having an insignificant failure probability due to material failure, the strength value of 490 MPa could be used for the first material, while only 250 MPa could be used for the second. Hence, for the same maximum stress during application, the cross-sectional area with the second material would need to be almost twice that of the first, thus increasing weight and cost.

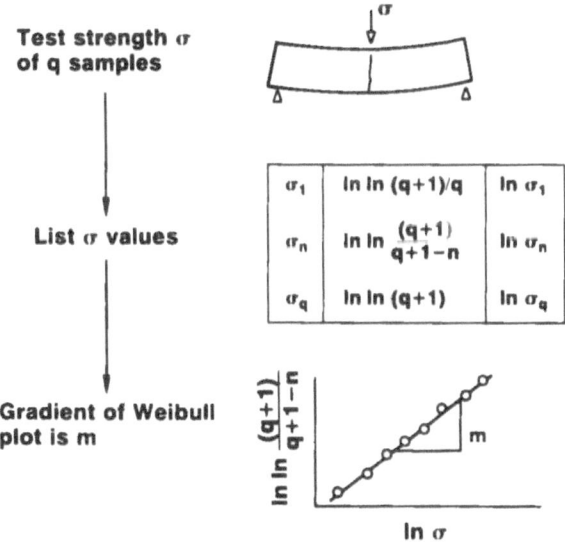

FIG. 2 Schematic representation of a Weibull modulus determination. (From Ref. 1.)

The variability in mechanical properties for ceramic materials that was mentioned earlier can now be described as a low Weibull modulus.

This discussion concerning several important mechanical properties of materials has been brief. Other properties important for specific applications have not been mentioned at all at this point. It is hoped that the treatment has provided sufficient background (or review) for those not very familiar with this area to facilitate assimilation of information on mechanical properties throughout this book. In specific instances throughout the book where other properties not discussed here are mentioned, the significance of the property will be described briefly. It is beyond the intended scope of this book to discuss the theory and application of mechanical property testing in detail.

III. GENERAL COMPARISON OF MECHANICAL PROPERTIES OF MATERIALS

In this section, properties of monolithic ceramics and other materials will be discussed in a more quantitative manner. The treatment draws heavily from the compilations of M. F. Ashby [2], who has assembled material property charts comparing many classes of materials. Table 1 indicates the classes and members of each class. The list is quite representative of materials used in load bearing and decorative applications, as well as elastomers ("rubbery" materials) and foams. Note that "engineering composites" refers only to polymer matrix composites, and not to metal matrix composites, or to ceramic matrix composites, the subject of this book. The term "engineering ceramics" is synonymous with the term "advanced ceramics" described earlier; Ashby's terminology will be used in this section. Also, he uses the term "porous ceramics" in preference to the synonymous term "traditional ceramics" discussed earlier.

TABLE 1.   Material classes and members of each class.

Engineering alloys (The metals and alloys of engineering)	Aluminum alloys	Al alloys
	Lead alloys	Lead alloys
	Magnesium alloys	Mg alloys
	Nickel alloys	Ni alloys
	Steels	Steels
	Tin alloys	Tin alloys
	Titanium alloys	Ti alloys
	Zinc alloys	Zn alloys
Engineering polymers (The thermoplastics and thermosets of engineering	Epoxies	EP
	Melamines	MEL
	Polycarbonate	PC
	Polyesters	PEST

TABLE 1 (Continued)

	Polyethylene, high density	HDPE
	Polyethylene, low density	LDPE
	Polyformaldehyde	PF
	Polymethylmethacrylate	PMMA
	Polypropylene	PP
	Polytetrafluorethylene	PTFE
	Polyvinylchloride	PVC
Engineering ceramics (Fine ceramics capable of load-bearing application)	Alumina	Al_2O_3
	Diamond	C
	Sialons	Sialons
	Silicon carbide	SiC
	Silicon nitride	Si_3N_4
	Zirconia	ZrO_2
(The composites of engineering practice. A distinction is drawn between the properties of a ply—"Uniply"—and of a laminate—"Laminates")	Carbon fiber reinforced polymer	CFRP
	Glass fiber reinforced polymer	GFRP
	Kevlar fiber reinforced polymer	KFRP
Porous ceramics (Traditional ceramics, cements, rocks and minerals)	Brick	Brick
	Cement	Cement
	Common rocks	Rocks
	Concrete	Concrete
	Porcelain	Pcln
	Pottery	Pot
Glasses (Ordinary silicate glass)	Borosilicate glass	B-glass
	Soda glass	Na-glass
	Silica	SiO_2
Woods (Separate envelopes describe properties parallel to the grain and normal to it, and wood products)	Ash	Ash
	Balsa	Balsa
	Fir	Fir
	Oak	Oak
	Pine	Pine
	Wood products (ply, etc)	Wood products
Elastomers (Natural and artificial rubbers)	Natural rubber	Rubber
	Hard butyl rubber	Hard butyl
	Polyurethanes	PU
	Silicone rubber	Silicone
	Soft butyl rubber	Soft butyl
Polymer foams (Foamed polymers of engineering)	These include:	
	Cork	Cork
	Polyester	PEST
	Polystyrene	PS
	Polyurethane	PU

Source: Ref. 1.

Figure 3 shows the important relationship of Young's modulus versus density. Note that ranges shown for specific materials in Figure 3 are indicated by the smaller closed areas, while the larger closed areas represent the range for all the materials of each general class (including more materials than listed in Table 1). Note also that both axes represent a logarithmic scale, so that enormous ranges in properties are represented.

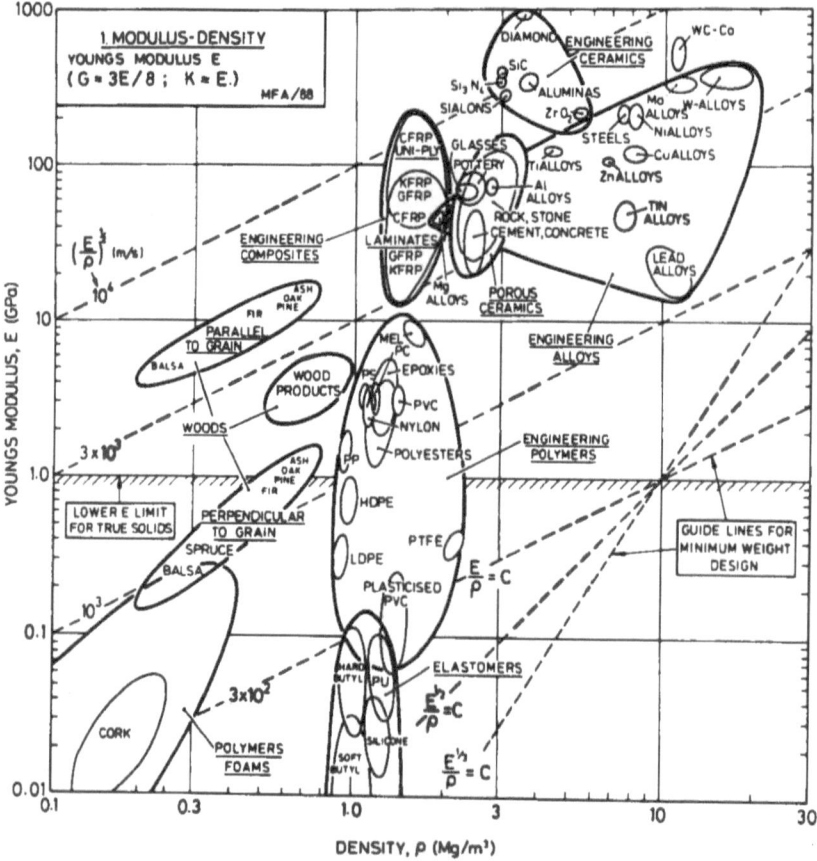

FIG. 3 Young's modulus, E, as a function of density, ρ, for various classes of materials. The heavy envelopes enclose data for a given class of materials. The diagonal contours show the longitudinal wave velocity. The guide lines of constant E/ρ, $E^{1/2}/\rho$, and $E^{1/3}/\rho$ allow selection of materials for minimum weight, deflection-limited design. (From Ref. 2.)

For many structural applications, Young's modulus, rather than strength, is the most important mechanical property. The Young's modulus of a material depends on both the stiffnesses of the individual chemical bonds and the number of bonds per unit area. Engineering ceramics, as a class, have the highest Young's moduli. There is some overlap with metals ("engineering alloys"), porous ceramics, and polymer composites. Since unreinforced polymers have considerably lower moduli, it is clear that the high moduli of the composites are due to the reinforcing fibers. Moduli for the foams, which are very porous, and elastomers, which contain tangled, long chain molecules, are lower than the calculated minima based on bond stiffness and density.

The density of a solid material depends upon the average atomic masses of its atoms or ions, their average size, and how they are packed. The major factor of the three in producing large density differences is the atomic mass, which varies by a factor of about 30 for elements that are stable solids at room temperature. Metals, as a group, have high densities because they contain heavy atoms packed relatively closely. Ceramics have lower densities because they contain light atoms such as oxygen, nitrogen, boron, and carbon. Polymers and other organic base materials have the lowest densities because they contain light atoms, mainly carbon and hydrogen, and tend to form linear or two- or three-dimensional networks that pack relatively poorly.

Strength versus density relationships are shown in Figure 4. Comparison of such diverse types of materials requires some compromise in the kinds of strength values. For metals, it is the "tensile yield strength," or the value in tension at which the metal begins to deform appreciably but does not catastrophically fracture. For polymer composites, it is the "tensile failure strength." For the elastomers, it is the "tensile tear-strength." For ceramics, it is the "compressive strength." The compressive strength of a ceramic is typically much greater than the tensile strength, while values are more nearly the same for metals. (The envelopes for the ceramic materials are shown as dashed lines to provide a reminder.)

Keeping in mind the limitation described above, engineering ceramics have the highest strengths of any materials. Of the other monolithic materials, engineering alloys have somewhat lower strengths, followed by engineering polymers. Reinforcement of the polymers increases strengths to the range of those of metals, at considerably lower densities.

Thus, as indicated qualitatively earlier, ceramics compare very favorably with other materials with respect to some key mechanical properties. However, as indicated qualitatively earlier and shown quantitatively in Figure 5, fracture toughness values are relatively low for ceramics, compared with engineering alloys or engineering composites. Most engineering alloys and engineering composites have toughness values greater than 20 MPa·m$^{1/2}$, which is often quoted as the minimum value required for conventional design of structures.

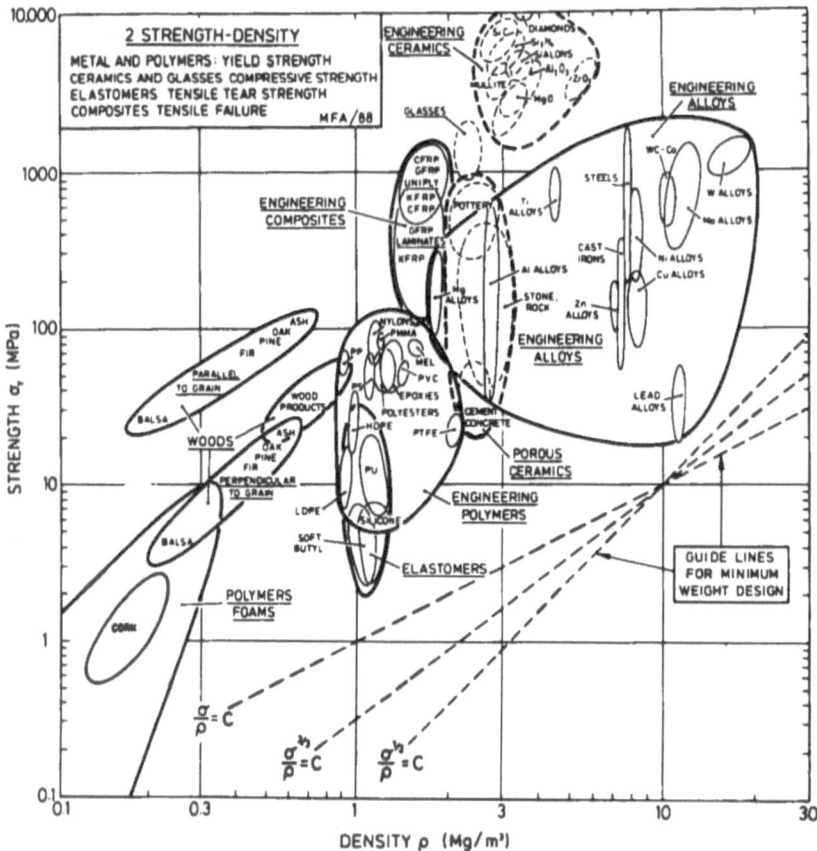

FIG. 4 Strength, σ, as a function of density, ρ, for various classes of materials (yield strength for metals and polymers, compressive strength for ceramics, tear strength for elastomers, and tensile strength for composites). The guide lines of constant σ^y, $\sigma^{y2/3}/\rho$, and $\sigma^{y1/2}/\rho$ are used in minimum weight, yield-limited design. (From Ref. 2.)

Engineering ceramics generally have values between 4 and 10 MPa·m$^{1/2}$, and many porous ceramics have values less than 1. The fracture toughnesses of engineering polymers are also lower than desirable for structural applications, but toughnesses are raised to the range of metals by incorporation of reinforcing fibers.

As described earlier, fracture toughness relates to the sensitivity of a material to a flaw. The greater sensitivity of a ceramic (or an unreinforced poly-

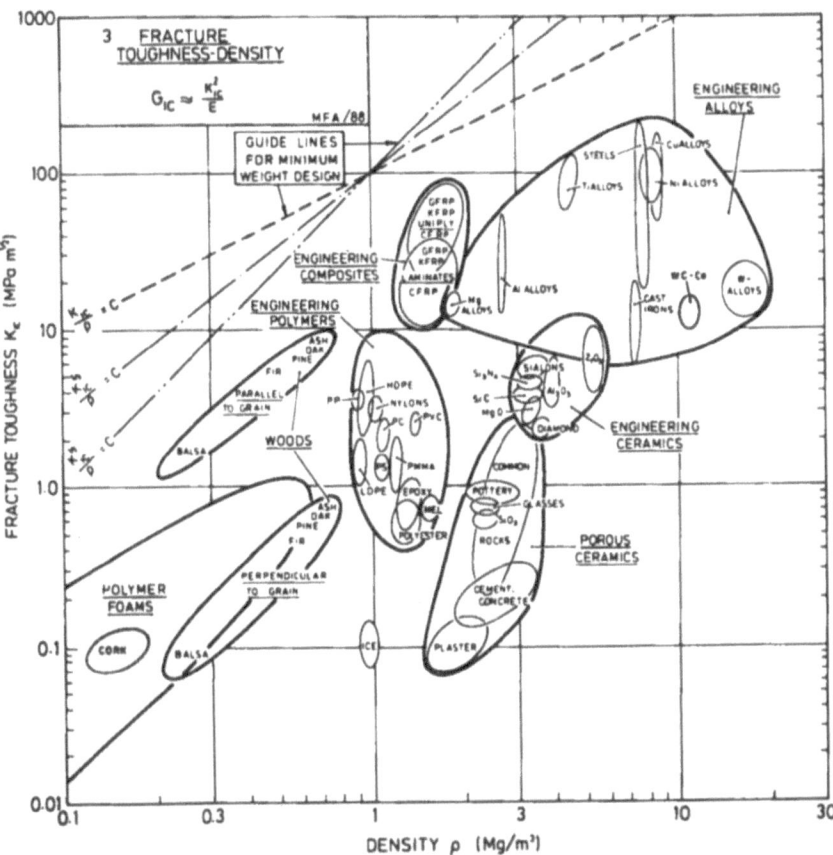

FIG. 5 Fracture toughness, K_{Ic}, as a function of density, ρ. The guide lines of constant K_{Ic}/ρ, $K_{Ic}^{2/3}/\rho$, and $K_{Ic}^{1/2}/\rho$ help in minimum weight, fracture-limited design. (From Ref. 2.)

mer) to a small flaw, relative to a metal, is due to the nature of the bonding. In metals, the "electron-sea model" describes a lattice of positive ions, with delocalized valence electrons. On application of a stress, atoms are able to shift positions relatively easily in the vicinity of a flaw and the flaw does not increase in size, or propagate. In other words, a metal tends to be ductile, or capable of being deformed appreciably without fracturing. In contrast, in a typical ceramic there are strong, highly directional covalent bonds that do not facilitate rearrangement of atoms in the vicinity of a flaw. Under mechanical stress, the flaw will tend to enlarge and lead to fracture.

Because of these fundamental differences, the energy required to propagate a crack in a metal is typically several orders of magnitude greater than that in a ceramic. From another perspective, a typical metal is susceptible to catastrophic cracking if it contains a flaw of the order of several centimeters in size, whereas a flaw of about 50 μm in size can cause failure of a ceramic.

Since, as we shall see later, comparisons of ceramic properties with properties of structural metal alloys are usually more relevant than those with polymers or reinforced polymers, Table 2 [1] compares some properties of ceramics and metals. This table encompasses some of the information shown graphically in previous figures, but includes the important measurement, Weibull modulus, which was not included in Ashby's review. In addition, tensile strengths, rather than compressive strengths, are given for the ceramics, so that a more direct strength comparison than derived from the strength versus density figure can be made.

Included in this table are properties for a typical cast iron, a typical steel, and a superalloy, Nimonic 90. Ceramics include the oxides, alumina and zirconia, and the non-oxides, silicon carbide and silicon nitride. Highest reported tensile strengths of all the ceramic materials exceed those of the typical cast iron and steel, and the highest reported values for the two oxides are comparable to those for the superalloy. Young's modulus values for all the ceramics exceed those for the alloys. Those for alumina and silicon carbide are about twice those for the alloys.

As indicated previously, a major shortcoming of the ceramics relative to the steel and superalloy are the low fracture toughness values. Typical values for the ceramics are one-tenth of those for the steel and superalloy. The upper value shown for zirconia is for "transformation toughened" zirconia. As will be indicated later, a shortcoming of this material is that the toughening mechanism is not effective at high temperatures. Weibull modulus values for the ceramics are less than one-tenth of those for the steel and superalloy.

TABLE 2.   Relative mechanical properties of some metals and ceramics.

Materials	Tensile strength (MPa)	Young's modulus (GPa)	Fracture toughness (MPa·m$^{1/2}$)	Weibull modulus
cast iron	170	175	5-13	10
steel	370	200	50-100	>100
Nimonic 90	1050	211	50-100	>100
Al_2O_3	276-1034	380	2.7-4.2	10
ZrO_2	600-1300	205	5-10	7-20
SiC	96-520	400	4.8	4-10
Si_3N_4	414-650	295	5.3	4-10

Source: Ref. 1.

Interestingly, the fracture toughness and Weibull modulus values of ceramics are similar to those for cast iron, which was used to a considerable extent as a structural material before the development of greatly improved metallic alloys. With continued research and development on monolithic ceramic processing, it is possible that monolithic ceramics having improvements in properties relative to present ceramics that are comparable to those of structural steels relative to cast iron will someday be available. However, in the near future, the development of reinforced ceramics for applications not suitable for metals or other structural materials is more promising.

IV. COMPARISON OF MATERIALS AT HIGH TEMPERATURES

Although the previous section gave an indication of clear advantages and disadvantages of ceramic materials relative to other engineering materials, the important factor of temperature was not considered. In general, ceramics have much higher melting or decomposition temperatures than the other types of materials. As a rough approximation, a material cannot be used for structural purposes at a temperature exceeding two-thirds of its melting or decomposition temperature, in °C. (A more conservative rule is that a material can be used at a temperature, expressed in °F, that equals its melting point in °C, or only up to about 55% of its melting point.) Even the most temperature resistant polymer base materials are not able to be used at temperatures exceeding about 400°C, so unreinforced or reinforced polymers can be ruled out for very high temperature structural applications. Only metals (including alloys and intermetallic compounds), ceramics, and carbon (which was included in Ashby's analysis only in the form of diamond and, implicitly, in fiber reinforced polymers) are candidates for high temperature structural use.

Figure 6 gives melting or decomposition temperatures as well as densities for a number of ceramics, metals, and carbon. Only ceramic materials having melting or decomposition temperatures above 1900°C are included. Using a conservative rule-of-thumb, all of these ceramics could be used structurally above 1000°C. The metals shown include those used as base metals for high temperature structural alloys, as well as selected other high temperature metals.

Nickel and titanium, used in high temperature aerospace alloys, are more temperature limited than the ceramic materials shown. In addition, many of the ceramic materials have considerably lower densities than nickel and titanium. Higher melting metals such as tungsten and tantalum are very dense.

Carbon has the optimum combination of "melting point" (carbon actually sublimes, rather than melts) and density. However, use of unprotected carbon in an oxidizing atmosphere at a temperature greater than about 400°C results in rapid conversion to carbon dioxide and carbon monoxide.

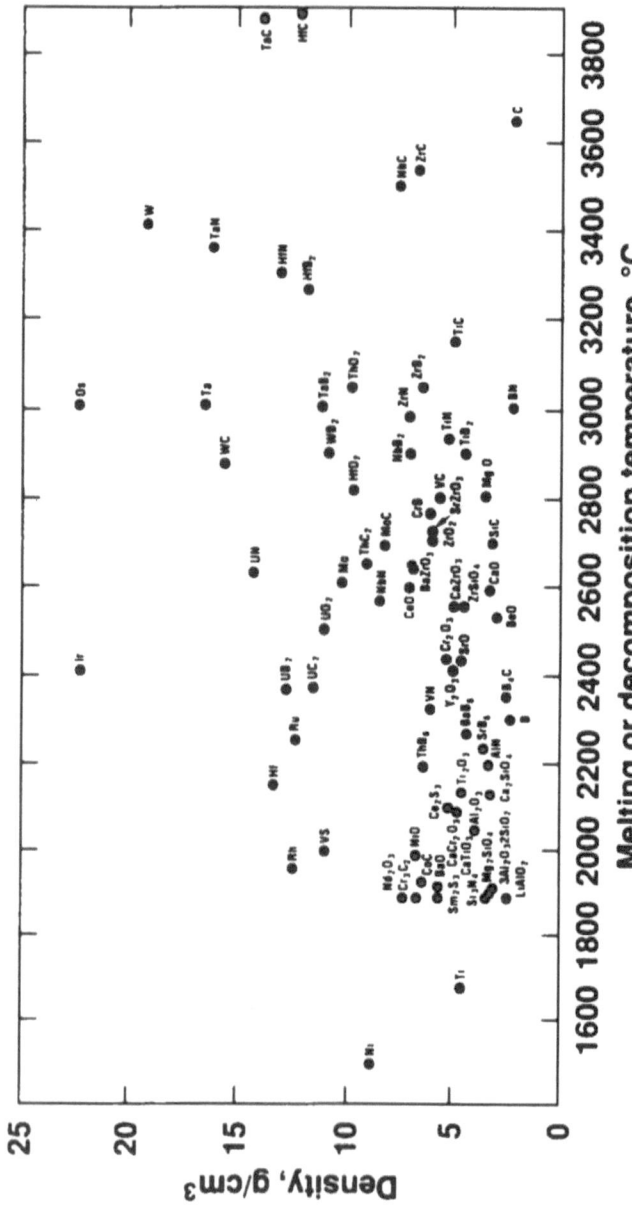

FIG. 6 Densities and melting or decomposition temperatures of selected ceramic and other materials. For many applications, the combination of a low density and a high melting or decomposition point is desirable.

V. GENERAL CHARACTERISTICS OF REINFORCED CERAMICS

Properties attainable by reinforcement of ceramics will be discussed in more detail in subsequent chapters, but a few comments on properties are warranted at this point. Reinforcement of a ceramic with continuous fibers can result in a strain-to-failure response that resembles that of a structural metal. For example, Figure 7 [3,4] compares the flexural stress-strain behavior of a typical monolithic ceramic, the same type of material reinforced with continuous fibers, and a common aluminum alloy. The stress-strain curve for the unreinforced silicon carbide is typical for a ceramic. The material is strong and stiff, but can tolerate little strain before fracturing. The reinforced ceramic has a much greater strain-to-failure than the unreinforced ceramic and the stress-strain curve is similar to that of the aluminum alloy. The greatly increased area under the stress-strain curve can be equated to increased toughness.

Ceramics reinforced with discontinuous fibers or whiskers can have increased toughness also, although not to the extent possible with continuous fibers. Even less toughening is generally available by particulate reinforcement. For example, Figure 8 [5] shows analytical modeling for one of a number of mechanisms for toughening, crack deflection. In this mechanism, a crack is forced to deflect around the reinforcing phase, requiring more energy for propagation, and hence, toughening the material. The calculations show that spheres (which can represent particle additions) provide much less toughening than rods (which can represent whisker or fiber additions) or discs (which can represent platelets). Although not shown in this figure, toughening is also predicted to increase with increasing rod aspect ratio (length divided by diameter).

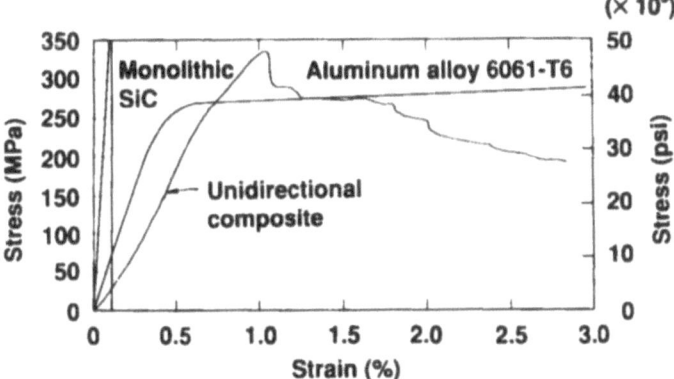

FIG. 7 Comparison of stress-strain curves for monolithic silicon carbide, fiber reinforced silicon carbide, and an aluminum alloy. (Ceramic curves from Ref. 3; aluminum alloy curve plotted from data in Ref. 4.)

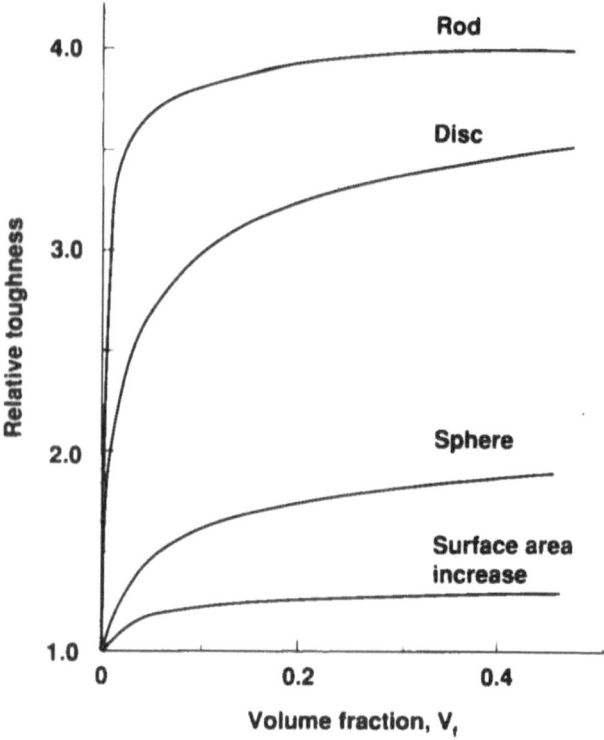

FIG. 8 Theoretical predictions of composite toughening by reinforcements of various shapes. (From Ref. 5.)

A special case of a particle-containing ceramic is a "transformation-toughened" ceramic such as partially stabilized zirconia (PSZ). In this material, precipitates of the high temperature tetragonal phase are retained at ambient temperature by rapidly quenching from a high temperature. When the material undergoes a tensile strain at a crack tip, transformation of tetragonal to mono-clinic zirconia can occur, with a volume increase of 5-6% due to a density difference between the two phases. On transformation and expansion, a radial compressive stress is introduced in the matrix, thereby opposing the external tensile stress. Hence, the material is toughened by increasing the tensile stress required for continued crack propagation.

Although toughness is increased appreciably by this mechanism, it is not effective at the high temperatures associated with many of the applications to be

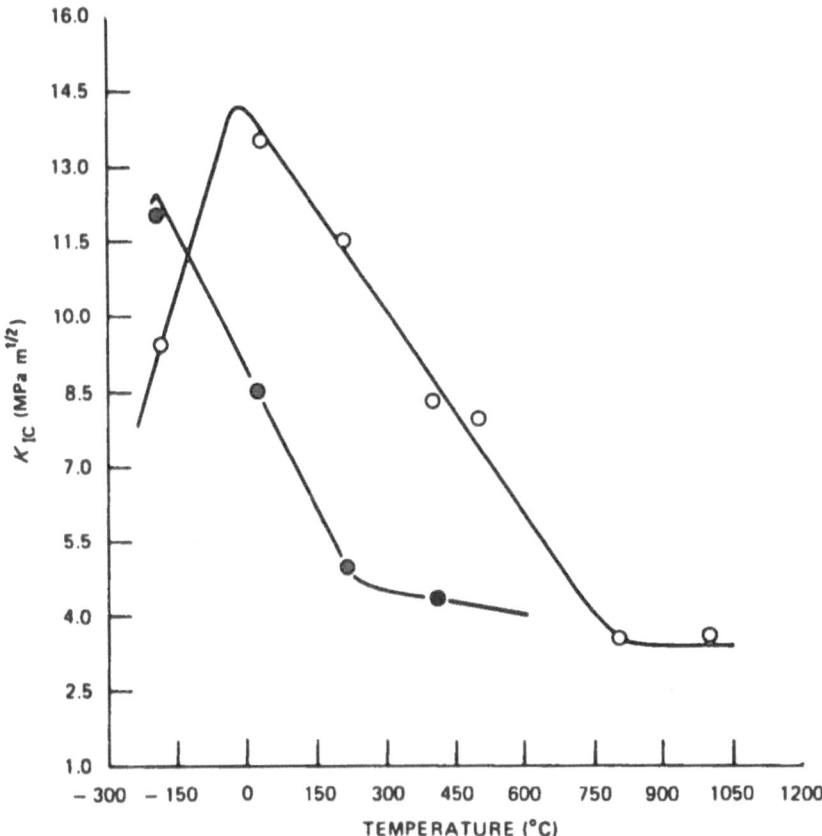

Fig. 9 Fracture toughness as a function of temperature for two partially stabilized zirconia materials. (From Ref. 6.)

described. For example, Figure 9 [6] shows fracture toughness as a function of temperature for two typical transformation-toughened zirconias. Although the toughness at room temperature for each material was over 10 MPa·m$^{1/2}$, the toughness dropped to less than half this value by about 175°C for one material and by about 750°C for the other material. It can be noted that the toughening effect is maximum at a specific temperature for a given material. This maximum is determined by the temperature at which spontaneous transformation of the tetragonal phase to the monoclinic phase begins to occur, which is a function of the tetragonal-phase precipitate size.

VI. SUMMARY

A variety of ceramic materials compare very favorably with other structural materials in terms of strength, Young's modulus, and usefulness at high temperatures. Densities of ceramics are also lower than those of most structural metals, which is advantageous in many applications. However, ceramic materials inherently have low fracture toughness and poor reliability.

Toughness and reliability of a ceramic material can be increased in a number of ways. Incorporation of particulate material (generally another ceramic material) into the structure can improve toughness and reliability to a limited extent. An appreciable toughness increase can be achieved for zirconia or alumina by a special type of particle addition, tetragonal zirconia, which can decrease stress intensity at a crack tip by a phase transformation. However, toughening by this mechanism is generally limited to relatively low temperatures.

Further increases in toughness and reliability which, under appropriate conditions, can be retained at very high temperatures can be accomplished by whisker or, especially, continuous fiber reinforcement. The mechanisms by which fibers and whiskers can toughen ceramics are described in the following chapter.

REFERENCES

1. Alford, N. McN., Birchall, J. D., and Kendall, K., Engineering ceramics—the process problem, *Materials Science and Technology*, vol. 2, 329-336 (1986).
2. Ashby, M. F., On the engineering properties of materials, *Acta Metall.*, vol. 37, no. 5, 1273-1293 (1989).
3. Caputo, A. J., Lackey, W. J., and Stinton, D. P., Development of an newer, faster process for the fabrication of ceramic-fiber reinforced ceramic composites by chemical vapor deposition, *Ceram. Eng. Sci. Proc.*, vol. 6, no. 7-8, 694-706 (1986).
4. Anon., *Aluminum Standards and Data 1979*, The Aluminum Association, Inc., Washington, DC, 1979.
5. Buljan, S. T., and Sarin, V. K., Silicon nitride-based composites, *Composites*, vol. 18, no. 2, 88-106 (1987). (citing the PhD thesis of K. T. Faber, University of California, Berkeley, CA, 1982).
6. Becher, P. F., Swain, M. V., and Ferber, M. K., Relation of transformation temperature to the fracture toughness of transformation-toughened ceramics, *Journal of Materials Science*, vol. 22, 76-84 (1987).

3

Toughening Mechanisms in Ceramic Matrix Composites

I. BACKGROUND

A major goal of reinforcing ceramics with fibers or whiskers is to increase fracture toughness. By understanding the mechanisms contributing to fracture toughness increases, systematic rather than trial-and-error attempts to develop tougher materials can be made.

Considerable work has been carried out on toughening mechanisms in both continuous fiber reinforced and whisker reinforced ceramics. To fully describe the state-of-the-art of ceramic composite toughening theory, a detailed mathematical treatment is required. However, a useful non-mathematical description of the most important toughening mechanisms can be provided for the reader interested in a general overall understanding of ceramic composites. Therefore, this will be the approach taken here. For readers desiring a more comprehensive, mathematical treatment of the subject of toughening mechanisms, review articles such as those by Evans and Marshall [1] and Becher [2] are recommended.

This chapter is limited to a description of toughening mechanisms. Other information related to composite toughening will be included in a chapter on mechanical testing of composites. Additionally, toughness values obtained for specific composites will be reported in chapters dealing with composite fabrication.

II. MAJOR TOUGHENING MECHANISMS IN CERAMICS REINFORCED WITH CONTINUOUS FIBERS

The toughening mechanisms operable in ceramic matrix composites depend upon a number of microstructural parameters affected by the fiber, matrix, and interface properties. The following description of some of these parameters draws heavily from a treatment by Osmani et al.[3] Figure 1 provides an excellent series of illustrations for describing several of the most important toughening mechanisms operating in ceramic matrix composites. Figures 1a-1h will be discussed in this section. (Figures 1i and 1j will be discussed in a later section on toughening by whiskers and discontinuous fibers.)

Figure 1a depicts a small volume element of ceramic material reinforced by uniaxially aligned fibers, the arrows representing a tensile stress in the direction of the reinforcing fibers. As a preliminary comment, it should be mentioned that in contrast to reinforced metals or polymers, where strain to fracture of the matrix material is usually considerably greater than that of the reinforcing fibers, strain to fracture for the fibers and matrix materials for a ceramic composite are more nearly the same. Although both the matrix and fibers have relatively low strains to failure in a ceramic matrix composite, strain to failure for the most useful reinforcing fibers is still somewhat greater than that of the matrix, for composites produced to date. Therefore, the following analysis assumes a greater strain to failure for the fibers. It is assumed further that the fibers are chemically or physically bonded to the matrix and that there is a critical stress required for debonding. Finally, it is assumed that even after debonding, there is a frictional resistance to sliding between the matrix and debonded fibers.

With the assumption of a greater strain to failure for the fibers, increasing tensile stress will cause microcracking of the matrix material (Figure 1b). Near such a crack, the fibers are loaded to a greater extent than previously, since the matrix no longer shares the load. If the critical stress for debonding between the fibers and matrix is much greater than the fiber strength, the fibers will break normal to the tensile stress without debonding at the interface (Figure 1c). In this case, the fracture energy is simply the weighted average of the surface energies for the fibers and the matrix. Little or no toughening occurs since both the fibers and matrix have low fracture toughnesses and a crack can readily propagate through both phases.

If, on the other hand, the critical stress for debonding the fibers is less than the fiber strength, debonding occurs before fiber fracture (Figure 1d). With the continuation of a tensile stress on the composite, debonding area increases while the fibers elongate, and energy is dissipated by friction between the debonded fibers and the matrix. Thus, several mechanisms of toughening of ceramics by continuous fibers should already be apparent. Energy is dissipated in the process of debonding fibers, the bridging fibers resist crack propagation, and energy is dissipated during frictional resistance to debonded fibers sliding in the matrix material.

An interesting phenomenon is illustrated in Figures 1e and 1f. Although variability of a typical ceramic material (quantified in terms of the Weibull modulus) is higher than desired for a structural material, as described earlier, ceramic composite toughening is enhanced if the fiber Weibull modulus, with respect to variation in strength along its length, is not particularly high. If the fiber Weibull modulus is high, the fibers will invariably break in an area where

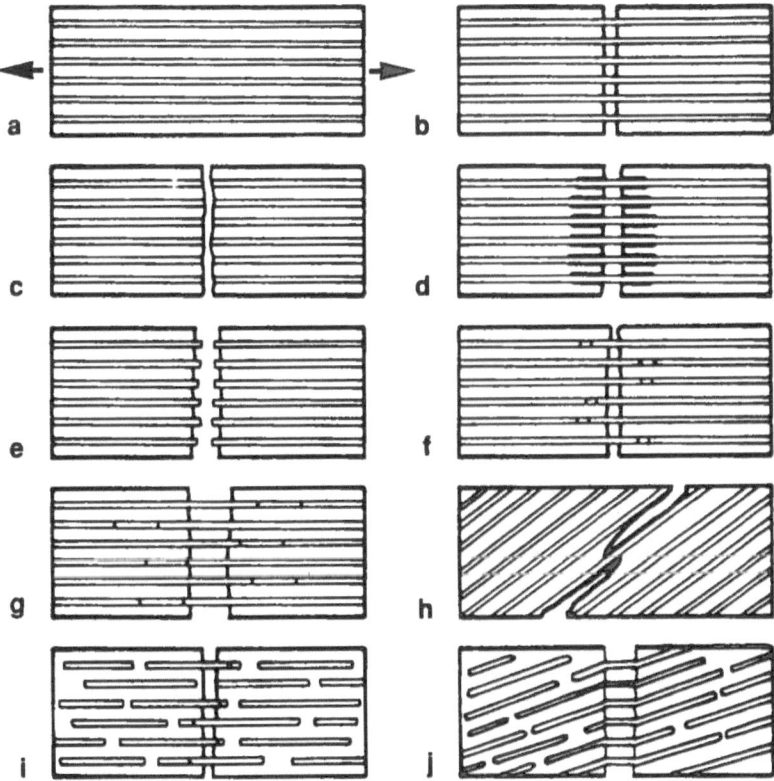

FIG. 1 Schematic representations of fracture mechanisms in fiber reinforced ceramics. The individual figures are discussed in the text. (From Ref. 3.)

the matrix has cracked because the load on the fibers in this region is greater than that where the matrix helps support the load (Figure 1e).

If the Weibull modulus of a fiber is small, with respect to strength variation along its length, the probability of breaking outside the matrix crack is significant. That is, the strength of a fiber in a matrix crack area can be sufficient to resist the higher load in this area, while the strength in an adjacent area, where the load is shared with the matrix, can be insufficient to resist the resulting lower fiber load (Figure 1f). In this case, separation requires fiber pullout from the matrix. Frictional resistance to fiber pullout dissipates energy and is, therefore, another important toughening mechanism. For continuous fiber reinforced ceramics, the observation of broken fibers extending from fracture surfaces has become a routinely used qualitative indicator of successful toughening.

Analysis of toughening mechanisms in ceramic composites containing uniaxially aligned fibers where the tensile stress is not parallel to the fiber axes (Figure 1h) is less straightforward. In this case, a crack tends to propagate parallel with the fibers. Hence, the overall crack behavior is not just "mode I" (propagation normal to the direction of stress), but also "mode II" (propagation parallel to the direction of stress) and "mixed mode". Although most of the same mechanisms described for the case when tensile stress is along the fiber axes are possible in this case also, the interpretation is more complex. It is obvious also that toughening of two- or three-dimensional fiber orientations cannot be explained in as straightforward a manner as for a uniaxial fiber orientation.

The general subject of debonding of fibers and frictional sliding of fibers from the matrix will be considered in more detail in a later chapter. However, some general comments are appropriate here. It is obvious that any chemical reactions that have occurred at the interface during composite fabrication or subsequent exposure to elevated temperatures and/or reactive atmospheres can lead to strengthening or weakening of the interfacial bonds. Bond strengthening caused by chemical reaction is well documented.

Debonding of unbroken fibers at a matrix crack would be expected to be aided by contraction in fiber cross section due to the Poisson's ratio effect, although this is not well documented. (Poisson's ratio is the ratio of the transverse contraction per unit dimension of a bar of uniform cross section to its elongation per unit length, when subjected to a tensile stress.) In addition, an interface subjected to an internal radial tensile stress would be expected to debond more readily than one without an internal stress, or, especially, one with a radial compressive stress.

A radial tensile stress occurs when the use temperature of the composite is lower than the fabrication temperature and the coefficient of thermal expansion of the fibers is greater than that of the matrix material. It occurs also when the use temperature is higher than the fabrication temperature and the coefficient of thermal expansion of the matrix material is greater than that of the fibers; but this is a less common situation. A radial compressive stress at the interface results from the opposite of either of the two situations given above.

After debonding and fracture of a fiber, the frictional resistance to sliding can be affected by several factors. Although the expected Poisson's ratio effect would not now be applicable, the radial or compressive stress factors are still applicable. In addition, roughness of the fiber can affect sliding.

III. MAJOR TOUGHENING MECHANISMS IN WHISKER OR DISCONTINUOUS FIBER REINFORCED CERAMICS

Figure 1i represents reinforcement by whiskers or discontinuous fibers, hypothetically uniaxially aligned. In this case, the relative magnitudes of whisker or fiber length, strength, and the interfacial bond strength influence the composite behavior. If the whisker or fiber strength is lower than the debonding stress, no toughening occurs, as in the analogous case with continuous fibers. If the debond stress is lower than the whisker or fiber strength but the whisker or fiber length is shorter than the typical debond length, minimal toughening occurs. However, if the debond stress is lower than the whisker or fiber strength and the fiber length exceeds the typical debond length, significant toughening can occur, both by fiber bridging and frictional resistance to sliding.

In Figure 1j, whisker or discontinuous fiber reinforcement with a tensile stress not along the whisker or fiber axes is illustrated. There is an increased probability of fiber fracture along the line of matrix cracking because of a bending force on the fibers.

In general, the specific situations shown in Figures 1i and 1j are not entirely realistic for whisker or discontinuous fiber reinforcement, since the reinforcing phase will tend to have a more random orientation. There can be some preferred orientation in one direction or in a plane, depending upon the method of fabrication, but not to the extent possible with continuous fiber reinforcement.

In a detailed theoretical and experimental study of whisker reinforced ceramics, Evans and coworkers [4] determined that whiskers not aligned in the direction of the tensile stress typically fail by bending and do not contribute to toughening by the pullout mechanism. Hence, they concluded that toughness increase of a ceramic material by incorporation of randomly oriented whiskers is limited, relative to toughness using continuous fiber reinforcement. Nonetheless, useful toughening by bridging of whiskers or discontinuous fibers prior to fracture and by pullout of fibers properly oriented to the direction of stress has been demonstrated, as will be shown later.

IV. OTHER TOUGHENING MECHANISMS IN CERAMIC COMPOSITES

It is believed that the mechanisms described provide the bulk of the toughening offered by whisker reinforcement and, especially, continuous fiber reinforcement of ceramics. A number of other toughening mechanisms have been proposed for fiber, whisker, and particulate reinforced ceramics. However, verification and modeling of many of these proposed mechanisms are limited. In addition, the requirements for some of these potential mechanisms run counter to those required for the more important, better documented mechanisms described above. Nonetheless, for a more complete treatment on the subject of toughening, a number of these mechanisms will be discussed briefly.

A useful summary on toughening in ceramic composites has been compiled by Gac.[5] He includes eight proposed toughening concepts. Those already discussed are matrix microcracking, crack bridging, fiber pullout, residual stress, and transformation toughening. Additional proposed mechanisms include modulus transfer, crack arresting (or blunting), crack bowing, and crack deflection.

The modulus transfer mechanism involves transferring the applied load from a lower modulus matrix to the higher modulus fibers to achieve a strain uniformity throughout the composite. A strong, non-slipping interface between fibers and matrix is required. Toughening can result from this concept since an increase in local driving force is necessary to propagate a crack throughout the composite. It will be noted that the requirements for this method of toughening run counter to those for the bridging and pullout mechanisms, which are well established as causes for significant toughness increases.

The crack arresting/blunting mechanism, at one extreme, might be considered an extreme case of fiber bridging, except that the reinforcing phase has sufficient strength and toughness to completely resist fracture, while, at the other extreme, it is proposed that toughening might occur due to a pore or hole which removes stress concentration at the crack tip. In the latter case, toughening might in fact occur, but undoubtedly with an adverse effect on strength.

Crack bowing involves the presence of particles, fibers, or whiskers in the path of a propagating crack, causing the crack to bow between the particles. This causes the stress intensity along the bowed segment of the crack to decrease, while producing a corresponding increase in the stress intensity at the particle, fiber, or whisker. Bowing increases until the particle, fiber, or whisker fractures, at which point crack advance continues. Since a requirement for this mechanism is a strong bond between the reinforcing phase and the matrix, it appears to be incompatible with some of the more important toughening mechanisms for fiber or whisker reinforced composites. However, the term "strong bond" used in conjunction with this mechanism is not well defined. If the interfacial bond is strong enough to cause some bowing of matrix cracks, but debonding occurs on continuation of stress, allowing the other mechanisms described above to operate, it can be considered to be compatible with these mechanisms.

The crack deflection mechanism is similar to the bowing mechanism but involves a non-planar crack front, such as in a composite containing randomly oriented whiskers or fibers. Some results calculated using a fracture mechanics model of this mechanism were shown earlier to provide an illustration of the greater potential of fibers or whiskers, relative to particles, for ceramic toughening. Like the crack bowing mechanism, the crack deflection mechanism might be compatible with the mechanisms described in a previous section. Initial propagation of a crack in the matrix might be retarded by crack deflection; as stress continues, debonding, crack bridging, and pullout can occur.

V. SUMMARY

The major toughening mechanisms in reinforced ceramics are rather well understood. Under the application of a sufficient tensile stress, matrix microcracking occurs, followed by partial debonding of fibers or whiskers, which dissipates some energy. This is followed by bridging of fibers or whiskers across matrix cracks, and additional dissipation of energy by frictional resistance to pullout of fibers that fracture some distance away from a matrix crack.

Toughening of whisker reinforced ceramics is less effective than toughening of continuous fiber reinforced ceramics. The toughening mechanisms described are most effective when the stress is parallel to fiber or whisker axes. In a typical whisker reinforced ceramic, the whiskers are more-or-less randomly oriented. In addition, the relatively short lengths of whiskers might decrease the effectiveness of the bridging mechanism. Several other toughening mechanisms might be operable, although these appear to be less important than debonding, bridging, and pullout.

REFERENCES

1. Evans, A. G., and Marshall, D. B., The mechanical behavior of ceramic matrix composites, *Acta Metall.*, vol. 37. no. 10, 2567-2583 (1989).
2. Becher, P. F., Microstructural design of toughened ceramics, *J. Am. Ceram. Soc.*, vol. 74, no. 2, 255-269 (1991).
3. Osmani, H., Rouby, D., Fantozzi, G., and Lequertier, J. M., Toughness, microstructure and interface characteristics for ceramic-ceramic composites, *Composites Science and Technology*, vol. 37, 191-206 (1990).

4. Campbell, G. H., Rühle, M., Dalgleish, B. J., and Evans, A. G., Whisker toughening: a comparison between aluminum oxide and silicon nitride toughened with silicon carbide, *J. Am. Ceram. Soc.*, vol. 73, no. 3, 521-530 (1990).

5. Gac, F. D., Is there anything of practical value hidden amongst the composite toughening theories?!—a Jim Mueller perspective, LA-UR-90-2431, 1990, Los Alamos National Laboratory, Los Alamos, NM.

4

Mechanical Testing of Ceramic Matrix Composites

I. BACKGROUND

Improving toughness or other mechanical properties of ceramics by fiber or whisker reinforcement must, of course, be demonstrated by testing. Several types of mechanical tests were described earlier. This chapter presents a more in-depth look at testing. Although mechanical test procedures for monolithic materials, including ceramics, have been well developed and standardized over the years, testing of fiber reinforced materials generally requires modification of existing test methods or development of new methods. Since the technology of polymer matrix composites is more mature than that of ceramic matrix composites, current mechanical test development for ceramic matrix composites draws considerably from the testing used for polymer matrix composites. However, just as test requirements for composites differ from those of monolithic materials, test requirements among different types of composites differ also.

II. MECHANICAL TESTING OF CONTINUOUS FIBER REINFORCED CERAMICS

A. Flexural Versus Tensile Testing of Reinforced Ceramics

In mechanical testing of any material, a test corresponding to a single type of stress, such as a pure tensile or compressive stress, is generally favored over a test corresponding to a combination of stress types. Interpretation of a test involving a single stress type is more straightforward, and the information is usually more readily used for structural design purposes. However, in the case of monolithic ceramic materials the flexural strength test (also termed the "modulus of rupture"), which involves both tensile and compressive stresses, is often used when tensile strength or, less often, compressive strength are of actual interest.

A simple schematic of a three-point flexural test was given in Chapter 2, Figure 2. (The pointed arrows in this figure are not intended to imply that the points of contact are narrow, which would increase the possibility of localized failure; rather, circular cross-sectional contacts (cylinders) are usually used in the test apparatus.) The test specimen is subjected to a compressive stress at the face in contact with the center cylinder and a tensile stress at the face in contact with the outer cylinders. In another type of flexural strength test, the four-point test, the single center contact is replaced by a pair of separated center contacts. Tensile and compressive stresses are characteristic of this variation also.

There are several reasons why flexural strength testing has become so widely used for monolithic ceramics. Although tensile strength of a ceramic material is often the property of major interest for an application, tensile strength testing is considerably more difficult to carry out on a ceramic than it is on a ductile material such as a structural metal. Because of the inherently low fracture toughness of a monolithic ceramic material, specimen gripping without inducing failure in the grip region is difficult. Test results when failure is in the grip region, rather than in the center span, are generally considered unacceptable. Hence, many tests might have to be run to generate a limited amount of acceptable data.

With a ductile material, reducing the cross section of the test specimen in the center span, relative to the grip area, is the common method of increasing the probability of specimen failure in the center span. However, this is not entirely satisfactory for ceramics since the relatively low toughness values increase the likelihood of failure-causing defects at areas where the cross-sectional area reduces appreciably.

Another reason for predominance of flexural testing over tensile testing for ceramics is that flexural testing is more easily carried out at the high temperatures of interest for many ceramics applications. Flexural test fixtures adequate for use at temperatures of 1000°C or higher are readily available. For tensile testing, two choices are available. One alternative is testing with the grips in the furnace. In contrast to the relatively simple fixtures required for flexural testing, the more complicated fixtures required for tensile testing are subject to problems arising from high temperatures. For example, differences in thermal expansion coefficients between different fixture materials and between fixture materials and the test specimen can present problems. In addition, reactions of the fixture materials at elevated temperatures can occur.

The other alternative, having only the center span of the test specimen at the temperature of interest, with the ends of the specimen and the grips outside the furnace, is unsatisfactory from some aspects also. Since there is only a limited length of the test specimen at the desired temperature, failures that are not in this zone are unacceptable. Machining the specimen to reduce the cross-sectional area in the portion that will be exposed to the temperature of interest is complicated by the fracture toughness problem, as outlined above.

Another consideration in using grips outside the furnace in an elevated temperature tensile test is the length of specimen required. While a specimen length as short as 10 cm is possible with hot grips, a specimen approaching 50 cm in length is more appropriate when the grips are outside the furnace. This can be a major problem for an experimental material when fabrication equipment limitations or cost considerations preclude the use of large specimens.

Because compressive strengths for most monolithic ceramic materials are considerably greater than tensile strengths, failure of a flexural test specimen will most frequently begin at or near the face subjected to tensile stress, that is, the face exposed to the outer contacts. Hence, the routine use of an easier-to-conduct flexural strength test, in preference to a tensile strength test, has generally resulted in a satisfactory indication of relative tensile strengths of ceramic materials, both at room temperature and at elevated temperatures.

However, use of flexural strength testing for fiber reinforced ceramics is much less satisfactory. As with a monolithic ceramic, a failure-causing crack will often originate at the tensile face. However, rather than propagating more-or-less normal to the long direction of the specimen, shear cracks along the length of the sample between fiber layers are likely because of low interlaminar shear strengths relative to compressive or tensile strengths for these composites. Delamination cracks are also typical. The resulting multiply-cracked specimen effectively acts as a thinner specimen when some fiber layers are fractured and delaminated, so that a strength calculation based on the original area is no longer applicable. The same process is iteratively repeated until sample failure, or until the sample bends to such an extent that further movement is restricted by the fixture.

Since the failure depends upon the relative tensile and interlaminar shear strengths, data generated from the test is of limited use for a fundamental analysis or from a structural design use. Although the extent of shear cracking can be reduced by increasing the span-to-depth ratio of the specimen, this presents other problems in the analysis.

In continuous fiber reinforced ceramics having a low density matrix, a matrix that is significantly microcracked during fabrication, or which has very weak bonds between the fibers and matrix, the failure-producing crack can originate from compressive rather than tensile stress, which further complicates the fracture analysis.

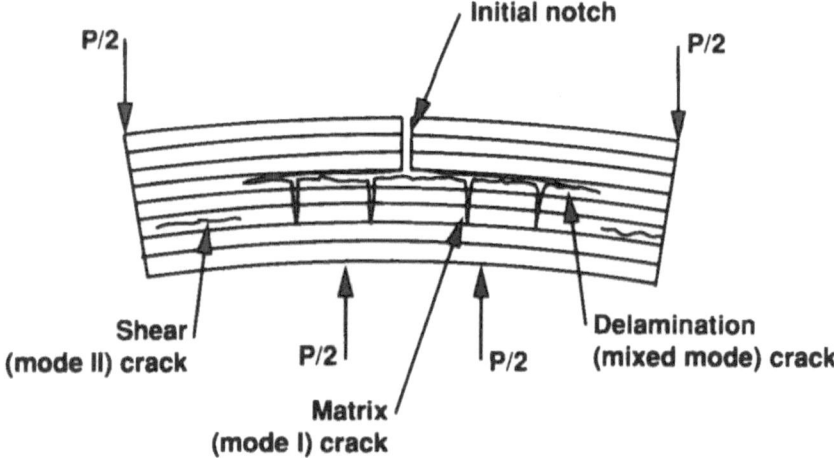

FIG. 1 Schematic illustration of failure modes in a continuous fiber reinforced ceramic in flexure. (From Ref. 1.)

A schematic illustration of fracture behavior during flexural testing has been given by Evans and Marshall (Figure 1).[1] The illustration is representative of a prenotched specimen, as would be used in a fracture toughness test, but it can be used to illustrate the behavior of an unnotched specimen in a conventional flexural strength test, if the assumption is made that the initial notch in the figure is actually a crack that initiated at the upper surface, under tensile stress, and has propagated through several fiber layers. If the toughening criteria described earlier were met, the matrix crack would have resulted in debonding of fibers, fiber bridging, fiber breakage at weak points (typically outside the matrix crack zone), and frictional resistance to sliding of broken fibers through intact matrix. (In that case, the crack at the upper surface of the schematic diagram would be shown more appropriately with fibers protruding from and receded into the matrix.)

At some point during the application of stress, delamination cracks occur. If sufficiently extensive, the effective cross section of the specimen is reduced in this area. That is, the neutral axis (balanced tensile and compressive stresses) moves toward the lower face in the diagram. Additional Mode I cracks can propagate from the delaminated area, and the crack pattern becomes increasingly more extensive. In addition, shear stress generated during flexural testing produces shear cracks in the matrix. Therefore, as described above, the measured strength depends upon a complex reaction to a complex combination of stresses, and is not particularly useful from a fundamental analysis and design standpoint.

Despite the interpretation problems noted for flexural testing of continuous fiber reinforced ceramics, the reader will become aware in subsequent chapters of this book that flexural strength data for characterization of various fiber reinforced materials is much more extensive than tensile strength data (or compressive or shear strength data). The ease of flexural testing relative to tensile testing has induced researchers to use this test extensively for characterization of continuous fiber reinforced ceramics. As materials become more mature and better structural design information is required, characterization by flexural testing will undoubtedly become less widely used.

Use of flexural strength testing for characterizing whisker reinforced ceramics is not as objectionable as for continuous fiber reinforced ceramics, although separate tensile and compressive testing is more desirable.

B. Flexural Strength Test Methodology

Both three-point and four-point bending tests with a variety of span lengths, span-to-depth ratios, and specimen dimensions have been used in flexural strength testing. In general, there are greater shear stresses in three-point testing.

A flexural test method for ceramic materials in general (not specifically composites) is being developed by the American Society for Testing and Materials (ASTM), through Committee C-28.[2] The test is adopted from an earlier U. S. Army test (Figure 2). Elevated temperature testing is being addressed also. Elevated temperature flexural testing does not differ in principle from room temperature testing, but it requires more attention to fixture materials of construction. The fixture must retain adequate mechanical properties and have adequate corrosion resistance at high temperatures in the test atmosphere. In addition, thermal expansion of the fixture materials must be taken into account.

C. Tensile Strength Test Methodology

Considerations in establishing a useful tensile test include specimen design, machining procedures and tolerances, alignment, optimal strain rate, specimen gripping design, and type of load frame. A related consideration is method of measuring strain in the specimen.

The load frame can be mechanically or servohydraulically driven. Of most general use is a machine capable of monotonic loading for fast fracture determinations and capable of cyclic loading for fatigue studies. Good quality universal testing machines used for testing metal or polymeric materials are useful for ceramics as well. Strain rates have not been standardized for ceramic composites.

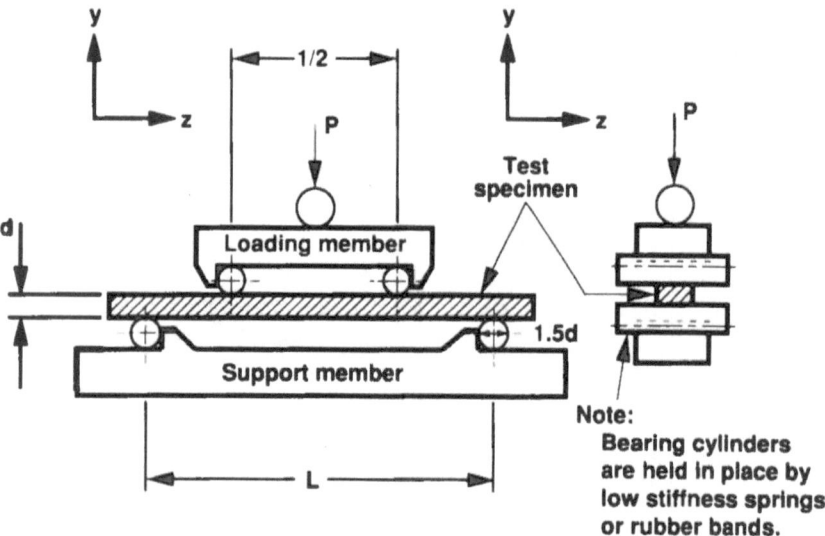

FIG. 2 Schematic representation of a flexural strength test fixture and specimen. (From Ref. 2.)

Specimen design, machining, and gripping are interrelated. Test configurations used for other composite materials such as polymer matrix composites and carbon/carbon composites serve as useful starting points, although adaptations to make them more appropriate for ceramic matrix composites are under development. Although modifications are generally based on scientific principles, trial-and-error plays a significant role.

A test configuration utilizing a specimen of uniform cross section with tabs attached by epoxy cement that has been found suitable for room temperature testing has been described by Lewis et al [3] (Figure 3). The specimen requires no complex machining, so the possibility of flaws caused by machining or by a complex sample configuration is minimized. On the other hand, the uniform cross section enhances the possibility for failure at the grips, where stress is concentrated, rather than in the center section. Failure of the epoxy bond is another potential problem. For elevated temperature testing, such a specimen would be practical only if a long specimen with the grips outside the hot zone were used, since satisfactory high temperature substitutes for epoxy have not yet been identified.

In order to increase the likelihood of failure in the center section and to eliminate cemented joints, tapered specimens with a reduced cross section at the center are preferred. Figure 4 schematically shows one type of specimen used

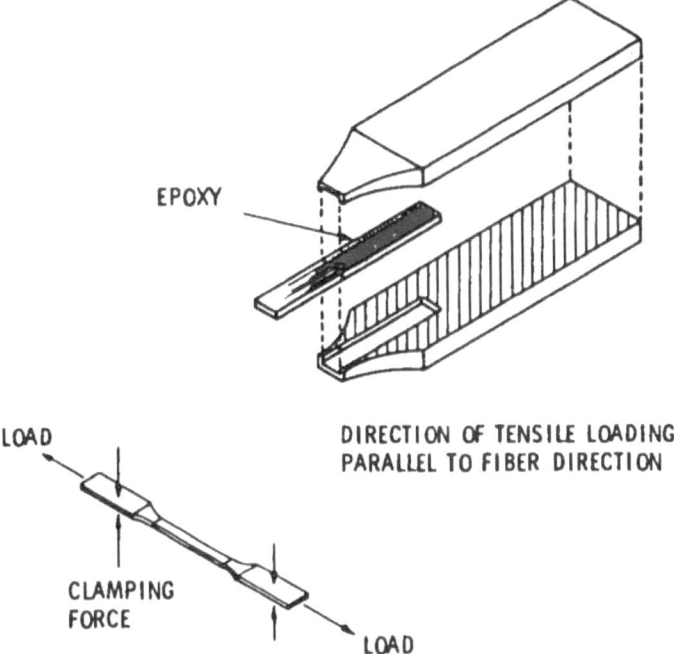

EPOXY

LOAD

DIRECTION OF TENSILE LOADING
PARALLEL TO FIBER DIRECTION

CLAMPING
FORCE

LOAD

FIG. 3 Schematic representation of a tabbed tensile test specimen suitable for fiber reinforced ceramics. (From Ref. 3.)

by Larsen et al.[4] With this type of specimen, elevated temperature testing with the grips in the hot zone is possible. The circles shown in each tab area represent holes for pinning of the specimen to the grips. The use of pins rather than frictional gripping depends on the ability of the specimen to support the load at the pins. Unfortunately, in tests with this type of configuration for composites with uniaxially aligned fibers, the load could not be supported by the specimen at the pins. Only with a cross-plied composite (which will be described in a subsequent chapter on fiber architecture) was this configuration satisfactory. Shear splitting caused by the rather abrupt contour of the transition region was also a problem.

Specimens with a much less abrupt change in cross section and use of a grip involving a metal wedge block were tested successfully by Mandell et al (Figure 5).[5] In general, less abrupt transition zones require increased specimen lengths, which can be a problem both from a test material availability standpoint and from the standpoint of furnace size if hot grips are used.

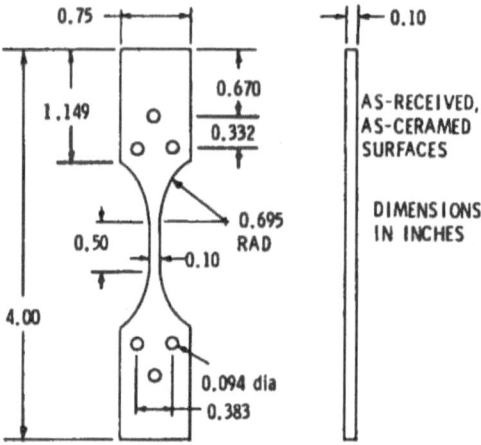

FIG. 4 Schematic representation of a pinned tensile test specimen suitable for fiber reinforced ceramics. (From Ref. 4.)

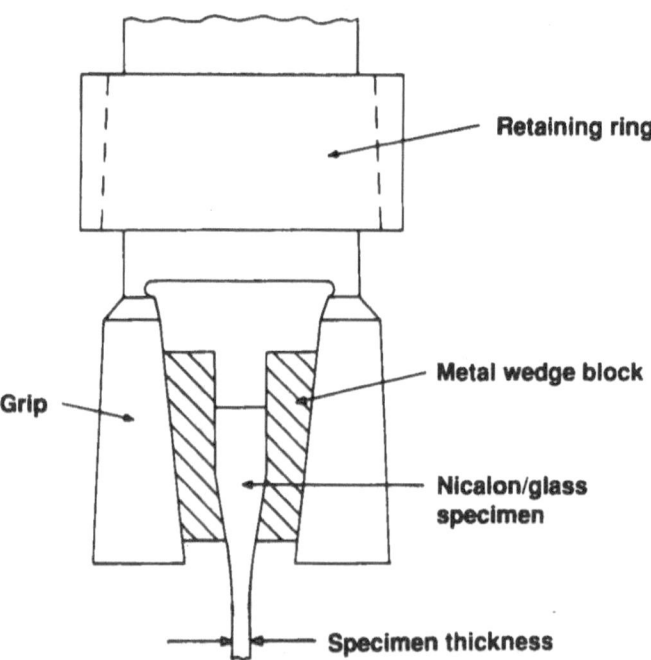

FIG. 5 Schematic representation of a wedge block fixture and specimen suitable for tensile testing of fiber reinforced ceramics. (From Ref. 5.)

Machining and cleaning procedures for specimens can affect the size and distribution of surface flaws and, in some cases, subsurface damage. A procedure found satisfactory at the Naval Research Lab (NRL) has been reported.[3] The as-fabricated piece is first rough machined using a high speed, 150-grit diamond saw, with a coolant to remove heat and debris. Typically, the sample is then rough ground using a 180-grit wheel, with feed depths of 0.012-0.12 mm, depending on the material. Particular attention is paid to the specimen axis relative to the fiber axis, since misalignment of only a few degrees will result in a considerable change in the measured tensile strength. The bulk piece is then sectioned using a 150-grit diamond saw to yield specimens about 0.5 mm larger than final dimensions. The specimens are ultrasonically cleaned in distilled water and alcohol or acetone, and are then air dried for at least one hour at 100°C.

A significant contribution to tensile testing of ceramic matrix composites is a self-aligning grip system suitable for use at high temperatures developed at Oak Ridge National Lab (ORNL) and being marketed by Instron Corporation under the trade name, Super Grip. Since the unwanted introduction of bending forces during tensile testing can significantly affect results, the ORNL grip system has a flexible coupling utilizing eight small hydraulic pistons that help eliminate bending stresses in the specimen.

Figure 6 [6] is a schematic representation of the original ORNL version and a modified version of the grip system. In the figure, the label (1) represents the hydraulic housing containing the piston assemblies that provide equal distribution of load to the metal pull rod assembly (2). A metal plug through which the load is transmitted is represented by (3), and a self-aligning tapered split collar by (4). In the modified version, the test specimen (5) is held in metal grips (6) that are connected to metal adaptors (7) that are held in the tapered split collar.

A furnace designed for tensile testing at up to 1600°C has been marketed by Instron in conjunction with Super Grip. Grips and furnace systems available from other equipment suppliers have been reviewed.[7] Some of the systems can be used at temperatures over 2000°C.

An important consideration in tensile testing of ceramic composites is measurement of specimen extension, or strain, during testing so that a stress-strain curve can be generated. At room temperature, strain gages adhesively attached to the specimen are convenient and accurate. Clip-on extensometers can be used also, but if attached too tightly, bending of thin tensile specimens can occur. If attached too loosely, slippage can be a problem.

At highly elevated temperatures, attachment-type gages or extensometers become more difficult to use successfully. Some of the systems offered by suppliers have provisions for contact extensometers. At moderately high temperatures, quartz is an attractive extensometer material because of its low coefficient

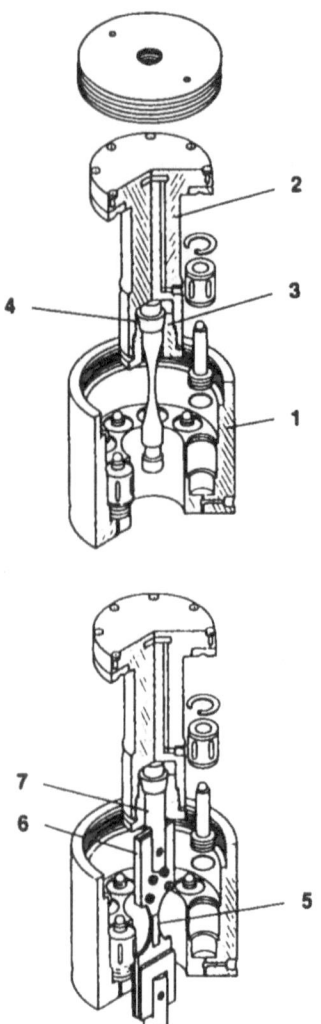

FIG. 6   Schematic illustration of ORNL-developed fixtures for elevated temperature tensile testing of ceramics. (From Ref. 6.)

of thermal expansion. A nickel-base superalloy, Haynes 214, has been used also. At temperatures above about 1000°C, aluminum oxide or silicon carbide can be used, and for temperatures over about 1600°C, carbon/carbon composite extensometers are marketed for use in vacuum or inert gas atmospheres. A capacitive strain-gage extensometer is marketed by Instron for use at up to

1600°C in conjunction with the furnace and grip system referred to earlier. This extensometer is reported to incorporate a nonslip design that requires very little operating force to deflect. A resolution of 0.1 mm at up to the maximum operating temperature has been reported.

As test temperature increases, noncontact extensometers become more attractive. A number of methods utilizing laser light are under development. Researchers at the University of Dayton Research Institute have used a device termed an optical speckle interferometer to measure uniaxial tensile strain of silicon carbide at temperatures up to 1400°C.[2] At ORNL, an optical method using interference fringes produced by diffraction of laser light passing through a narrow slit has been developed. The device is capable of instantaneous measurements for dynamic loading conditions, or an intermittent mode for static, slow-strain-rate, or creep measurement conditions. Another laser light device has been used for creep deformation measurements at the National Institute of Standards and Technology (NIST).

D. Compressive Strength Test Methodology

Compressive strength test development for continuous fiber reinforced ceramics has been limited and relatively few compressive strength results have been reported. Test methods have generally been adaptations of those used for polymer matrix composites. One test, developed at the Illinois Institute of Technology Research Institute (IITRI), uses trapezoidal wedge grips and end support. Test specimen can be either straight sided or tapered. Axial alignment is assured by guide pins in the lower fixture and roller bushings in mating holes in the upper fixture.

Values measured using other compressive strength test methodology will not necessarily agree closely with those obtained using this technique. Compressive failures of composites can involve fiber buckling and fracture, matrix crushing and shear failures, and delamination. It is likely that test fixture and specimen details affect the relative importances of these factors, and, hence, measured strength values.

E. Shear Test Methodology

Even more variation exists in shear tests for continuous fiber reinforced ceramics than for the other tests described above. Shear testing has been reviewed by Larsen.[8] The most appropriate test from a theoretical standpoint is torsion of a thin walled tube. However, most continuous fiber reinforced ceramics fabricated to date have been plates, so this test is not feasible. Perhaps the easiest to perform test is the short beam shear test, which is simply a flexural test with a short span-to-depth ratio. However, the deficiencies inherent to the ordinary flexural strength test are true for this version as well.

Another type of test for measuring in-plane shear is a tensile test conducted on a specimen prepared so that the fiber alignment is 10° off-axis. A test for use on composites with cross-ply fiber alignment is a tensile test with the specimen prepared so that the plies are oriented at 45° angles to the long axis. Another shear test is the Iosipescu test, which utilizes a specimen with 90° notches at the midpoint of each face and loading fixtures that induce a shear stress in the notch area.

F. Toughness Test Methodology

Toughness testing has been briefly described earlier. As shown (Chapter 2, Figure 2), a flexural specimen with a pre-introduced flaw (a notch) on the face subject to tension is used to determine K_{Ic}. This test is commonly referred to as the "single edge notched beam" (SENB) test. A more refined version of this test is the "chevron notch test," in which the notch is machined in such a man-

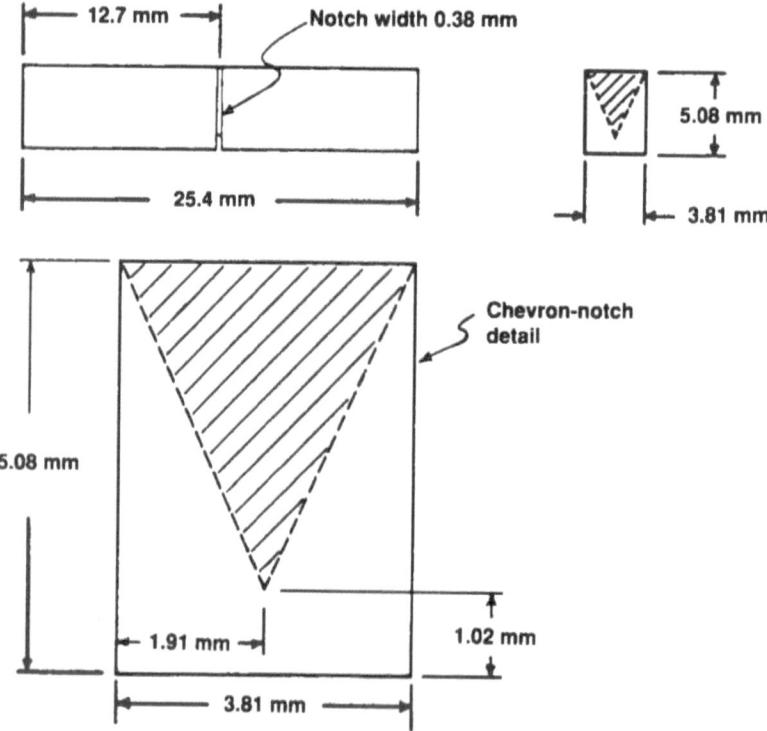

FIG. 7 Schematic representation of the chevron notch test for fracture toughness of ceramics. (Ref. 9.)

ner (in the shape of a chevron) that helps direct crack propagation and, hence, makes test interpretation more straightforward. One version of a test specimen is illustrated schematically in Figure 7.[9] Other tests for toughness, including some based on damage caused by an indenter, have been used also. For example, a Vickers hardness tester, which has a pyramid-shaped diamond indenter, can be used. The median crack length caused by the indenter is measured and K_{Ic} is calculated using a formula involving the crack length, Young's modulus, Vickers hardness, and applied load.

A fundamental problem with determination of K_{Ic} for continuous fiber reinforced ceramics by notched flexural specimens and, probably, by other tests as well, is the complex mode of fracture. Some investigators have challenged the validity of these tests for ceramics reinforced with continuous fibers. Nonetheless, tests of these types have been conducted on many of the materials to be described in this book, and available values will be reported.

Another method of indicating toughness of continuous fiber reinforced ceramics is measurement of the area under a stress-strain curve. In contrast to monolithic ceramics, which have very little area under the curve because of their very low strain-to-failure values, continuous fiber reinforced ceramics can have relatively large areas.

III. MECHANICAL TESTING OF WHISKER REINFORCED CERAMICS

Many of the problems outlined with regard to mechanical testing of continuous fiber reinforced ceramics are not applicable to whisker (and discontinuous fiber) reinforced ceramics. Whisker reinforced ceramics can be considered to be more like monolithic ceramics than continuous fiber reinforced ceramics, from a mechanical testing standpoint. Perhaps the major consideration is the extent of whisker alignment. Depending upon fabrication conditions, the whiskers can be oriented randomly, or might have a preferred orientation in a plane, or even in one direction. In the latter two cases, the probability of anisotropic properties, as with continuous fiber reinforcement, must be considered in specimen preparation and testing.

IV. OTHER PERSPECTIVES ON MECHANICAL TESTING

This chapter was not intended to provide an in-depth treatment of mechanical testing, but, rather, to supply sufficient information to provide an understanding of the problems encountered in testing ceramic matrix composites and to give a general idea of the state-of-the-art. In later chapters dealing with specific composite systems, considerable mechanical property information will be given. As suggested in this chapter, relatively little test standardization for ceramic matrix composites has yet occurred. Test values obtained using a specific test method will not necessarily agree with those obtained by a different method. Hence, in most cases only semi-quantitative comparisons, at best, can be made among the composite materials to be described.

Test standardization is progressing. Eventually, direct quantitative comparisons of ceramic composite materials by well established and accepted test methods will be possible. In the interim, it would be of benefit it there were more exchange of materials between workers in this field, or supplying materials to a single organization for testing. To some extent, the latter is beginning to occur in conjunction with bid and proposal activities on contracts for further material development for various advanced aerospace projects. However, it is questionable to what extent this information will be generally available, since suppliers may have the option of limiting the dissemination of test information on their materials and, in addition, general dissemination of information with regard to these advanced aerospace projects is often restricted by government regulations.

Finally, it can be noted also that despite the shortcomings of flexural testing for continuous fiber reinforced ceramics, flexural strength will be the most commonly cited mechanical property for many of the composite materials to be described later in this book. The researchers involved were surely aware of the limitations of flexural testing for continuous fiber reinforced ceramics. However, the simplicity of test fixtures and test specimens, both at room temperature and at elevated temperatures, and the need for relatively little material for testing have simply tended to make flexural testing more attractive than other characterization tests. It is not unreasonable to assume that early ceramic composite material development was advanced more effectively by using a less-than-ideal test from a design and analysis standpoint, but straightforward from an experimental standpoint, than by using tests that are more fundamentally sound, but not nearly as well developed experimentally.

V. SUMMARY

The relative simplicity of flexural testing has made flexural strength the most common mechanical property measured on ceramic matrix composites, despite serious shortcomings in the use of such test results for continuous fiber reinforced ceramics. Various versions of tensile, compression, and shear tests for ceramic matrix composites are under development, as are fracture toughness tests. High temperature test fixtures are becoming more available. Results will be most useful when various material systems are tested using identical test methods.

REFERENCES

1. Evans, A. G., and Marshall, D. B., The mechanical behavior of ceramic matrix composites, *Acta Metall.*, vol. 37, no. 10, 2567-2583 (1989).
2. Geiger, G., Advancements in mechanical testing of advanced ceramics, *Ceramic Bulletin*, vol. 69, no. 11, 1794-1800 (1990).
3. Lewis, D., Bulik, C., and Shadwell, D., Standardized testing of refractory matrix/ceramic fiber composites, *Ceram. Eng. Sci. Proc.*, vol. 6, no. 7-8, 507-523 (1985).
4. Larsen, D. C., Stuchly, S. L., and Bortz, S. A., In *Metal Matrix, Carbon, and Ceramic Matrix Composites*, NASA Conference Publication 2406, 1985, 313-334.
5. Mandell, J. F., Grande, D. H., and Edwards, B., Test method development for structural characterization of fiber composites at high temperatures, *Ceram. Eng. Sci. Proc.*, vol. 6, no. 7-8, 524-535 (1985).
6. Sheppard, L. M., Tensile testing of ceramics, *Advanced Materials and Processes*, vol. 131, no. 5, 11-15 (1987).
7. Baxter, D. F., Jr., Tensile testing at extreme temperatures, *Advanced Materials and Processes*, vol. 139, no. 2, 22-30 (1991).
8. Larsen, D. L., and Stuchly, S. L., The mechanical evaluation of ceramic fiber composites. In *Fiber Reinforced Ceramic Composites*, K. S. Mazdiyasni (Ed.), Noyes Publications, Park Ridge, NJ, 1990, 182-221.
9. Bar-on, I., Tuler, F. R., and Roman, I., Comparison of analytical and experimental stress-intensity coefficients for chevron V-notched three-point bend specimens. In *Chevron-Notched Specimens: Testing and Stress Analysis*, ASTM Special Technical Publication 855, J. H. Underwood, S. W. Freiman, and F. I. Baratta (Ed.), 1983, 98-111.

5

Fibers and Whiskers for Ceramic Reinforcement

I. BACKGROUND

Fibers or whiskers having high strengths and Young's moduli are required for useful reinforcement of ceramics. In addition, chemical compatability between the fibers and matrix are required. Also, for elevated temperature applications, the fibers must be acceptably thermally stable at the application temperature. Low fiber density is obviously an advantage. Finally, cost (or, in the case of experimental fibers, projected cost based on commercial production) must not be prohibitive for intended applications.

Although metal and polymeric fibers have been available for many years, these fibers do not meet most of the requirements outlined above for ceramic reinforcement. Hence, development of reinforced ceramics has concentrated on use of carbon fibers and on ceramic fibers and whiskers. The development of industrial scale production processes for high quality continuous carbon fibers preceded processes for continuous ceramic fiber production. Even though carbon fibers do not have the capability of extended high temperature use when exposed to air, they are useful for ceramic reinforcement for high temperature applications if adequately protected from air by the matrix or coatings. Since the more recent development of a variety of continuous ceramic fibers and ceramic whiskers, emphasis has been on these materials for ceramic reinforcement.

This chapter will cover continuous carbon and ceramic fibers and ceramic whiskers, with regard to manufacture, composition, and properties.

II. CONTINUOUS CARBON FIBERS

A. History

Use of carbon fibers dates back over 100 years.[1-3] Thomas Edison filed a patent on carbon fibers in 1879. Carbon fibers were used as filaments in electric light bulbs prior to the introduction of tungsten filaments. However, these carbon fibers, prepared from precursors based on cellulose, were not sufficiently strong to be attractive from a reinforcement standpoint.

A number of workers have played roles in the more recent development of strong, high modulus carbon fibers. In the late 1950s, researchers at Wright-Patterson Air Force Base in the USA investigated fibers prepared using viscose rayon, and the process was developed further at Union Carbide Corp. A. Shindo in Japan prepared carbon fibers from polyacrylonitrile (PAN) in 1961. His fibers had Young's moduli averaging about 170 GPa. Subsequently, workers at Rolls Royce Ltd. in Britain found that modulus of a PAN-derived fiber could be increased dramatically, to about 600 GPa, by stretching the fiber during processing. In 1965, S. Otani in Japan made pitch based fibers for the first time. In the intervening years, a great deal of work on high performance carbon fibers produced from PAN, rayon, and pitch has been carried out.

B. General Characteristics of Carbon and Carbon Fibers

Carbon is among the materials having the greatest high temperature stability; it sublimes at about 3500°C. There are two well-known crystalline forms of carbon, diamond and graphite. In diamond, the carbon atoms are arranged in a three-dimensional array, with each atom having four covalent bonds, separated by 109.5° angles, to other carbon atoms. As the result of its strong three-dimensional covalent bonding, diamond is the hardest material known.

The structure of carbon that has been utilized in fiber production is the graphite structure. Carbon in graphite form is extremely anisotropic, with strong covalent bonds in layer planes but only weak bonds between planes. In the planes, each carbon atom is bonded to three other atoms. The bonds are at 120° angles, which results in a hexagonal array. As a consequence of the weak bonding between planes, distance between successive planes is relatively large, and graphite has a density of 2.3 g/cm^3, much lower than that of diamond, 3.5 g/cm^3. The graphite form of carbon is shown schematically in Figure 1.[4]

Graphite is strong and has a very high Young's modulus in the direction of the planes, but weak with a very low modulus normal to this direction. The

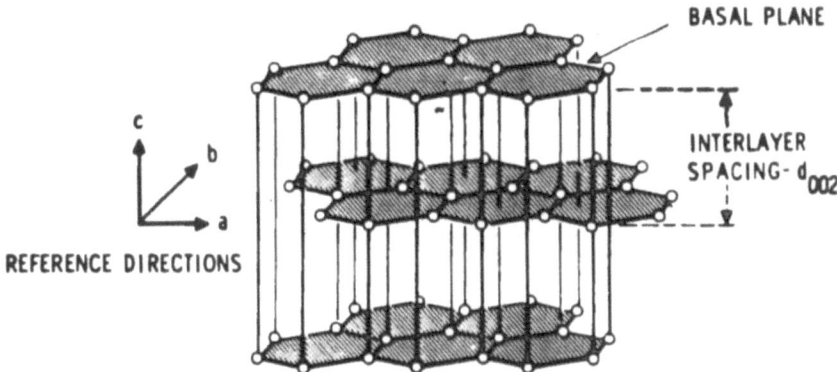

FIG. 1 Schematic representation of the structure of graphite. (From Ref. 4.)

ease with which graphite planes can move relative to other planes makes graphite an excellent solid lubricant. Values for mechanical properties of graphite are retained, or increase, at temperatures over 2000°C in an inert atmosphere. However, graphite readily oxidizes in air at temperatures over about 400°C.

Although another form of carbon, amorphous carbon, is sometimes described, amorphous carbon may be considered as the graphite form of carbon that is characterized by the following: small crystallite sizes, structural disorder which results in a larger average interplanar distance than the minimum possible for graphite, misorientation of crystallites relative to one another, and cross-linking among crystallites.

More recently, carbon molecules such as those containing sixty carbon atoms arranged in a "soccer ball" configuration (termed "buckminsterfullerine," after Buckminster Fuller, inventor of the similarly configured "geodesic dome") have been considered to be a new form of carbon. However, bonding is of the graphite type, although the bonds are highly distorted.

In order to produce high strength fibers having the graphite form of carbon, the layer planes must be oriented in the direction of the fiber axis. This is true for all the types of carbon fibers that have been commercially developed. Differences in fiber properties depend on how well the planes are oriented in the fiber direction, the orientation of the plane axes in the radial direction, the degree of cross-linking among crystallites, and the presence of impurities, voids, and other flaws. The Young's modulus, in particular, depends upon the first factor. If the planes are in somewhat of a "zigzag" configuration relative to the fiber axis, applying a stress will result in straightening of the planes and, hence, modulus will be lower than for a fiber in which the planes are well ori-

ented in the fiber direction. Tensile strength, on the other hand, is more dependent upon flaws. Hence, processing conditions that maximize strength are not
necessarily the same as those that maximize modulus.

C.   Carbon Fibers Derived from Polyacrylonitrile (PAN) Fibers

Acrylonitrile is a derivitive of ethylene, $H_2C=CH_2$, in which one of the hydrogen atoms has been replaced by a -CN, or nitrile, group. Using an appropriate
catalyst such as a peroxide, the acrylonitrile monomers can be polymerized into
long chains of H_2C-CHCN units.[5] Polyacrylonitrile (PAN) is dissolved in a
solvent such as N,N-dimethylformamide and spun into threads of "acrylic
fibers." Spinning consists of forcing the solution through a spinneret, which
can contain thousands of tiny holes. Individual filaments are combined into
"tows." Figure 2 [4] schematically shows a spinneret. Fiber diameters of about
10 μm are typical. PAN fibers used for carbon fiber production are produced
specially for that purpose.

To prevent melting (and, hence, destruction) of PAN fibers during heat
treatment to convert the fibers to carbon fibers, the fibers must be preoxidized,
or stabilized, at 200° to 300°C in an oxidizing atmosphere.[1-3] This promotes
cross-linking among chains and raises the melting point. The basic polymeric
structure is modified, and oxygen-containing groups such as C-OH, C=O, and
C-O are formed. Ring type structures form at some locations along the chain,
and some dehydrogenation of the C-C backbone structure occurs. Providing
preoxidized fibers that yield carbon fibers with good properties has entailed

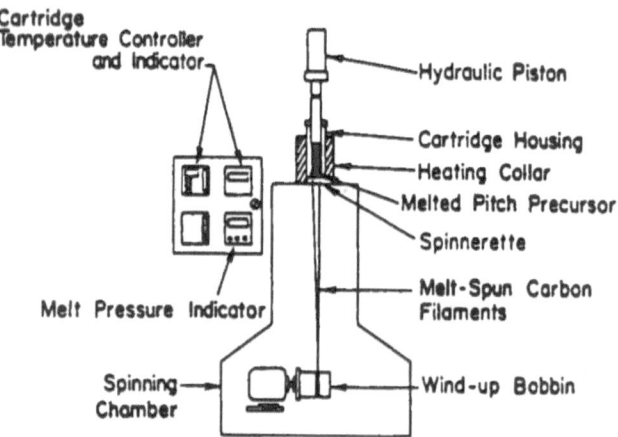

FIG. 2   Schematic illustration of a spinneret used for fiber production. (From
Ref. 4.)

considerable work on optimizing temperature, soak time, and atmosphere. Inadequate stabilization or, on the other hand, excessive formation of ring structures in the chains (over-oxidation) have both been found to result in inferior carbon fiber properties.

Relaxation of dipole reactions among the strongly polar nitrile groups, and some of the reactions mentioned above, result in shrinkage of the fibers during preoxidation and consequent loss of the good linear orientation of the original PAN chains. The poor orientation results in fibers having a low modulus. Hence, modulus is greatly improved by stretching the fibers during this process. Stretching of several hundred percent can be carried out. Excessive stretching can cause some bond rupturing, leading to reduced tensile strength.

Between 300° and 600°C, most of the chemical reactions leading to carbon fibers take place, with evolution of water, ammonia, and hydrogen cyanide. Hydrocarbons in the form of tar may be evolved also, which is undesirable because it results in elimination of carbon atoms from the chain structure, leaving defects, and can result in sticking together of fibers. Tar formation is minimal with properly stabilized fibers and a relatively slow heating rate. By 1000°C, the fibers are mostly carbon. The yield of carbon fibers from PAN fibers is up to about 55%. Fiber density is typically about 1.8 g/cm^3. Production of PAN-derived fibers (along with production of pitch-derived fibers, to be described below) is illustrated in Figure 3.[4] A surface treatment to affect fiber-matrix properties in a composite is often applied. A "size" to aid

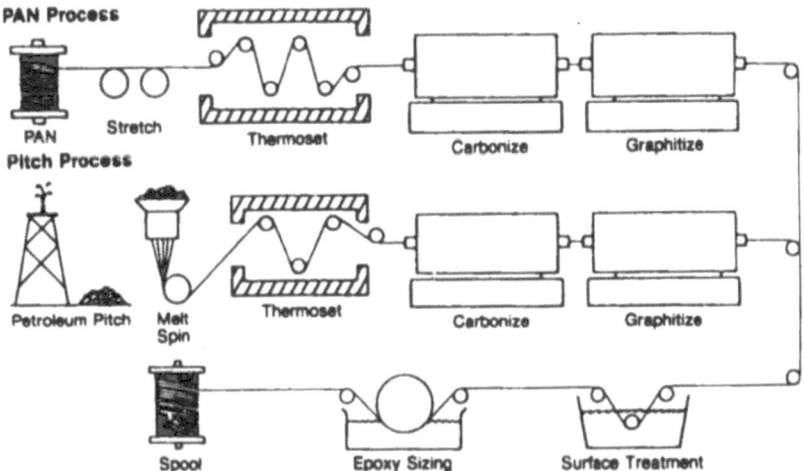

FIG. 3 Schematic illustration of production processes for PAN derived and pitch derived carbon fibers. (From Ref. 4.)

in fiber handling is invariably added also. Although an epoxy size is indicated in the figure, other materials may be used as well.

After heat treatment at 1000°C, the fiber molecular structure is based on the graphite-type structure. However, because of extensive and complex cross linking between polymer chains during the early stages of carbonization, the crystallites are not large and not well oriented, except for a generally good orientation of planes along the fiber axis. In addition, many fibers have one type of arrangement in an outer layer and a different arrangement in the core. One characterization for graphite is L_c, the average height of a crystallite. L_c is typically about 10 Å for a carbon fiber, compared with L_c values of over 1000 Å for natural graphites.

The graphite-like character of fibers is progressively increased by heat treatments at increasing temperatures. Although temperatures approaching 3000°C are considered "graphitizing temperatures" for carbon, L_c values for typical carbon fibers remain below 100 Å at 3000°C. Fibers stretched during high temperature heat treatment may attain an L_c value of 200 Å, which is still far below that for natural graphite. Hence, even fibers heat treated at 3000°C are more appropriately referred to as carbon fibers rather than graphite fibers.

The maximum heat treatment temperature depends upon the properties desired for a particular application, with the added cost of high temperature heat treatment as a consideration. For example, typical fibers heat treated only to 1000°C have tensile strengths of about 2.0 GPa and moduli of up to 170 GPa. When fibers are heated to about 1500°C, tensile strength can increase to over 3.5 GPa and modulus to about 275 GPa. When heat treatment at 2500°C is carried out, strength decreases to less than 2.8 GPa (probably due to introduction of elongated pores due to shrinkage of crystallites, which is a consequence of increasing graphitization), while modulus increases to as high as 480 GPa. It is possible also to further stretch the fibers at about 2000°C to increase modulus to as high as 700 GPa, at a greatly increased cost.

D. Carbon Fibers Derived from Rayon Fibers

Viscous rayon is a synthetic polymeric fiber derived from a natural polymeric material, cellulose, which constitutes the cell membranes of the higher plants.[5] Cellulose consists of multiply-linked sugar molecules, which are made up of only carbon, hydrogen, and oxygen. Molecular weights can range from one to two million. Pure cellulose is often obtained from cotton by dewaxing with an organic solvent and then reacting with hot sodium hydroxide solution. Rayon manufacture is based on the reaction of hydroxyl groups in cellulose with carbon disulfide in the presence of sodium hydroxide. Fibers are spun from a viscous solution, "viscose," in which the cellulose has been degraded to an average molecular weight of about 70,000.

In the process of converting rayon fibers to carbon fibers, desorption of water is the main reaction up to 150°C.[1-3] Above this temperature, the desired reaction is removal of -OH groups, but tar formation can also occur. Catalysts can be used to facilitate -OH removal.

Yield of carbon fibers from rayon fibers is relatively low, about 30%. In addition, mechanical properties are inferior to those of PAN-derived fibers. A tensile strength of 0.7 GPa and a Young's modulus of 70 GPa are typical. In contrast to PAN-derived fibers, it is not possible to stretch at low temperatures to improve properties. However, in common with PAN-derived fibers, it is possible to stretch rayon-derived fibers during heat treatment at about 2000°C to improve strength to about 2.8 GPa and modulus to 550 GPa. Because rayon-derived fiber properties are generally inferior to PAN-derived fiber properties, without the costly high temperature stretching process, use for structural applications has been more limited.

E.   Carbon Fibers Derived from Pitch

Pitches derived from coal tars or petroleum residues are thermoplastic materials having relatively high carbon contents. They are composed of carbon-containing molecules having a broad range in molecular weight. Most of the compounds are aromatic (benzene base) molecules that tend to convert to the graphite-type carbon structure during heat treatment. Pitches can readily be spun into fibers suitable for conversion into carbon fibers. As with PAN-derived fibers, preoxidation at 200° to 300°C is required to prevent melting prior to decomposition into carbon. Untreated pitches have an isotropic arrangement of molecules, which results in fibers with relatively little orientation of graphitic planes in the direction of the fiber axis. Because of the weak forces between planes in the graphite-type carbon structure, strength and modulus of such a fiber are poor.

When coal tar or petroleum base pitches are heated at above about 350°C, spherical shaped liquid crystals containing layers of aromatic molecules separate from the isotropic pitch. These are termed "mesophase." This mesophase material is highly anisotropic. When temperature is increased, these mesophase spheres grow and coalesce, producing a highly anisotropic material. Fibers spun from mesophase pitch using a high extrusion pressure have a highly oriented molecular structure along the fiber axis, obviating the necessity for stretching during preoxidation.

As with the polymer-derived fibers, a carbonization temperature of about 1000°C is required, and the fibers can be heat treated further up to about 3000°C. Yield of carbon from mesophase can exceed 75%. Density is typically about 2.0 g/cm^3. A tensile strength of about 3.5 GPa and a Young's modulus of 400 GPa have been attained with fibers carefully prepared to minimize flaws. Heat treatment at 3000°C has produced fibers having moduli of about 700 GPa.

F. Commercial Availability of Carbon Fibers

Carbon fibers are readily available commercially from a number of suppliers. These differ appreciably in parameters such as tensile strength, Young's modulus, strain-to-failure (a function of strength and modulus), number of filaments per tow, and cost. A survey conducted in the mid-1980s [6] reported 17 manufacturers of high performance fibers and 74 fiber grades. The largest number of manufacturers (7) were in the USA. Other countries represented were Japan (4), Germany (2), Great Britain (2), France (1), and Israel (1).

The commercially available carbon fibers can be classified into Ultra High Modulus, High Modulus, Intermediate Modulus, Very High Strength, and High Strength categories. Specific modulus (modulus/density) values among commercial fibers differ by a factor of three and specific strength (strength/density) values differ by a factor of three and one-half. Percent strain-to-failure (100 × strength/modulus) values are as high as 2.0, which is appropriate for toughening of ceramic materials by the mechanisms discussed earlier. The number of filaments per tow ranges from less than 1000 to over 100,000. Price can vary by about two orders of magnitude (from roughly $10 per pound to roughly $1000 per pound) between relatively low performance fibers and premium fibers.[3]

III. CONTINUOUS CERAMIC FIBERS

Continuous ceramic reinforcing fibers have more recently become available for use in reinforcing ceramics. Because of the oxidation problem with carbon fibers, most recent work on continuous fiber reinforced ceramics has been with ceramic fibers. These include both oxide and non-oxide base fibers. Some of the fibers are made from a single phase, relatively pure material, while others are multiphase materials. A number of fabrication methods have been used.

A. Fibers Formed by Polymer Decomposition

Several types of continuous ceramic fibers are produced by a method analogous to production of carbon fibers. Polymer melts or solutions are spun into fibers and then heat treated to convert the polymeric fibers into ceramic fibers. By far the most developed fiber of this kind is Nicalon™, a silicon carbide base fiber. Because of its prominence in the field of continuous ceramic fibers, emphasis in this section will be on Nicalon.

1. NICALON™ FIBERS

a. Technical Basis for Nicalon Fiber Development

In a number of publications, beginning in 1975, S. Yajima and coworkers at Tohoku University in Japan reported the successful preparation of silicon car-

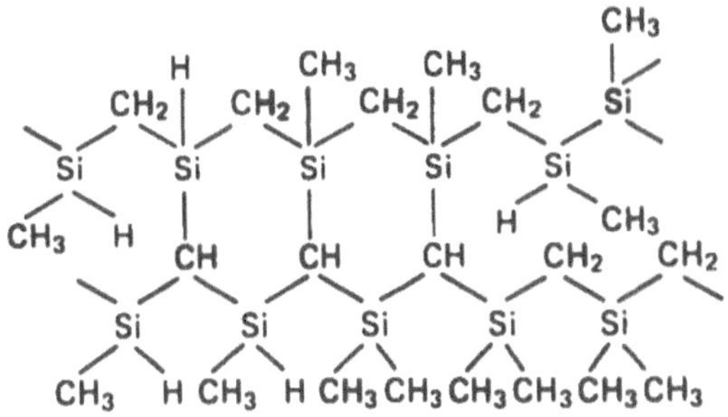

FIG. 4   General representation of a section of a polycarbosilane molecule. (From Ref. 7.)

bide base fibers by spinning and pyrolysis of a polymer termed polycarbosilane, which has alternating silicon and carbon atoms in its chain structure. A general representation of a section of a polycarbosilane molecule is shown in Figure 4.[7] The procedures described by Yajima and coworkers formed the basis for eventual commercialization, but, undoubtedly, changes have been made over the years.

Polydimethylsilane was produced by reacting methyldichlorosilane, $(CH_3)_2SiCl_2$, with sodium in xylene.[8] The xylene was removed by suction filtration and excess sodium was removed by reaction with methanol. On addition of water, a white powder was obtained. This powder, polydimethylsilane, was washed with additional water and then with acetone, prior to final drying in vacuum at 150°C.

This material was further polymerized under argon at temperatures of 450° to 470°C for 14 h in an autoclave. Starting pressure was 100 kPa. Pressure increased during the process and final pressure depended on the temperature. The product in each case was a yellowish-brown viscous substance. Products were dissolved in n-hexane and the solution filtered. Solvent was removed by evaporation and a fraction boiling at up to 280°C under 0.13 kPa pressure was removed. The product after removal of the distillate fraction was a yellowish-brown glassy polycarbosilane-containing material.

As with other polymers, the term polycarbosilane does not refer to a single material but to a range of materials of varying molecular weights and structural details. Hence, the specific chemical and physical properties of the polycarbosilane products in this work and in other work by Yajima and coworkers were affected by reaction conditions.

Infrared spectra for the polycarbosilane produced by polymerization at 470°C showed absorptions at 600 to 920 cm^{-1} (Si-CH$_3$ bonds), 1020 and 1355 cm^{-1} (Si-CH$_2$-Si bonds), and 2100 cm^{-1} (Si-H bonds), as well as some less well characterized absorptions. Number-average molecular weights of the polycarbosilane products ranged from 1250-1750, depending upon polymerization temperature. Chemical analysis indicated a product containing 50.0 wt% Si, 37.9 wt% C, 1.0 wt% O, and 6.6 wt% H. Based on the chemical analysis, the empirical chemical formula for the polycarbosilane product from polymerization at 470°C was $SiC_{1.77}O_{0.03}H_{3.70}$. Various mechanisms were proposed by Yajima and coworkers for conversion of the Si-Si chain linkage in polydimethylsilane to the Si-C chain linkage in polycarbosilane and for molecular weight increase of polycarbosilane.

Fibers were obtained by melt-spinning at about 350°C from the nozzle of a capillary drawn out from the bottom of a quartz tube.[9] Melted polycarbosilane was kept under nitrogen gas to avoid oxidation. Fibers were wound on a 50 cm diameter drum placed 50 cm below the nozzle. Drawing speed was about 200 m/min.

Fibers were cured in air using an upheat rate of 30°C per hour to 190°C and a 30 minute hold time at that temperature. This curing process was necessary to produce cross-linking of polycarbosilane chains, which in turn prevented melting and, hence, destruction of the fibers when heat treated to produce the silicon carbide base fibers. This is analogous to the process for PAN-derived and pitch-derived carbon fibers. Gravimetric and differential thermal analysis showed that an exothermic reaction, with a weight gain, started at about 150°C. Chemical analysis indicated a product of 44.4 wt% Si, 31.0 wt% C, 5.3 wt% H, and 15.1 wt% O, corresponding to an empirical formula of $SiC_{1.63}H_{3.34}O_{0.61}$.

Oxygen content was about 15 times that in the uncured material. Based on infrared absorption results, the mechanism of curing was postulated to be the oxidation of Si-H and $Si-CH_3$ bonds to produce Si-O-Si and Si-O-C bonds.

The cured fibers were heat treated using an upheat rate of 100°C per hour at various temperatures up to 1300°C and held at the maximum temperature for 60 minutes. Heat treatment was done both under vacuum and in nitrogen gas flowing at a rate of 100 cm^3 per minute.

In the first stage of fiber heat treatment, up to about 550°C, it was concluded on the basis of infrared absorption analysis that cross-linking by condensation occurred, along with dehydrogenation. Infrared absorption bands corresponding to various hydrogen related bond motions decreased.

From about 550° to 850°C, thermal decomposition of side chains such as Si-H, $Si-CH_3$, and C-H in $Si-CH_2-Si$ were postulated to occur, since infrared absorption bands corresponding to Si-H and $Si-CH^3$ almost totally disappeared. A new band due to Si-O stretching began to occur also.

Above 850°C, conversion to an inorganic material was completed. The infrared spectra of fibers heat treated at 1300°C were very similar to those of commercial silicon carbide. However, as will be described later, the heat-treated fibers are far from stoichiometric silicon carbide.

b. Commercialization of Nicalon Fibers

In the late 1970s, this process for silicon carbide base fiber production was carried out on a pilot scale.[10] Multifilament tows about 0.5 m long and containing 4000-5000 fibers were produced. Commercial production by Nippon Carbon Co., Ltd. in Japan began in 1981. Marketing and distribution in the USA is by Dow Corning Co. A general overall scheme for the process is outlined in Figure 5, although the accuracy of this schematic flow diagram is problematic, since the process is proprietary.

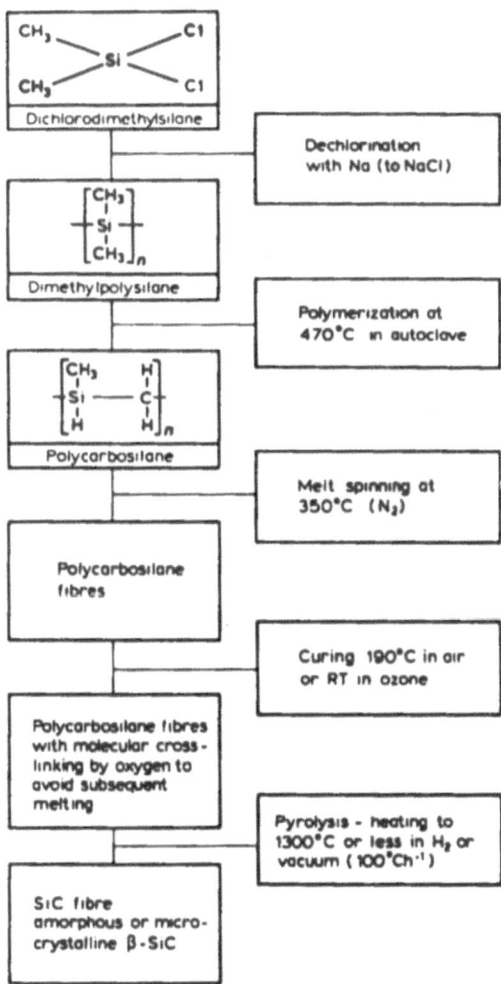

FIG 5. Schematic flow diagram for Nicalon™ fiber production. (From Ref. 10.)

An improvement by Yajima and coworkers over the process scheme described above that has been cited [11] as a major factor in commercialization is the use of polyborodiphenylsiloxane to bring about the rearrangement of a linear polysilane into a linear polycarbosilane at ambient pressure, rather than using batch processing in an autoclave.

A large proportion of the early laboratory fibers had diameters below 10 μm, but average value for commercial Nicalon is about 15 μm. There is a distribution in average diameter among fibers and variations in diameter along the length of an individual fiber can be observed also. Fiber density is about 2.6 g/cm^3. Figure 6 shows Nicalon fibers.

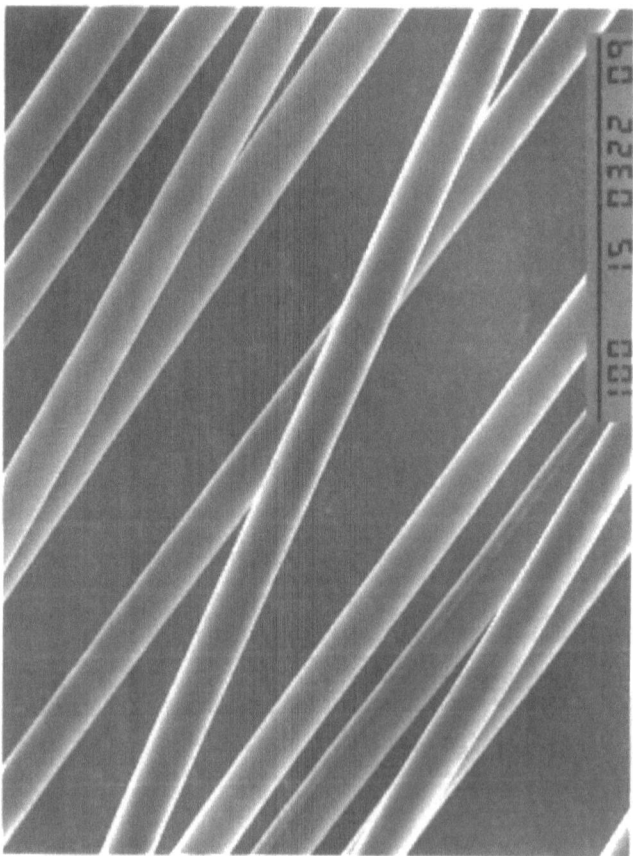

FIG. 6   Micrograph of Nicalon™ fibers. (Courtesy of Dow Corning Corp.)

c. Structure of Nicalon Fibers

From the chemical composition given previously, it is apparent that Nicalon fibers are not at all close to stoichiometric silicon carbide. Numerous structural studies have been carried out since the inception of the commercial process. Although understanding of the structural details has been refined during this period, the basic understanding was worked out relatively soon after commercialization.[10]

A number of investigations into the structure of Nicalon fibers have been conducted. The results of several of these [12-15] are summarized in this section. X-ray diffraction and transmission electron microscopy indicate Nicalon fibers to be microcrystalline with a major phase of randomly oriented cubic β-silicon carbide and a minor amorphous phase. Crystallite diameter is about 2.5 nm. The amorphous material is silicon dioxide and free carbon. Equivalent molar composition calculated for one lot of commercial Nicalon fibers, assuming that the entire oxygen content is present in silicon dioxide, corresponded to 61% silicon carbide, 35% silicon dioxide, and 3.5% free carbon.

Nicalon fibers are currently supplied in both a "standard grade" and a "ceramic grade." The oxygen content of ceramic grade is approximately 10 wt%, while that of standard grade is nearly twice as high. Microstructural details are also reported to differ.

d. Room Temperature Mechanical Properties of Nicalon Fibers

As a preface to discussing properties of Nicalon fibers, some comments should be made on the general subject of ceramic fiber mechanical property values. Many of the fibers to be described are in relatively early stages of development, and properties can be expected to differ due to both evolutionary process changes and due to less-than-perfect quality control. Even the most mature of the fibers, Nicalon, probably changed somewhat over the years, particularly during early years of commercial production.

In addition to variations in properties that can be expected from the fibers themselves, other variations can be expected because of the lack of standardization of fiber property testing. Factors such as handling and the lengths of the fibers tested have documented effects, as will be shown. In addition, test details such as strain rate and degree of success in avoiding bending stresses undoubtedly influence results. Strengths of individual fibers calculated from tests on fiber tows can also differ from strengths determined on individual fibers.

Because of the factors outlined above, reported results on the various types of ceramic fibers can differ appreciably. The selected data that will be shown are representative values, and values given for the same type of fiber will often differ. No attempt was made to assemble all data on fiber testing and make a critical assessment. The results given are believed sufficient to provide a general understanding of room temperature properties and property retention at elevated temperatures for available fibers.

Sawyer et al [13] tested ceramic grade Nicalon fibers described by the manufacturer as having an average tensile strength of 2.8 GPa and average modulus of 200 GPa, as determined on tests using fiber tows. They used single fiber test method ASTM-3379, "Test Method for Tensile Strength and Young's Modulus for High Modulus Single Filament Materials." Sawyer et al found strengths ranging from 0.3 GPa to over 3.6 GPa for individual fibers. They found a large dependence on fiber length. This dependence is related to a larger probability for a failure-producing flaw as length is increased, and is typical for fibers and whiskers in general. Mean strength was about 1.0 GPa with a 25-cm gage section, about 2.4 GPa with a 2.5-cm gage section, and over 3.1 GPa with a 0.3-cm gage section.

Scanning electron microscopy along with other techniques permitted the analysis of the types of flaws causing fracture. Low tensile strengths (about 0.8 GPa or lower) generally resulted from surface or internal contaminants. Granular defects and surface defects generally caused failure in the range of 2.4-2.8 GPa. Above about 2.8 GPa, tensile failure initiated within the fiber.

Andersson and Warren [10] also identified several types of flaw populations, causing fracture in different strength regimes, as well as an effect of fiber length. Figure 7 shows data for 1.0 and 10.0-cm gage sections. The data for the 10.0-cm length clearly shows the presence of two types of fracture causes. The designation, "m," represents the Weibull modulus. The defect size shown at the top of the figure was calculated from an equation relating strength to fracture toughness and defect size. This figure also indicates how reported tensile strength values can vary due to fiber handling. Fibers that were damaged during size removal and during removal from tows failed more commonly at lower stresses than carefully handled fibers having the same gage length.

In other work on room temperature properties of Nicalon fibers, Pysher et al [16] found an average tensile strength of about 1.9 GPa for individual fibers of ceramic grade Nicalon having a 7.5-cm gage length. This value is in reasonable agreement with those described above for the longer gage sections. Their value of about 185 GPa for the Young's modulus agrees reasonably well with the manufacturer's value, cited in work described above.

e. Elevated Temperature Properties of Nicalon Fibers

Production of ceramic matrix composites usually entails temperatures in excess of 1000°C. In addition, ceramic matrix composites are envisioned for applications in which temperatures exceed those tolerable by structural metals or polymer composites. For these reasons, stabilities of reinforcing fibers at elevated temperatures are of considerable importance.

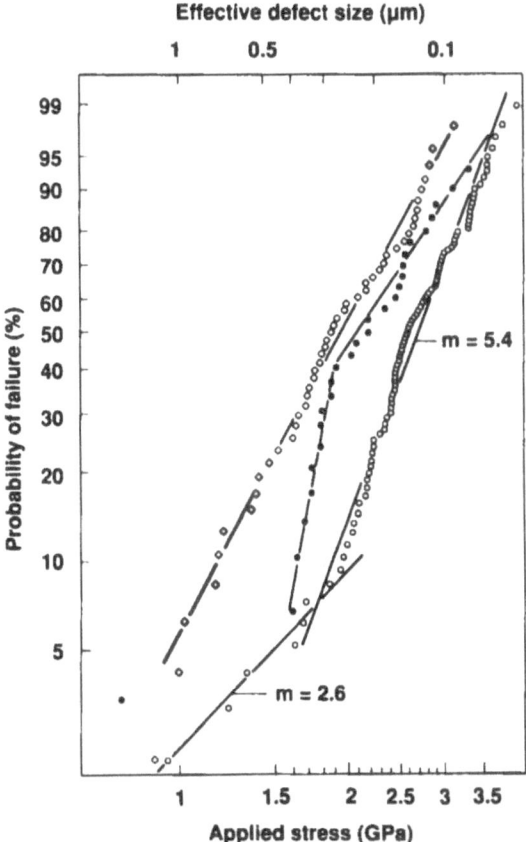

FIG. 7 Two-parameter Weibull plots for Nicalon™ fibers. (Open circles indicate 1-cm gage section, fibers carefully handled prior to testing; closed circles indicate 1-cm gage section, fibers damaged during size removal; diamonds indicate 10-cm gage section, fibers carefully handled.) (From Ref. 10.)

Strengths of Nicalon fibers measured at high temperatures or measured at room temperature after heat treatment at elevated temperatures have been reported by a number of investigators. For example, ceramic grade Nicalon fiber tensile strength at room temperature is shown in Figure 8 as a function of heat treatment time and temperature in air and in an inert atmosphere.[7] At 800°C, strength was not reduced in argon, even after 100 h. At 1000°C, strength reduction after 100 h was minimal. At 1200°C, strength was reduced

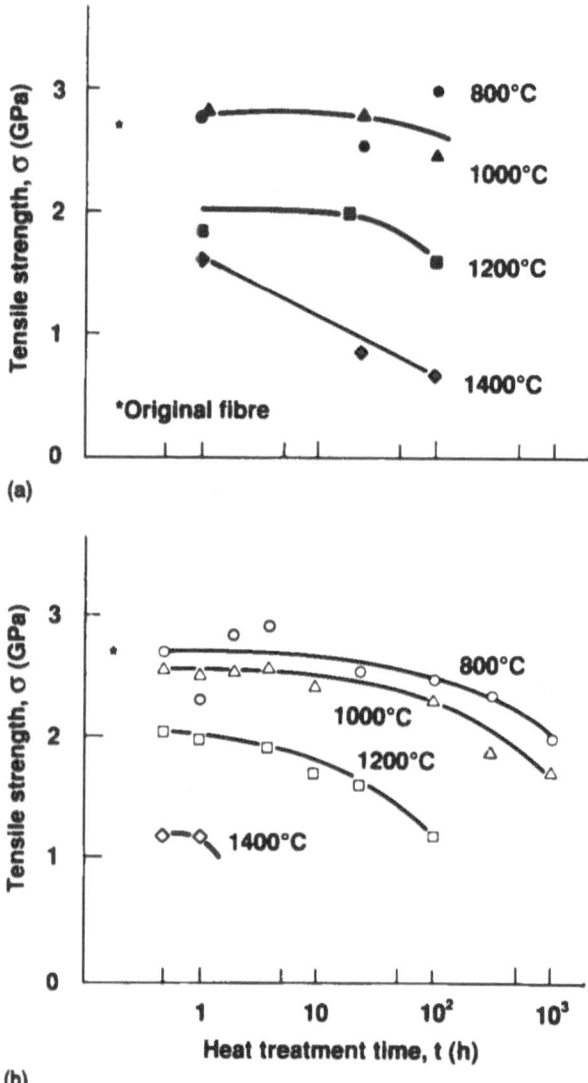

FIG. 8 Room temperature tensile strengths of Nicalon™ fibers after various heat treatments in argon (a) and air (b). From Ref. 7.

considerably after only 1 h, although there was not much additional reduction even after 100 h. At 1400°C, strength decreased considerably with increasing heat treatment time.

In air, only minor strength reductions occurred at 800° or 1000°C in up to 100 h, but deterioration became more significant at times longer than 100 h. At 1200°C, strength was considerably reduced in 1 h, and further decreased with increasing heat treatment time. Strength was reduced to about 40% of the room temperature strength after 0.5 h at 1400°C.

Clark et al [12] determined tensile strengths of both standard grade and ceramic grade Nicalon fibers after heat treatments at 1000°, 1200°, and 1400°C in wet air and in argon for 12 h. In some contrast to the work cited above, strength was reduced appreciably, from about 2.4 GPa to 1.4 GPa, after treatment at 1000°C. (As pointed out earlier, differences in both fibers and test details may contribute to such differences in reported test values.) Strength was reduced to about 0.5 GPa after treatment at 1400°C. Standard grade Nicalon fibers had a lower room temperature strength, about 2.0 GPa and deteriorated with increasing heat treatment temperature to a value of 0.4 GPa.

In wet air, strengths of both types of fibers decreased appreciably with increasing heat treatment temperature. However, the ceramic grade fibers maintained a strength advantage at all temperatures.

Pysher et al [16] determined tensile strengths and elastic moduli of ceramic grade Nicalon at elevated temperatures in air after a short time. The original room temperature strength of close to 2.0 GPa was retained at 800°C, but decreased to about 1.5 GPa at 1000°-1200°C, and to about 1.0 GPa at 1250° and 1300°C. The original Young's modulus of about 190 GPa decreased only slightly, to about 175 GPa at 800°C, and then decreased almost linearly with increasing temperature to a value of about 100 GPa at 1300°C.

Andersson and Warren also reported a decrease in the Weibull modulus, from about 4.0 to 2.5, on heating Nicalon fibers at 1200°C or higher.

Thus, although reported results are not completely consistent, it is obvious that a considerable decrease in strength and modulus of Nicalon fibers occurs when tested or heat treated at 1200°C or higher, and in some studies, at 1000°C. Tensile strength is retained somewhat better in an inert atmosphere than in air.

f. Mechanisms for Nicalon Fiber Degradation at Elevated Temperatures

Studies have been conducted on determining mechanisms for the reduction in strength of Nicalon fibers at elevated temperatures. Elemental analysis, optical microscopy, scanning electron microscopy, X-ray diffraction, thermal gravimetric analysis, mass spectrometry, and other analytical techniques have been used. As with mechanical property determinations, there are some inconsistencies in the findings. However, major phenomena are well established.

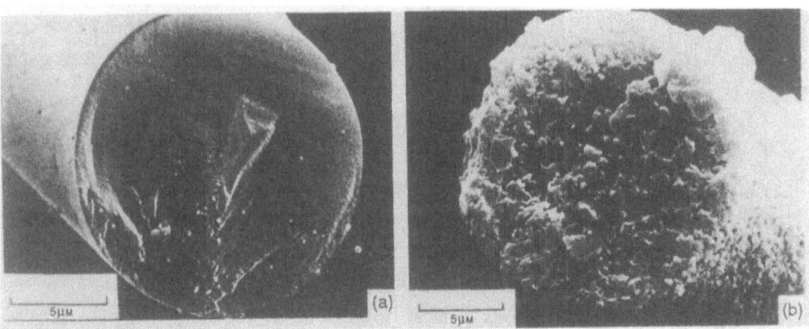

FIG. 9 Scanning electron micrographs of fracture surfaces of Nicalon™ fibers heat treated under conditions not resulting in appreciable grain growth (a), and under conditions producing extensive grain growth (b). (From Ref. 17.)

Many investigators have found that the fibers convert from a largely amorphous or microcrystalline material to a more obviously crystalline material, and that grain growth occurs. Grain growth is not unexpected, since this is a common phenomenon for ceramic materials subjected to high temperatures. Figure 9 [17] shows a fracture surface for a ceramic grade Nicalon fiber that has been heat treated under conditions not leading to appreciable grain growth (and thus resembling an as-produced fiber) and a fracture surface for a fiber that has undergone extensive grain growth.

It has also been generally found that both carbon monoxide (CO) and silicon monoxide (SiO) are evolved during heat treatment. Thermodynamic calculations [18] confirm that fibers containing silicon, oxygen, and carbon are inherently unstable at elevated temperatures and change in composition with time. Carbon monoxide can derive from reaction between silicon dioxide and carbon. In air, direct reaction of free carbon at the fiber surface with oxygen is a possibility. Silicon monoxide and carbon monoxide can both be produced by reaction of silicon carbide with silicon dioxide. It is believed that the evolution of carbon monoxide and silicon monoxide causes voids, which serve as flaws, and contributes to enlargement of existing surface flaws, thus reducing tensile strength.

Clark et al [12] found differences between standard grade and ceramic grade Nicalon fibers. The lower-oxygen-content ceramic grade fibers were more stable. On heating in argon, the standard grade fibers underwent extensive growth of β-silicon carbide grains, while the ceramic grade showed microstructural and compositional stability and remained finer grained. Both types reacted with oxygen when heated in air, to form an outer silicon dioxide layer. Cracking and spalling of this layer can contribute to strength reduction.

g. High Temperature Stability of Nicalon Fibers in Elevated and Reduced Pressures and in Other Environments

Determinations of strength retention and structural changes at elevated and reduced pressures and in atmospheres other than air and argon have proven to be useful. Jaskowiak and DiCarlo [17] heat treated standard and ceramic grade Nicalon fibers at temperatures ranging from 1000° to 2200°C under vacuum and under argon gas pressures of 0.1 and 138 MPa. A shift of extensive weight loss to a higher temperature under the high argon pressure, relative to the low argon pressure or vacuum, suggested that the high pressure suppressed evolution of carbon monoxide and silicon monoxide (Figure 10). Grain growth also followed the same trend as weight loss, i.e. accelerated grain growth was shifted to a higher temperature under high pressure.

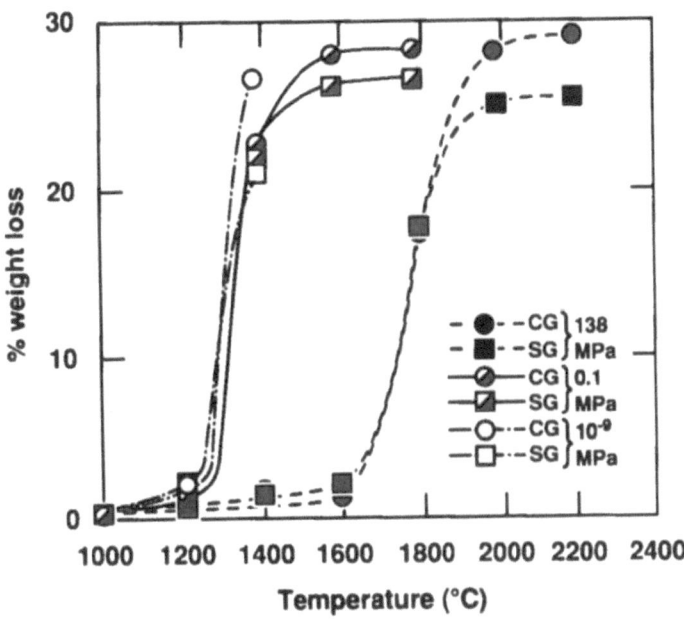

FIG. 10 Weight loss as a function of heat treatment temperature for Nicalon™ fibers in various atmospheres. (CG indicates ceramic grade fibers; SG indicates standard grade fibers.) (From Ref. 17.)

Room temperature tensile strength results (Figure 11) showed better strength retention with high argon pressure, intermediate strength retention with low argon pressure, and poorest retention with vacuum. Although both grain growth and gas evolution accompanied strength reduction, a comparison of weight loss, grain growth, and tensile strength results indicated that tensile strength reduction accompanied extensive weight loss, but preceded grain growth. Hence, it was inferred that flaw generation or enlargement due to gas evolution, and not flaw generation during grain growth, is the major factor causing decreased strength of Nicalon fibers at elevated temperatures. Microstructural analysis supported the hypothesis of growth of existing surface flaws due to gas evolution.

Although contributing to understanding of mechanisms of Nicalon fiber degradation at elevated temperatures, the fibers treated under high argon pressure were not permanently improved. Fibers initially treated at 1200° and 1600°C under a 138 MPa argon atmosphere were subsequently heat treated at 1400°C in 0.1 MPa of argon. These fibers had the same weight loss and low strength as the as-received fibers treated at 1400°C in 0.1 MPa of argon.

Bender et al [19] compared room temperature tensile strengths of ceramic grade Nicalon fibers after heat treatment in nitrogen or argon and after heat treatment in nitrogen with the fibers packed in carbon powder. Strengths were appreciably higher for the fibers packed in carbon powder. For example, with a heat treatment temperature of 1600°C, fibers treated in nitrogen or argon retained only 5% of their original strength, whereas fibers packed in carbon and treated in nitrogen retained about 30%.

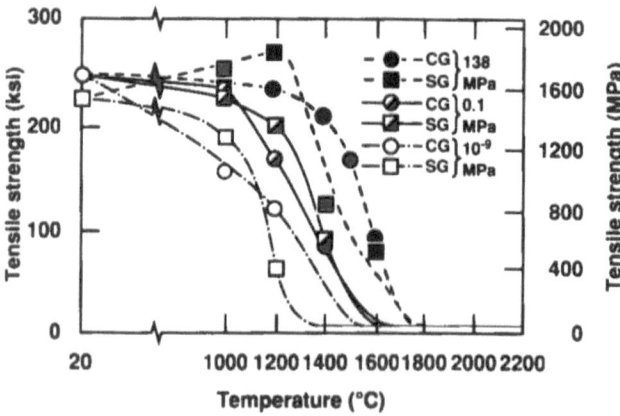

FIG. 11 Room temperature tensile strength as a function of heat treatment temperature for Nicalon™ fibers in various atmospheres. (CG indicates ceramic grade fibers; SG indicates standard grade fibers.) (From Ref. 17.)

The fibers heat treated in argon developed a fine porous, pinholed microstructure, with individual grains about 50 nm in diameter and surface pores up to 1.0 µm in diameter. Fibers heat treated in nitrogen had growths of silicon nitride whiskers on their surfaces. On the other hand, fibers packed in carbon and treated in nitrogen had a dense reaction layer about 30 nm thick on the surface. Electron microscopy indicated that the layer was α-silicon carbide. This layer, which apparently formed by reaction of evolving silicon monoxide with carbon powder, prevented further massive gas evolution.

Heat treatment in carbon monoxide had an even more favorable effect on retained room temperature tensile strength. Room temperature strengths of fibers heat treated in carbon monoxide were at least 30% higher than those of fibers treated in nitrogen at temperatures ranging from 1000° to 1600°C. Room temperature strengths of fibers treated at 1600°C averaged about 1.4 GPa, compared with near zero strength for fibers treated in nitrogen.

Various analyses indicated that the original fiber composition was essentially maintained during heat treatment in carbon monoxide, although grain size increased (from about 2 nm in the original fibers to 5-10 nm at 1600°C). Apparently, the carbon monoxide partial pressure during heat treatment was greater than the equilibrium carbon monoxide partial pressure for the reaction of carbon and oxygen in the fibers. As a result, pores that can act as flaw sites were not generated.

This work supported the finding of Jaskowiak and DiCarlo that gas evolution rather than grain growth is most responsible for high temperature deterioration of Nicalon fibers. Whether the fibers resulting from the heat treatments of Bender et al were permanently improved with respect to further heat treatment in air or inert atmosphere was not reported.

The results described in this section indicate that fiber properties at room temperature after heat treatment in air or an inert atmosphere, or properties at elevated temperatures in air or an inert atmosphere, might be quite different than properties at elevated temperatures in a composite, where the fibers can be exposed to a more complex gaseous environment, solid phases, and, perhaps in some cases, liquid phases.

2. TYRANNO™ FIBERS

a. Technical Basis for Tyranno Fiber Development

Development of continuous fibers containing silicon, titanium, carbon, and oxygen have been reported by Yamamura et al.[20] These fibers, having the trade name Tyranno, are produced by Ube Industries Ltd. in Japan. Tyranno fibers are in many respects similar to Nicalon, but are claimed to have superior high temperature performance.

Polytitanocarbosilane is produced by heating a mixture of polydimethylsilane and polyborodiphenylsiloxane (which serves as a catalyst) with a titanium alkoxide at about 340°C in nitrogen. In this process, Si-Si bonds in polydimethylsilane are cleaved and Si-H and Si-CH$_2$-Si bonds are formed. Simultaneously, condensation of Si-H bonds and cross-linking by the titanium compound occur. A curing process in air, at about 180°C, is carried out after melt spinning of the fibers. The cured fibers are then heated at about 1300°C to convert the polymer to a ceramic material. Fiber strength and Young's modulus maximize at this temperature. Fiber diameters are 8-12 μm.

b. Mechanical Properties of Tyranno Fibers

Typical room temperature tensile strength and Young's modulus are fairly similar to those for Nicalon fibers. Although some work has indicated that Tyranno fibers have better heat resistance than other fibers, due to grain growth inhibition by the addition of titanium, improvement with respect to Nicalon fibers has not been demonstrated in more recent work. For example, Bender et al [19] reported on strength retention of Nicalon and Tyranno fibers at room temperature after heating in nitrogen for 3 h. Original strength for Nicalon fibers averaged close to 4.0 GPa and strength for Tyranno fibers averaged about 3.0 GPa. Slight reductions in strength were found for both types of fibers after treatment at 800°C. After treatment at 1100°C, strengths of both types of fibers were reduced to approximately the same level, about 2.0 GPa. Nicalon fibers retained this strength after treatment at up to 1500°C, before falling to near zero strength at 1600°C. On the other hand, strengths of Tyranno fibers continued to fall after treatments between 1200° and 1500°C, to a value near zero after treatment at 1500°C.

Strength retention of Tyranno fibers after heat treatment in nitrogen was improved moderately when the fibers were packed in carbon during the heat treatment, for temperatures between 800° and 1600°C. Treatment in a carbon monoxide atmosphere resulted in little strength reduction (and strength increase, in some cases) for heat treatment temperatures up to about 1400°C. A thin titanium carbide layer was found on the fibers after high temperature treatment in carbon monoxide. Whether the subsequent high temperature stability of Tyranno fibers in air or inert gas was improved by these treatments was not reported.

Testing at elevated temperatures in air by Pysher et al [16] also showed Nicalon fibers to have superior high temperature stability to Tyranno fibers. In contrast to the work reported above, Tyranno fibers were found to have a higher tensile strength before heating, about 2.5 GPa for Tyranno fibers and slightly less than 2.0 GPa for Nicalon fibers. Strengths of both types of fibers were reduced but were nearly equal to one another at temperatures between 800° and 1200°C. Values at 1200°C were about 1.5 GPa. At temperatures of 1300° and 1400°C, the Nicalon fibers had higher strengths. Strength at 1400°C was about 1.0 GPa for Nicalon fibers and less than 0.5 GPa for Tyranno fibers.

3. HPZ FIBERS

a. Technical Basis for HPZ Fiber Development

Another fiber more recently offered in commercial quantities, by Dow Corning Corporation, is HPZ fiber.[21] This fiber is amorphous with a typical composition of 57% silicon, 28% nitrogen, 10% carbon, and 4 % oxygen. HPZ fibers are produced from the polymer, hydridopolysilazane. Work on this fiber and some related fibers to be described below was supported by the Defense Advanced Research Projects Agency (DARPA) to provide improved polymer-derived silicon carbide or silicon nitride base fibers. Celanese Corporation was also involved in the development of these fibers. Details of the commercial process have not been given, but laboratory work on HPZ fibers is outlined below.

Polysilazanes are organosilicon polymers containing Si-N bonds. Hydridopolysilazane has been produced by the reaction of trichlorosilane, $HSiCl_3$, with hexamethyldisilazane, $[(CH_3)_3Si]_2NH$, in a series of steps with a maximum reaction temperature of 200°C or higher.[22] The product was a clear, colorless solid. A multitude of cyclization and branching alternatives during polymer formation prevented identification of individual molecular structures in the growing polymer. Based on elemental analysis, nuclear magnetic resonance, and chromatography data, a calculated empirical formula based on types of bonds was $[Si-H]_{39.7} [(CH_3)_3Si]_{24.2}[N-H]_{37.3}[N]_{22.6}$. Yield of product was typically greater than 95% of theoretical, based upon the starting trichlorosilane.

Fibers were produced from the molten polymer by spinning in an inert atmosphere and cured by exposure to trichlorosilane vapor at 70°C to cross-link the polymeric chains. Pyrolysis was carried out at 1200°C in nitrogen. A number of species were evolved as temperature was raised. Above 700°C, only methane and hydrogen were evolved. During pyrolysis, 25-30% weight was lost and density increased from about 1.0 to 2.3 g/cm³. Typical composition, by weight, was 60.0% silicon, 32.6% nitrogen, 2.3% carbon, and 2.2% oxygen. These carbon and oxygen contents differ substantially from those quoted for the commercially available fibers, probably reflecting changes required for practical process scale-up. Fiber diameters were 15-20 μm.

The structure of HPZ fibers is amorphous by X-ray diffraction. Studies by a variety of analytical techniques were all consistent with random bonding of silicon to carbon, nitrogen, and oxygen. Excess carbon is present in a microcrystalline graphitic structure very similar to that of pyrolytic carbon. The fibers are not fully dense, although no porosity is found using usual techniques.

b. Mechanical Properties of HPZ Fibers

Tensile strengths of up to 3.1 GPa and Young's moduli of up to 260 GPa were reported for laboratory-produced fibers. These are higher than those reported for the commercial fibers (1.9-2.6 GPa tensile strength; 180-235 GPa Young's modulus).[21] The commercial fibers are available with 500 or 1000 filaments per tow. Diameter is 10-12 µm.

High temperature properties or retained room temperature properties after high temperature heat treatment for HPZ fibers have apparently not yet been published in the open literature. The fact that much of the developmental work was sponsored by a defense-related agency (DARPA) has probably slowed publication of results in the open literature.

4. MPS FIBERS

As part of the DARPA-sponsored program leading to the development of the HPZ fibers, MPS fibers (named for their polymeric precursor, methylpolysilane) have been produced also.[23] Methylpolysilanes were produced from a mixture of methylchlorodisilanes. This polymer was melt spun, cross-linked (cured) and pyrolyzed in an inert atmosphere. By variations in the reagents, fibers ranging from carbon-rich through stoichiometric have been produced. Elemental compositions (by weight) have ranged from 52-75% silicon, 24-47% carbon, 0-0.1% nitrogen, 0.5-6.0% oxygen, and less than 0.1% hydrogen. A calculated composition of one fiber lot that was extensively characterized is 81 wt% silicon carbide, 12 wt% silica, and 7 wt% carbon.

The predominant band in infrared absorption spectra of MPS fibers is Si-C. There is also an indication of Si-O bonds and some residual C-H bonds. Other analytical techniques also indicate that silicon carbide is the major component and suggest some silicon oxycarbide species and, perhaps, C-Si-O species. X-ray diffraction indicates the presence of crystalline silicon carbide having a crystallite size of about 2 nm.

Typical tensile strength values were 1.7 GPa and Young's modulus values were 210 GPa for fiber diameters of 10-20 µm. Individual fibers had strengths as high as 2.8 GPa and Young's moduli of 240 GPa. Weight loss on heating in helium to over 1500°C was only 3%, compared to a loss of 18% with ceramic grade Nicalon fibers under the same conditions. Although this suggests improved high temperature mechanical properties for the MPS fibers, high temperature properties have apparently not been reported yet in the open literature.

5. MPDZ FIBERS

Another fiber from the DARPA-sponsored work is MPDZ (named for its precursor, methylpolydisilazane). This fiber has a typical nitrogen content between those of its companion fibers, HPZ and MPS. A typical composition, by weight, has been given as 47% silicon, 30% carbon, 15% nitrogen, and 3% oxygen.[24] Tensile strength range is reported as 18-21 GPa and Young's modulus as 175-210 GPa. Density is 2.3 g/cm^3 and fiber diameter is 10-15 µm.

6. OTHER NON-OXIDE FIBERS FROM POLYMERIC PRECURSORS

Other types of fibers derived from polymer precursors have mostly been produced in relatively small quantities, often by hand drawing from melts or solutions. The previously cited paper by Okamura [7] gives a general treatment of silicon carbide base ceramic fibers.

B. Silicon Carbide Fibers by Chemical Vapor Deposition

Another type of silicon carbide base fiber has been produced on a commercial scale using a totally different process, chemical vapor deposition (CVD). The CVD process involves deposition of a solid material onto a heated substrate from a gaseous precursor. In producing fibers by this method, a resistively heated core fiber serves as the substrate. The process is continuous, utilizing tube reactors with mercury seals. CVD has been used for producing boron fibers on a tungsten core for about 30 years, and the experience was utilized in developing a process for silicon carbide. (Since boron fibers are not deemed attractive for ceramic reinforcement, they will not be discussed.)

1. SCS FIBERS

a. Manufacturing Process

CVD-produced silicon carbide fibers, trade named SCS fibers, have been available for over 10 years from AVCO Specialty Products Division (now Textron Specialty Materials) in Lowell, MA. The SCS designation is followed by a numerical suffix, which designates the type of surface coating.

The fiber manufacturing process is very similar to a boron fiber process developed at AVCO.[25] For the silicon carbide fibers, a carbon filament having a diameter of 33 μm serves as the core material. A pyrolytic carbon coating about 1 μm in thickness is first deposited onto the core fiber to improve the surface for silicon carbide deposition. The core fiber is fed into a glass reactor through a mercury contact, which provides a gas-tight seal and facilitates electrical resistance heating of the fibers. Hydrogen, argon, and silane vapors are fed into the reactor at about 1300°C. Figure 12 is a schematic diagram of a process for SCS fiber production by CVD. Figure 13 shows a series of reactors of the type used for SCS fiber production.

Columnar grains of β-silicon carbide, oriented radially, are deposited onto the core fiber. With no fiber coating, tensile strength is about 2.1 GPa. The relatively low strength is due to exposed grain boundaries acting as stress concentrators. A thin coating of carbon deposited in the reactor during the final stage of fiber production to seal over the grain boundaries can nearly double the strength.

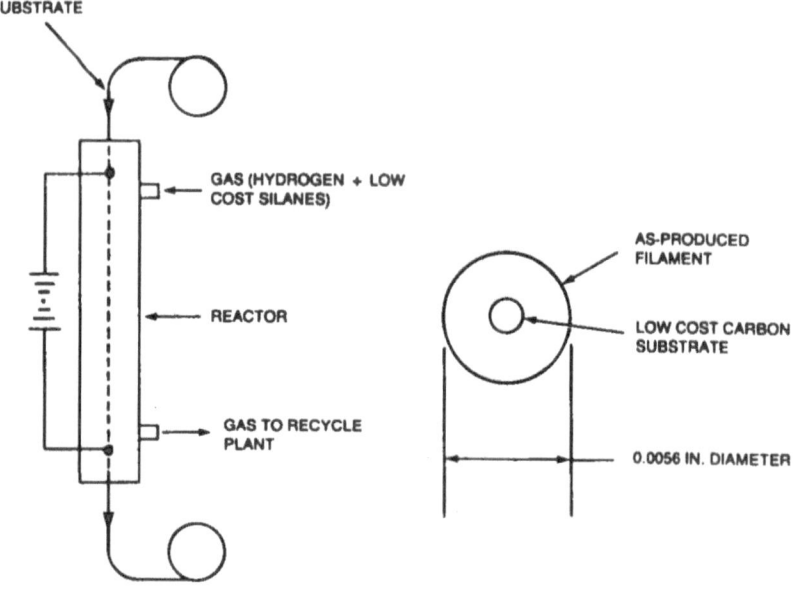

FIG. 12   Schematic illustration of the production process for SCS fibers. (Courtesy of Avco Specialty Materials/Textron.)

For the commercial fibers, more complex coatings of carbon and graded mixtures of silicon carbide and carbon are deposited onto the fiber, depending on the intended use. The fiber grade developed for ceramic reinforcement, termed SCS-6, has the structure shown in Figure 14.[26] This coating consists of a graded silicon carbide inner layer with an increasingly greater carbon content with increasing thickness. Eventually, a pure carbon layer of less than 0.5 μm is deposited. Finally, a graded coating with an increasing silicon carbide content is deposited. Total fiber diameter is 143 μm. Density is 3.0 g/cm³.

b. Mechanical Properties of SCS Fibers at Room Temperature

As with polymer derived fibers, mechanical property values reported for these CVD-produced fibers can differ, depending on both test method and material variation. For example, Foltz [25] reported a Young's modulus range of 400-414 GPa for SCS fibers (without identification of coating type). The tensile strength information is not clear. In the table in which the Young's modulus was reported, tensile strength range is listed as 8.4-9.1 MPa (which obviously

FIG. 13 Reactors of the type used for SCS fiber production. (Courtesy of Avco Specialty Materials/Textron.)

was a typographical error, with GPa intended). However, a figure in the same paper, showing range and average for one month's production of SCS-2 fibers, shows a strength range from about 2.0 GPa to 5.5 GPa, with a mean of about 4.0 GPa. Le Petitcorps et al [27] measured tensile strengths for two variations of SCS-6 fibers in which the outer coating differed, reporting average tensile strengths of 3.6 and 4.7 GPa and Young's moduli of 360 and 390 GPa for fibers designated SCS-6(O) and SCS-6(N), respectively. Weibull moduli were 12.5 and 16.0 for the two types.

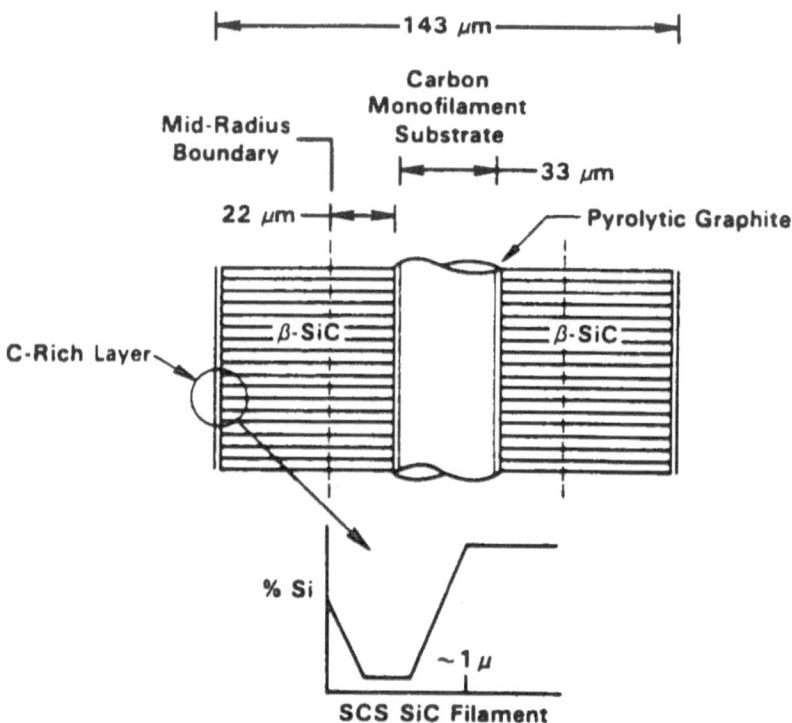

Fig. 14   Schematic illustration of the composition of SCS-6 fibers. (From Ref. 26.)

c. Properties of SCS Fibers at Elevated Temperatures or after Exposure to Elevated Temperatures

Foltz [25] reported a tensile strength of 1.0 GPa for short term tests at 1400°C in either air or argon and a strength of 3.1 GPa at room temperature after a 15-min heat treatment in air (citing R. Bhatt of NASA Lewis for the latter measurement). Ahmad et al [28] reported a short term strength of 2.0 GPa at 1093°C, 1.9 GPa at 1204°C, and 1.6 GPa at 1316°C.

The strengths at elevated temperatures or after elevated temperature exposures are higher than those of polymer derived fibers. This is reasonable, since the bulk of the SCS fiber is nominally pure silicon carbide, whereas the polymer derived fibers have a mixture of phases. However, a factor that will obviously contribute to degradation of SCS fibers during long term heat treatment in air is the oxidation susceptibility of both the carbon core and the carbon-rich coating.

2. SIGMA FIBERS

Another type of silicon carbide base fiber produced by CVD onto a core fiber is the Sigma fiber, produced by Berghof, in Germany. References to this material are as yet relatively sparse. Le Petitcorps et al [27] reported that this fiber is 100 μm in diameter, on a 13-μm tungsten core, with a mean tensile strength of 3.3 GPa and a mean Young's modulus of 390 GPa. Ko [29] reported a density of 3.4 g/cm^3, a room temperature tensile strength of 3.45 GPa, and a room temperature Young's modulus of 410 GPa.

C. Oxide Fibers

1. BACKGROUND

Oxide fibers would appear to be better candidates for reinforcement of ceramic matrix composites for high temperature applications in air than the fibers already discussed because of inherent oxidation resistance of oxide fibers (provided that the cations are in their highest oxidation state). However, strengths at high temperatures or after high temperature heat treatment of existing oxide fibers are degraded by other mechanisms, so that their potential has not yet been fully realized.

A number of oxide fibers have been available commercially for 10 years or longer. Commercially available oxide fibers of promise for high performance ceramic reinforcement are based on aluminum oxide (alumina), ranging from about 60 wt% alumina to greater than 99%.

2. FIBER FP

a. Manufacturing Process

A continuous, multi-filament polycrystalline α-alumina fiber known as Fiber FP had been produced since about 1970 by E. I. du Pont de Nemours & Co. Inc. In 1991, Du Pont ceased production of these fibers. The following description is, then, primarily of historical interest, although there are undoubtedly aspects of the technology that can be utilized in the future.

This fiber was made by spinning of an aqueous base slurry of alumina particles along with organic spinning additives.[30] (Slurry processing of ceramics will be described in more detail in a later chapter dealing with whisker reinforced ceramics.) The firing process was carried out in two steps, with the second step utilizing a flame to convert it to a dense α-alumina fiber. Mean fiber diameter is about 20 μm. Density is 3.9 g/cm^3.

b. Properties of FP Fibers

Fiber FP has a grain size of about 0.5 μm and a rough surface. For some purposes, Fiber FP was coated with a layer of silica. The coated version has a smoother surface and has a strength about 50% higher than uncoated Fiber FP. Presumably, this is due to reduction of surface flaws. Fiber FP is reported to have a room temperature tensile strength of 1.4 GPa, a Young's modulus of 380 GPa, and an elongation to breakage of 0.6%. These data were based on a 0.64-cm gage length; values did not change appreciably as gage length was increased to over 20 cm.

FP grade fibers held for two hours at elevated temperatures and tested at room temperature were found to retain their original tensile strength with heat treatment at 800°C, to be degraded only slightly after heat treatment at 1000°C, and to have a tensile strength of 1.0 GPa after heat treatment at 1300°C. Longer heat treatment times at 1300°C did not result in further degradation, relative to the 2-h heat treatment. The major change in fiber microstructure upon high temperature heat treatment is grain growth, which is probably accompanied by generation of surface and internal flaws.

Another group testing FP grade fibers at elevated temperatures measured room temperature strength at about 1.0 GPa [16], yet another example of different values determined for the same fibers due to differences in testing methods and, perhaps, different fiber lots. Strength was not reduced at 800°C, but decreased at temperatures above 800°C to a value of about 0.4 GPa at 1200°C. Young's modulus was measured as 300 GPa at room temperature. Modulus decreased to about 190 GPa at 1200°C.

3. PRD-166 Fibers

Another oxide fiber, trade named PRD-166, was also produced by DuPont. Production of it too was discontinued in 1991. This fiber is composed of polycrystalline α-alumina with about 20 wt% partially stabilized zirconia.[30] Yttria or other rare earth oxides were used to stabilize the tetragonal zirconia phase. The fabrication process was similar to that for Fiber FP. Room temperature strength is improved about 50% over Fiber FP, perhaps due to transformation toughening. Fiber diameter is about 20 μm. Density is 3.9 g/cm³. By scanning electron microscopy coupled with energy dispersive X-ray analysis, alumina grain size is about 0.5 μm, while zirconia grain size tends to be smaller, down to about 0.1 μm.

PRD-166 fibers were reported to have a room temperature tensile strength of 2.1 GPa, a Young's modulus of 380 GPa, and an elongation to breakage of 0.6%. These values were for a 0.64-mm gage length. Tensile strength decreased appreciably with increasing gage length (in contrast to Fiber FP). This was attributed to incorporation of flaws during the fiber-making process, which was less mature than the process for Fiber FP. Whether processing conditions improved in the intervening years (since 1985) is problematic.

After heat treatment at elevated temperatures and testing at room temperature, PRD-166 fibers retained their original strength after treatment at up to

1000°C. Tensile strength decreased only slightly after heat treatment at 1200°C, and it decreased to about 1.7 GPa after treatment at 1400°C.

Another group measuring strengths at elevated temperatures reported a lower room temperature tensile strength for PRD-166, 1.3 GPa.[16] Strength was about 1.2 GPa at 800°C, 0.9 GPa at 1000°C, and 0.5 GPa at 1200°C. Young's modulus was 280 GPa at room temperature, 300 GPa at 800°C, 220 GPa at 1000°C, and 150 GPa at 1200°C.

4. NEXTEL™ FIBERS

a. Manufacture and Characteristics of Nextel Fibers

Several types of fibers, trade named Nextel fibers, are produced by 3M Company. The fibers that have been used in ceramic reinforcement are trade named Nextel 320, Nextel 440, and Nextel 480. These fibers all consist of alumina, silica, and boric oxide, and are produced by a sol-gel method.[31] In the sol-gel process appropriate compounds are mixed to form a homogeneous solution (sol) that is then caused to gel by heat or catalysis. Further heat treatment converts the gel to a glass or ceramic. (The sol-gel process will be described in more detail in a later chapter on matrix formation.) As for most of the fibers previously described, the fibers are spun through multi-hole spinnerets into multi-filament tows.

Nextel fibers are oval in cross section, with the major axis up to twice the dimension of the minor axis. Diameter is reported by 3M as 10-12 μm. Presumably, this refers to the equivalent diameter of a circular fiber with the same cross-sectional area. Typically, there are about 750 fibers per tow. Nextel 312, the first of the three fibers developed, has a composition of 62% (by weight) alumina, 24% silica, and 14% boric oxide.[32] Density is about 2.8 g/cm^3. The major phases present are aluminum borate (9 Al$_2$O$_3$·2 SiO$_2$) and amorphous silica. Crystal size is reported to be less than 500 nm.

Nextel 440 and Nextel 480 both have compositions of 70% (by weight) alumina, 28% silica, and 2% boric oxide. Mullite (3 Al$_2$O$_3$·2 SiO$_2$), γ-alumina, and amorphous silica are the major phases in Nextel 440. The Nextel 480 fibers are more extensively crystallized during production, so that the only major phase is mullite. Both have densities of 3.1 g/cm^3.

b. Mechanical Properties

Room temperature tensile strengths are reported to be 1.7 GPa for Nextel 312 fibers, 2.0 GPa for Nextel 440 fibers, and 1.9 GPa for Nextel 480 fibers.[31] Corresponding Young's moduli are 150 GPa, 190 GPa, and 220 GPa.

Tensile strengths measured at room temperature after heat treatment at elevated temperatures in air, hydrogen, and vacuum are shown in Figure 15

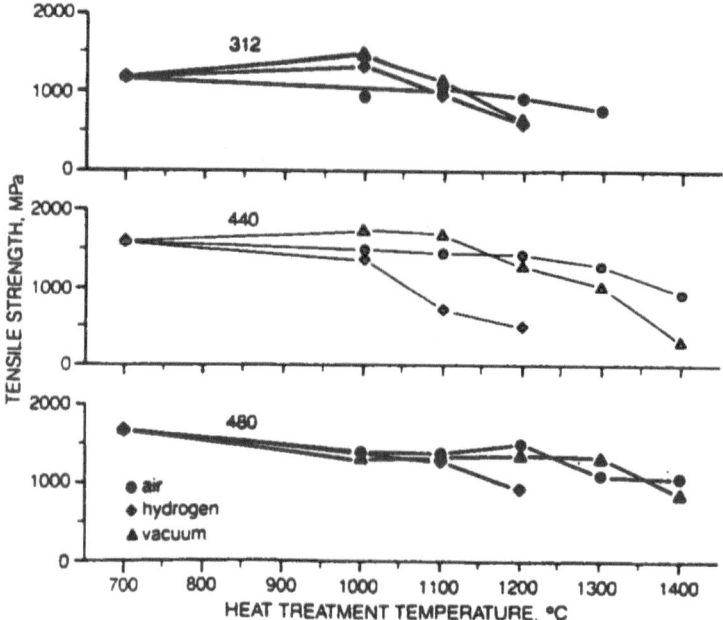

FIG. 15   Room temperature tensile strengths of Nextel™ fibers heat treated in several atmospheres as a function of heat treatment temperature. (From Ref. 32.)

[32]. Although there are apparent differences due to atmosphere, only the larger differences shown may be significant, since the authors reported a +/-20% uncertainty range at the 95% confidence level.

With all fibers, values after heat treatment at 700°C were lower than those shown for the fibers not heat treated. For Nextel 312 fibers, there were some indications of increased strengths after treatment at 1000°C, relative to those after treatment at 700°C. For temperatures above 1000°C, strength was reduced the most in a hydrogen atmosphere. Nextel 480 fibers had the best strength retention after high temperature treatment.

Figure 16 [32] shows tensile strength during short term heating at elevated temperatures in air. Nextel 312 fiber strength decreased appreciably at temperatures above 800°C. Nextel 440 fibers lost little strength at up to 1000°C. Nextel 480 fibers began to lose strength at 800°C, but strength retention at higher temperatures was better than that of Nextel 440 fibers. This is apparently due

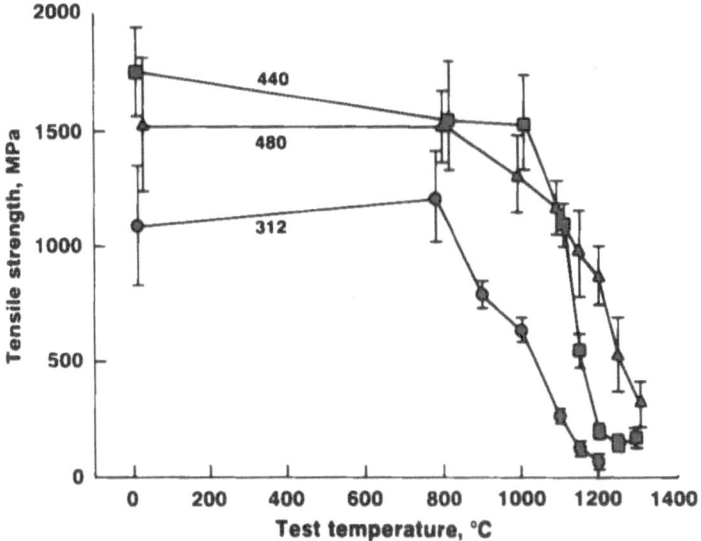

FIG. 16   Tensile strength as a function of test temperature for Nextel™ fibers. (From Ref. 32.)

to the more crystalline nature of Nextel 480 fibers, relative to Nextel 440 fibers, since plastic deformation began to be observed at the higher temperatures with Nextel 440 fibers, but not with Nextel 480 fibers.

Young's modulus values during short term tests at elevated temperatures are shown in Figure 17. The Nextel 480 fibers had the best modulus values at all temperatures, although modulus began to decrease at 1000°C.

Based on creep measurements on the three fibers at two stress loads, Nextel 480 fibers have the best creep resistance, followed by Nextel 440 fibers, and Nextel 312 fibers. The amount of creep permissible depends upon the application. If a limit of 1% per 100 h at one of the stress levels used, 69 MPa, is considered tolerable, the Nextel 480 fiber could be used to about 1190°C.

As for other fibers, different investigators have reported somewhat different mechanical properties. Pysher et al [16] reported a room temperature tensile strength of 1.5 GPa for Nextel 480 fiber, and a reduction to 1.2 GPa at 800°C. Room temperature Young's modulus was reported as 190 GPa, and modulus decreased to 125 GPa at 800°C.

5.   OTHER OXIDE FIBERS

Although a number of other oxide fibers are available, many have mechanical properties significantly inferior to those reported above, have poorer high tem-

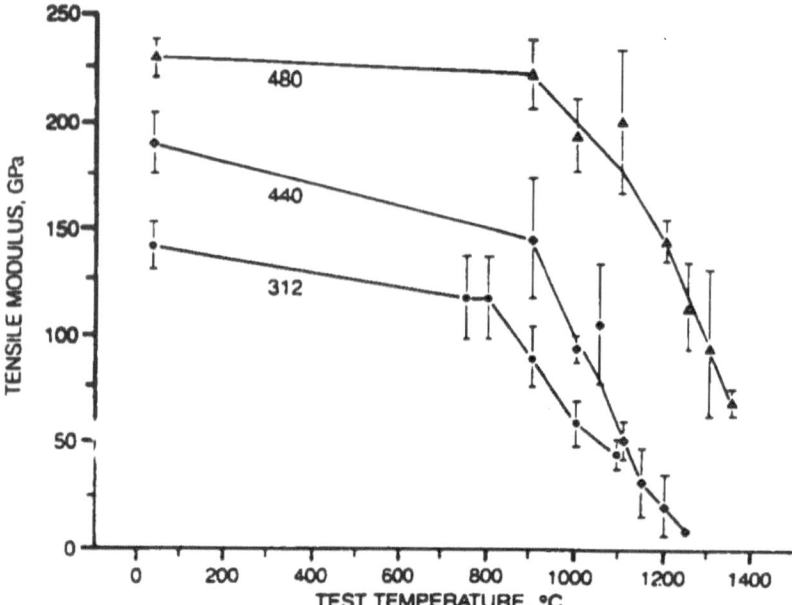

FIG. 17 Young's modulus as a function of test temperature for Nextel™ fibers. (From Ref. 32.)

perature stability, or are not available in continuous form. One exception is the Sumitomo fiber, which is about 85% (by weight) alumina and 15% silica.[31] This fiber has an average diameter of 17 μm and a density of 3.2 g/cm³. Room temperature tensile strength is reported as 1.8 GPa and Young's modulus as 210 GPa. Sumitomo fibers have apparently been investigated for use in ceramic reinforcement to a lesser extent than the other fibers described.

While not appreciably investigated for use in ceramic matrix composites, one other type of fiber will be mentioned for completeness. Single crystal sapphire alumina fibers (trade named EFG sapphire fibers) are grown from molten alumina by Saphikon, Inc., Milford, NH. Fiber growth is seeded by touching the liquid film atop a molybdenum die with a c-axis single crystal alumina fiber, and slowly pulling the fiber.

D. Comparative Mechanical Testing of Continuous Ceramic Fibers

As shown throughout this chapter, most of the results on mechanical properties of fibers reported by different investigators do not agree closely. Some of this

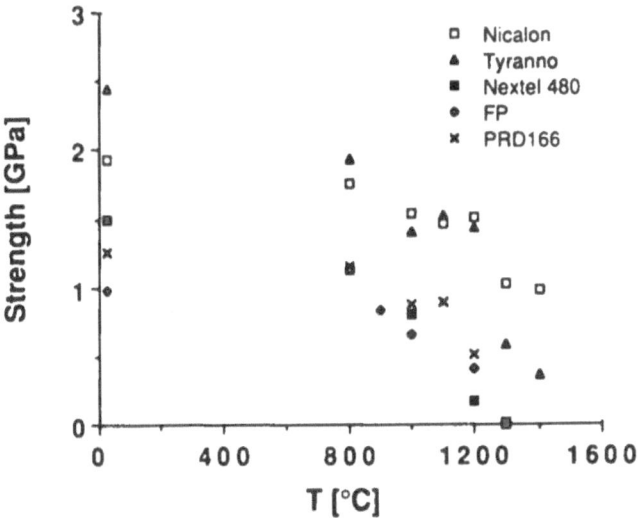

FIG. 18 Tensile strength as a function of temperature for several types of ceramic fibers. (From Ref. 16.)

variability is probably due to differences in different lots of fibers, while the remainder is due to different test method details. One work that has been cited throughout this chapter [16] involved testing of several of the types of fibers described under identical conditions, so that variability due to test method was eliminated. These comparative results would be expected to be much more valid than those obtained by plotting the independent results from different research groups, who used different experimental methods.

Figure 18 shows comparable tensile strength versus temperature data for five of the fiber types described above. General observations are that all fibers began to lose strength below 1000°C, that the non-oxide fibers were superior to the oxide fibers at the highest test temperatures, and that Tyranno fiber was inferior to Nicalon fiber at the highest test temperatures. Although variation due to test method was eliminated, it cannot be ruled out that some of these trends would differ if other lots of the fibers were tested. Testing of multiple lots of the same fibers would need to be conducted to make statistically valid comparisons.

E. Mechanical Testing of Fiber Tows

All of the mechanical data described above were for single fiber testing. Rather different behaviors at elevated temperatures or after heat treatment at elevated temperatures are found when fiber tows are tested. In general, strength retention is poorer and relative rankings of different fibers can change. There are several reasons for this. Strength is generally poorer because significant bending stresses are present due to contacts between the hundreds of fibers in a typical tow. Moreover, these contacts with other fibers can lead to failure-producing surface flaws. The potential for surface flaw generation is somewhat mitigated for as-produced fibers due to the "size" that is invariably applied to the fibers to facilitate handling and weaving. Treatment at temperatures above 200°-300°C thermally removes the size and accentuates the detrimental effects of fiber-to-fiber contact during application of stress.

For example, it has been found [34] that room tensile strength of Nicalon fiber tows decreased from 1.5 GPa in the as-received condition to only about 0.5 GPa after treatment at only 600°C in argon for one hour. Tow strength decreased further to 0.03 GPa after treatment at 1000°C, and to 0.002 GPa at 1300°C. Even less strength was retained in air at each temperature. These values are much poorer than those obtained for single fibers.

Sawko and Tran [35] reported low strength retentions during testing of fabrics at relatively low temperatures. They found severe degradation below 500°C for Nicalon fiber fabric, Nextel fiber fabrics, and a silica fiber fabric (Figure 19). Only one Nextel fiber fabric (Nextel 440) retained its as-fabricated strength to 1200°C, but the as-fabricated strength was considerably lower than those of the other fibers. Although the reasons for the large differences in strength retention were not discussed in Sawko and Tran's report, they may be due to such factors as variations in tow twist and in susceptibilities to surface damage among the various types of fibers.

Although it might be assumed that the tow or fabric tests give more realistic results, since the fibers are incorporated into matrices as tows or fabrics, this is not necessarily the case. In a good composite, the matrix material penetrates the fiber tows and surrounds individual fibers. Hence, there is a limited amount of fiber-to-fiber contact, which is responsible for the degradation demonstrated in the tow and fabric tests. In actual usage, fiber tensile strengths lie between the values measured on individual fibers and those measured on tows or fabrics, but might be closer to the values measured on individual fibers.

F. Cost Considerations for Continuous Ceramic Fibers

Reinforcing material cost is obviously an important consideration to the ceramic matrix composite designer. However, price information will not be detailed herein since current prices for the materials described above probably

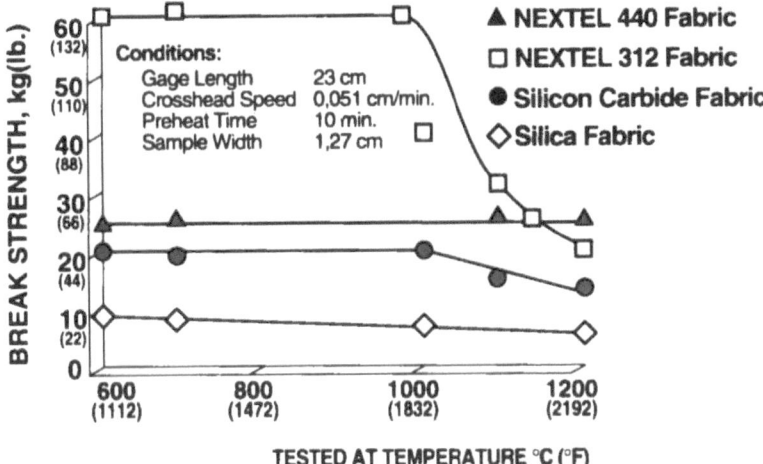

FIG. 19 Tensile strength as a function of temperature for several types of ceramic fiber tows. (From Ref. 35.)

do not reflect relative prices expected for ultimate large volume production. However, a few comments on fiber prices will be made. At this time, Nicalon fibers are the most widely used fibers in ceramic matrix composite development. A price quoted in 1990 for continuous ceramic grade Nicalon fibers was about $510/kg in lots of over 500 kg. Most other fibers are more expensive, but production has been lower than that of Nicalon. In the case of carbon fibers, price is dependent upon fiber quality. For example, price of low performance carbon fibers is about $20/kg, but highest performance fibers produced in small quantities can cost over $1000/kg. A cost projection for the various reinforcement materials when available in high volume production is beyond the scope of this book.

IV. CERAMIC WHISKERS

A. Background

Whiskers are fibrous single crystals, typically having cross-sectional diameters of less than 10 μm. Length can be as much as 10,000 times greater than the diameter. Because whiskers are single crystals and are, in most cases, produced with relatively few flaws, strengths and Young's moduli can be extremely high, approaching a significant fraction of those predicted from bond strengths. Compared with polycrystalline fibers having a similar composition, whiskers have significantly greater strengths and moduli. Strength of a silicon carbide whisker can be 10 times that of a silicon carbide fiber.[36]

Whiskers of a variety of materials have been produced by a number of methods. In this section, discussion will be limited to ceramic whiskers, mostly silicon carbide, since these have been the major whiskers used in ceramic matrix composites. Two processes, vapor-solid (VS) and vapor-liquid-solid (VLS) will be described. In the first method, atoms from the vapor phase are incorporated directly onto the whisker. In the second method, atoms from the vapor phase are incorporated into a liquid droplet on the whisker tip and are then deposited onto the whisker from the liquid.

Silicon nitride whiskers will be described briefly. A number of other ceramic whiskers have been produced, but will not be mentioned here since investigations on use in ceramic matrix composites are as yet very limited.

B. Silicon Carbide Whiskers Produced by the VS Process

Although details are not readily available due to proprietary considerations, most silicon carbide whiskers available today are believed to be produced by vapor-solid processes. A classic method for this type of production, using rice hulls as the raw material, was patented by I. B. Cutler at the University of Utah. As reported by Lee and Cutler [37], rice hulls are a unique by-product from rice milling. Large quantities are available from rice mills located throughout the rice producing regions of the United States. The unique characteristic of rice hulls is their high ash and silica content. About 15 wt% of the hulls consists of an ash that is predominantly silica.

Rice hulls were heated at 900°C in the absence of air and then coked further in carbon monoxide at 1350°C, yielding an intimate mixture of amorphous silica and carbon, roughly 50 wt% silica and 50 wt% carbon. In some experiments, prior to heat treating the hulls were soaked in an iron-containing solution and then treated with ammonium hydroxide to fix the iron within the organic (cellulose) structure, for determination of any catalytic effect.

Heating at about 1500°C resulted in a high yield of silicon carbide production. Iron was found to catalyze the reaction, and reaction rate increased linearly with increasing surface area of the iron. It was concluded that the reaction proceeded by way of the gas phase by dissociation of silica and subsequent reaction with carbon. In this original work, about 10% of the silicon carbide was in the form of whiskers, the remainder being particles about 0.1 μm in diameter. It is likely that processing conditions to increase whisker yield have been developed since then.

Figure 20 [38] is a schematic of silicon carbide whisker commercial production from rice hulls. After grinding, the rice hulls are heated (coked) in the

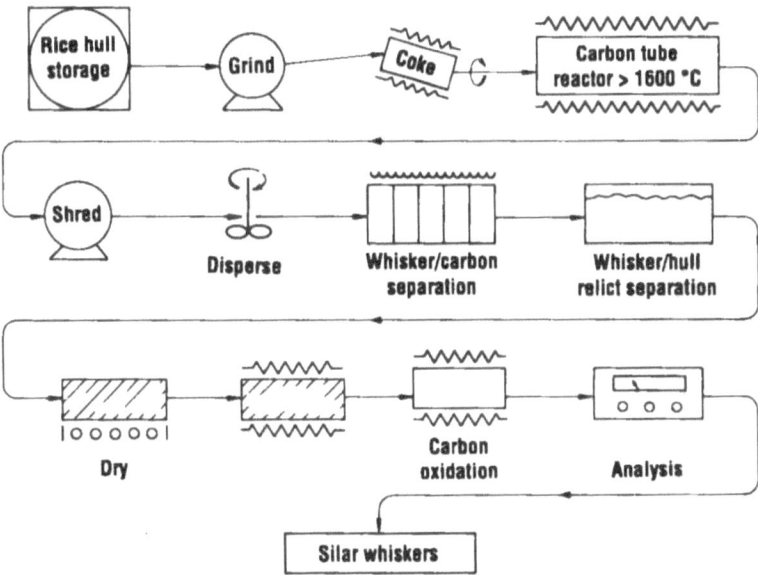

FIG. 20 Schematic flow diagram of silicon carbide whisker production from rice hulls. (From Ref. 38.)

absence of air at about 700°C to remove volatile materials and convert remaining carbon compounds to elemental carbon. The coked rice hulls are then heated at 1500° to 1600°C for about one hour in a nitrogen or ammonia atmosphere to produce silicon carbide whiskers. Silicon carbide particles as well as whiskers are produced and some free carbon remains. Wet separation processes are used to separate whiskers from particles, and excess carbon is removed by oxidation in air at about 800°C.

C. Silicon Carbide Whiskers Produced by the VLS Process

A vapor-liquid-solid process for silicon carbide whisker production has been developed at Los Alamos National Laboratory (LANL).[39] Whiskers produced by the VLS process tend to have diameters larger than those produced by the rice hull process, and also have high length/diameter ratios (aspect ratios).

Figure 21 is a schematic representation of the VLS process. As a first step, a catalyst material is distributed on a compatable substrate. The substrate is then heated to a high enough temperature to melt the catalyst. Typically, the catalyst reacts with the substrate to form liquid drops of catalyst and craters in

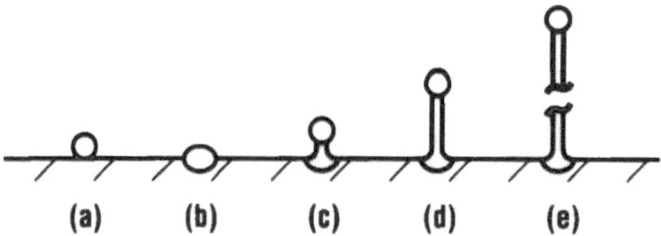

FIG. 21 Schematic representation of a vapor-liquid-solid process for silicon carbide whisker production, showing a solid metal catalyst particle (a), melting and crater formation (b), nucleation and initial growth of a whisker (c), and continued whisker growth (d) and (e). (From Ref. 39.)

the substrate. The appropriate gases are then admitted to the system and the catalyst material is enriched in silicon and carbon atoms. Eventually, the catalyst droplets become supersaturated, and a whisker is nucleated. The catalyst particle is lifted off the substrate surface by the growing whisker.

An appropriate catalyst is an alloy containing manganese, cobalt, and nickel, which can be obtained as an atomized powder and brushed onto the substrate from an organic suspension. Carbon is supplied initially from the substrate and continuously by methane in the gas stream. Silicon is incorporated into the catalyst droplet from silicon monoxide vapor, which is produced *in situ* by the reaction of silicon dioxide and carbon at an elevated temperature.

Figure 22 is a schematic diagram of a pilot-scale reactor that has been used at LANL. Substrate plates are graphite. The silicon monoxide generators are porous brick impregnated with carbon and silicon dioxide. Metal catalyst particles are applied to the growth surfaces. The process gases (hydrogen, nitrogen, methane, and carbon monoxide) are introduced through a bottom plenum. The separation between the plates is about 3.2 cm. Whisker growth is limited by the distance at which opposing growth leads to intermixing (about 1.6 cm). Longer whiskers have been grown by removing some plates.

D. Properties of Silicon Carbide Whiskers

1. MORPHOLOGY AND SURFACE CHEMISTRY

Karasek et al [40] measured morphology and surface chemistry of a number of available silicon carbide whiskers. Morphology varied considerably. In general, diameters and lengths of individual whiskers of a particular grade had a large range. For example, diameters of individual whiskers of one grade ranged from 0.2 to 5 μm and lengths ranged from 5 to 50 μm. With many whisker grades, diameter varied along the length. An extreme example of this is a "shish

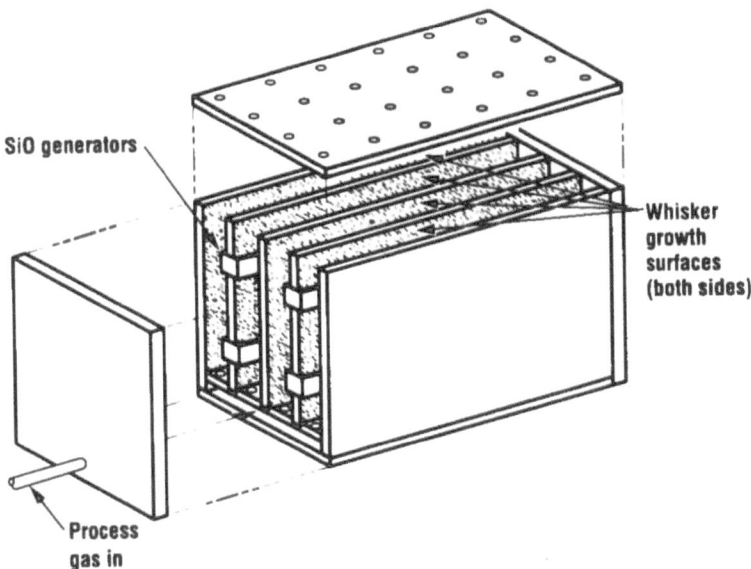

FIG. 22 Schematic illustration of a pilot-scale reactor used at LANL for production of silicon carbide whiskers by a vapor-liquid-solid process. (From Ref. 39.)

kebab" appearing fiber (alternately increasing and decreasing in diameter over short lengths). With some grades, branched and bent whiskers were found. Figure 23 [41] shows examples of different whisker morphologies.

VLS whiskers were generally straight and long, with diameters ranging from 0.25 to 5 μm and many lengths over 100 μm. Variation in diameter along the length was observed. Metallic catalyst impurities were seen at the end of some whiskers, consistent with the production scheme outlined previously.

Using X-ray photoelectron spectroscopy, whiskers having four basic types of surface chemistries were identified. One type had a high surface oxygen content, and the oxide resembled silicon dioxide. Another type had a low surface oxygen content, and the oxide resembled a glass containing silicon, oxygen, and carbon. The third type had a carbonaceous hydrocarbon surface, and the last type had a high surface oxygen level that resembled the silicon-oxygen-carbon-containing glass.

Whisker surface chemistry can be important from several aspects. As indicated in an earlier chapter, the interfacial bond strength between the rein-

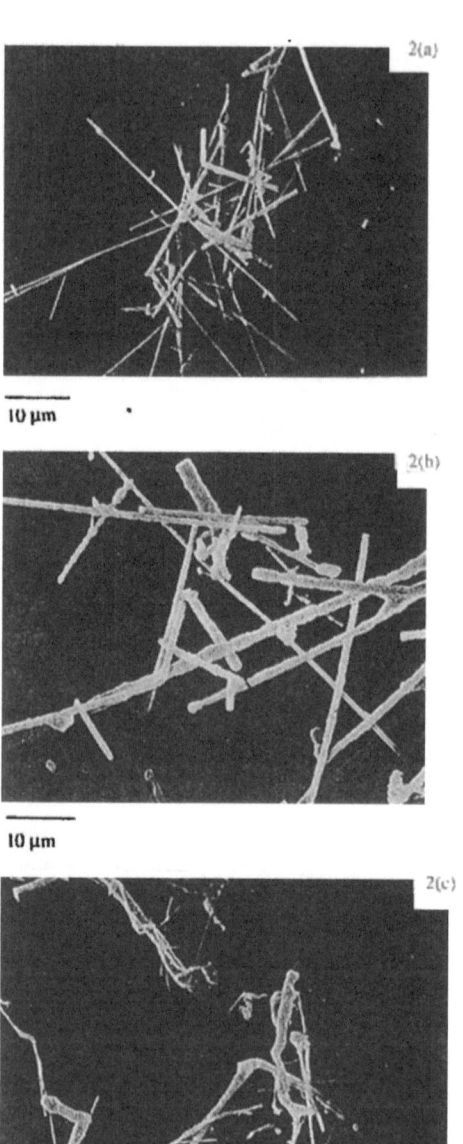

FIG. 23 Scanning electron micrographs of silicon carbide whiskers from three suppliers. (From Ref. 41.)

forcing material and the matrix plays a major role in the mechanical performance of a composite. The nature of the whisker surface can appreciably influence the interfacial strength. In addition, whisker reinforced ceramics are ordinarily produced starting with slurries of whiskers and matrix particles. The whisker surface chemistry will have an effect on the dispersibility of the whiskers in the slurry.

Whiskers from a number of suppliers had surface impurities. These impurities included calcium, cobalt, iron, and nickel. The presence of the impurities suggested that catalysts were used in the processes. Whisker samples also included various levels of debris, indicating that the separation processes outlined above were not totally successful in removing particulate material, or were not used. One type of whisker with a low surface oxygen content had a high surface fluorine level, suggesting that a hydrofluoric acid etch had been used to remove silicon dioxide on the surface.

2. MECHANICAL PROPERTIES

Mechanical properties of whiskers are obviously difficult to determine because of their small size. However, since lengths and diameters of the VLS whiskers are comparatively large, it has been possible to measure properties. The whiskers were carefully aligned and cemented into grips of a micro-tensile tester.[39] After fracture, scanning electron microscopy was used to determine cross-sectional area. Strain was determined by the relative displacement of two drops of epoxy cemented onto the whiskers, which permitted calculation of the Young's modulus.

Average tensile strength for forty whiskers with equivalent circular diameters of 4-8 μm and 5 mm lengths was 8.4 GPa and the average modulus was 578 GPa. Tensile strength ranged from 2 GPa or less to over 23 GPa. This large variation was attributed to a distribution of the type and size of flaws. Range in Young's modulus was also considerable, but this may have been due largely to measuring error with the small gage length.

In other determinations, whiskers up to 75 mm in length were tensile tested. Broken fragments were then retested and the process was continued until lengths of broken fibers were too short to test. Strengths of the original whiskers were 2-4 GPa, and strength increased with decreasing lengths of broken whiskers until plateaus at approximately 16 GPa were reached at about 25 mm gage lengths. Analysis of the data suggested two failure regimes. Low strength failures (below 10 GPa) typically occurred at 5 mm separation from observable defects such as appendages or internal cavities. Flaws causing failures at about the 16 GPa level could not be identified.

The most valid comparison of mechanical properties of these silicon carbide whiskers with those of polycrystalline silicon carbide fibers would appear to be with SCS fibers, since Nicalon and other fibers produced from polymeric precursors are far from stoichiometric silicon carbide and contain amorphous phases. Average strength of SCS fibers is about 4 GPa, which is less than half that reported for the VLS whiskers. Average Young's modulus of the SCS fibers is about 400 GPa, or roughly two-thirds of that reported for the whiskers.

Whisker strength retention at elevated temperatures or after heat treatment at elevated temperatures does not appear to have been studied. Since silicon carbide whiskers are pure and essentially monocrystalline, retention of mechanical properties at high temperatures would be expected to be superior to those of the available continuous fibers.

E. Silicon Nitride Whiskers

Details of production processes or properties of silicon nitride whiskers are not as readily available as for silicon carbide whiskers. However, fairly detailed information on properties of silicon nitride whiskers from one supplier, Ube Industries, Ltd., was reported by Hirata et al.[42] Composition of elements other than silicon and nitrogen included 1% oxygen, 0.28% yttrium, 700 ppm iron, 10 ppm aluminum, and 10 ppm calcium. Whiskers were straight, with lengths of 1.7 to 20 μm, with an average of 3.7 μm. Diameters were 0.1 to 1.3 μm, with an average of 0.4 μm. Lattice constants by X-ray diffraction were similar to a silicon nitride standard.

F. Availability of Silicon Carbide and Other Whiskers

According to Milewski [36], the estimated total capacity of all whisker producers in the USA in 1989 was about 50,000 pounds per year, and there is considerable production capacity elsewhere, particularly in Japan. Many companies reported that they could double or triple their capacity within a six month period. Companies involved in whisker research, production, and/or distribution have varied from year to year. For example, it was reported that between February, 1988, and September, 1989, the number of companies involved in whisker research, production, and/or distribution increased slightly, from 26 to 28, but that eight companies had dropped out of this area during that time and ten companies had entered the area. The great potential of whisker reinforcement has obviously provided a strong driving force for entry into this area. On the other hand, technical difficulties and, perhaps, slowness of market development have caused others to leave the field.

Because of this rapid rate of change, a current list of whisker producers will not be given here. Only a few commercial sources will be mentioned. Advanced Composite Materials Co. (ACMC), which was formerly ARCO Chemical Co., has produced silicon carbide whiskers, trade named Silar whiskers, by the rice hull process for many years. These whiskers are used for ceramic reinforcement in commercial cutting tool bits, as will be covered in a later chapter. In addition, Tateho Chemical Industries and Tokai Carbon Co. have been whisker suppliers for quite a few years. A number of other Japanese companies are active in whisker production. These companies may use the rice hull or related processes. Several companies are attempting to scale up the VLS process also. Tateho Chemical Industries and Ube Industries, Ltd. have supplied silicon nitride whiskers.

In 1989, Milewski reported silicon carbide whisker prices of $600 to $800/kg in small lots, and $450 to $500/kg in large lots; a more quantitative description of lot size was not given.

V. SUMMARY

A variety of carbon fibers, continuous ceramic fibers, and ceramic whiskers are available for use in ceramic matrix composites. Carbon fibers are the most well developed of these fibers and whiskers, and are available in a number of types having appreciably different properties and prices.

Although carbon fibers have been successfully used in ceramic matrix composites, their susceptibility to oxidation at elevated temperatures makes them less than ideal for this purpose. Several silicon carbide base fibers, produced by polymer spinning and decomposition and by chemical vapor deposition have been developed. The current CVD-produced fibers are limited by their large diameters. Oxide fibers of various types are also available. Since oxide fibers should be more resistant to oxidation than non-oxide fibers, they are, perhaps, more promising reinforcements than non-oxide fibers. However, current oxide fibers suffer from about as much high temperature degradation as non-oxide fibers.

Most available ceramic whiskers are silicon carbide produced by vapor/solid methods. These whiskers differ appreciably in their size, surface chemistry, and morphology. A vapor/liquid/solid process for silicon carbide whisker production is also under development. Silicon nitride whiskers are also available in some quantity, and other ceramic whiskers have been produced on a smaller scale.

REFERENCES

1. Donnet, J. B., and Bahl, O. P., Carbon fibers. In *Encyclopedia of Physical Science and Technology*, vol. 2, Academic Press, Inc., 1987, 515-539.
2. Chawla, K. K., *Composite Materials*, Springer-Verlag New York Inc., New York, NY, 1987, 6-57.
3. Volk, H. F., Carbon fibers and fabrics. In *Kirk-Othmer Encyclopedia of Chemical Technology*, Third Edition, vol. 4, H. F. Mark, D. F. Othmer, C. G. Overberger, and G. T. Seaborg (Ed.), John Wiley & Sons, New York, NY, 1978, 622-628.

4. Buckley, J. D., Carbon-carbon, an overview, *Ceramic Bulletin*, vol. 67, no. 2, 364-368 (1988).

5. Noller, C. R., *Chemistry of Organic Compounds*, Second Edition, W. B. Saunders Co., Philadelphia, PA, 1957.

6. Lovell, D. R., A comparison of available carbon fibers. In *Carbon Fibers: Technology, Uses and Prospects*, Noyes Publications, Park Ridge, NY, 1986, 39-47.

7. Okamura, K., Ceramic fibres from polymer precursors, *Composites*, vol. 18, no. 2, 107-120 (1987).

8. Yajima, S., Hasegawa, Y., Hayashi, J., and Iimura, M., Synthesis of continuous silicon carbide fibre with high tensile strength and high Young's modulus, *Journal of Materials Science*, vol. 13, 2569-2576 (1978).

9. Hasegawa, Y., Iimura, M., and Yajima, S., Synthesis of continuous silicon carbide fibre, *Journal of Materials Science*, vol. 15, 720-728 (1980).

10. Andersson, C.-H., and Warren, R., Silicon carbide fibres and their potential for use in composite materials. Part 1, *Composites*, vol. 15, no. 1, 16-24 (1984).

11. Emsley, E. F., Sharp, J. H., and Bailey, J. E., The fabrication of silicon carbide fibres by the polymeric precursor route. In *Fabrication Technology*, R. W. Davidge and D. P. Thompson (Ed.), British Ceramic Proceedings, No. 45, The Institute of Ceramics, Stokes-on-Trent, UK, 1990, 139-151.

12. Clark, T. J., Arons, R. M., Stamatoff, J. B., and Rabe, J., Thermal degradation of Nicalon fibers, *Ceram. Eng. Sci. Proc.*, vol. 6, no. 7-8, 576-588 (1985).

13. Sawyer, L. C., Arons, R., Haimbach, F., Jaffe, M., and Rappaport, K. D., Characterization of Nicalon: strength, structure, and fractography, *Ceram. Eng. Sci. Proc.*, vol. 6, no. 7-8, 567-575 (1985).

14. Ishikawa, T., Ichikawa, H., and Teranishi, H., Strength and structure of SiC fiber after exposure to high temperature. In *Symposium on High Temperature Materials Chemistry - IV*, The Electrochemical Society, Pennington, NJ, 1988, 205-217.

15. Johnson, S. M., Brittain, R. D., and Lamoreaux, R. H., Degradation of SiC fibers. In *Symposium on High Temperature Materials Chemistry - IV*, The Electrochemical Society, Pennington, NJ, 1988, 355-362.

16. Pysher, D. J., Goretta, K. C., Hodder, R. S., and Tressler, R. E., Strengths of ceramic fibers at elevated temperatures, *J. Am. Ceram. Soc.*, vol. 72, no. 2, 284-288 (1989).

17. Jaskowiak, M. H., and DiCarlo, J. A., Pressure effects on the thermal stability of silicon carbide fibers, *J. Am. Ceram. Soc.*, vol. 72, no. 2, 192-197 (1989).

18. Griel, P., Thermodynamic calculations of Si-C-O fiber stability in ceramic matrix composites, *Journal of the European Ceramic Society*, vol. 6, 53-64 (1990).

19. Bender, B. A., Wallace, J. S., and Schrodt, D. J., Effect of thermochemical treatments on the strength and microstructure of SiC fibres, *Journal of Materials Science*, vol. 26, 970-976 (1991).

20. Yamamura, T., Ishikawa, T., Shibuya, M., Hisayuki, T., and Okamura, K., Development of a new continuous Si-Ti-C-O fibre using an organometallic polymer precursor, *Journal of Materials Science*, vol. 23, 2589-2594 (1988).

21. Anon., Improved processing of ceramics and composites, *Ceramic Bulletin*, vol. 70, no. 6, 687 (1991).

22. LeGrow, G. E., Lim, T. F., Lipowitz, J., and Reaoch, R. S., Ceramics from hydridopolysilazane, *Ceramic Bulletin*, vol. 66, no. 2, 363-367 (1987).

23. Lipowitz, J., LeGrow, G. E., Lim, T. F., and Langley, N., Silicon carbide fibers from methylpolysilane (MPS) polymers, *Ceram. Eng. Sci. Proc.*, vol. 9, no. 7-8, 931-942 (1988).

24. Mah, T.-I., Mendiratta, M. G., Katz, A. P., and Mazdiyasni, K. S., Recent developments in fiber-reinforced high temperature ceramic composites, *Ceramic Bulletin*, vol. 66, no. 2, 304-317 (1987).

25. Foltz, T. F., SiC fibers for advanced ceramic composites, *Ceram. Eng. Sci. Proc.*, vol. 6, no. 9-10, 1206-1220 (1985).

26. Shetty, D. K., Pascucci, M. R., Mutsuddy, B. C., and Wills, R. R., SiC monofilament-reinforced Si_3N_4 matrix composites, *Ceram. Eng. Sci. Proc.*, vol. 6, no. 7-8, 632-645 (1985).

27. Le Petitcorps, Y., Lahaye, M., Pailler, R., and Naslain, R., Modern boron and SiC CVD filaments: a comparative study, *Composites Science and Technology*, vol. 32, 31-55 (1988).

28. Ahmad, I., Hill, D. N., and Hefferman, W., SiC filament as reinforcement for high temperature superalloy matrices, *Proc. 1st Int. Cong. on Composite Mater.*, TMS, Warrendale, PA, 1976, 85-102.

29. Ko., F. K., Preform fiber architecture for ceramic-matrix composites, *Ceramic Bulletin*, vol. 68, no. 2, 401-414 (1989).

30. Romine, J. C., New high-temperature ceramic fibers, *Ceram. Eng. Sci. Proc.*, vol. 8, no. 7-8, 755-765 (1985).

31. Sowman, H. G., and Johnson, D. D., Ceramic oxide fibers, *Ceram. Eng. Sci. Proc.*, vol. 6, no. 9-10, 1221-1230 (1985).

32. Holtz, A. R., and Grether, M. F., High temperature properties of three Nextel ceramic fibers. Paper presented at the 32nd International SAMPE Symposium and Exhibition, Anaheim, CA, April 6-9, 1987.

33. Johnson, D. D., Holtz, A. R., and Grether, M. F., Properties of Nextel 480 ceramic fibers, *Ceram. Eng. Sci. Proc.*, vol. 8, no. 7-8, 744-754 (1985).

34. Fareed, A. S., Fang, P., Koczak, M. J., and Ko, F. M., Thermomechanical properties of SiC yarn, *Ceramic Bulletin*, vol. 66, no. 2, 353-358 (1987).

35. Sawko, P. M., and Tran, H. K., Strength and flexibility properties of advanced ceramic fabrics, *SAMPE Quarterly*, vol. 17, no. 1, 7-13 (1985).

36. Milewski, J. V., Whisker reinforcements: current global producers and their products. Paper presented at the 21st International SAMPE Technical Conference, Atlantic City, NJ, September 25-28, 1989.

37. Lee, J.-G., and Cutler, I. B., Formation of silicon carbide from rice hulls, *Ceramic Bulletin*, vol. 54, no. 2, 195-198 (1975).

38. Rhodes, J. F., Rootare, H. M., and Peters, J. E., Whisker-reinforced ceramic composites, *Proc. Int. Conf. PM Aerosp. Mater.*, MPR Publ. Serv. Ltd., Shrewsbury, UK, 1988, 0-45.

39. Hurley, G. F., and Petrovic, J. J., Silicon carbide whiskers for composites - growth and properties. In *Advanced Composites: Proceedings of the Conference*, Dearborn, MI, December 2-4, 1985, ASM, Metals Park, OH, 1985, 207-212.

40. Karasek, K. R., Bradley, S. A., Donner, J. T., Yeh, H. C., and Schienle, J. L., Characterization of recent silicon carbide whiskers, *Journal of Materials Science*, vol. 26, 103-111 (1991).

41. Shih, C. J., Yang, J.-M., and Ezis, A., Processing and performance of several SiC whisker-reinforced Al_2O_3 matrix composites, *Materials and Manufacturing Processes*, vol. 5, no. 1, 35-49 (1990).

42. Hirata, Y., Nakagama, S., and Ishihara, Y., Dispersion and consolidation of silicon carbide whisker in aqueous suspension, *J. Mater. Res.*, vol. 5, no. 3, 640-646 (1990).

6

Fiber Architecture

I. GENERAL CONSIDERATIONS

For continuous fiber reinforced ceramics, the arrangement of the fibers, or fiber architecture, provides the structural backbone of the composite, analogous to the steel beams in a skyscraper.[1] By proper selection of the geometry of the fiber "beams," the structural performance of the composite can be altered to meet design requirements.

Fiber architecture can range from a simple uniaxial alignment to a highly complicated three-dimensional array. Much of the technology for producing preforms of fibers of interest for reinforcing ceramics has been adapted from textile technology, although there are greater limitations with high modulus, high performance fibers. For whisker reinforced ceramics, the term fiber architecture has less significance. The whiskers tend to be more or less randomly oriented, although orientation of the whisker axes can be biased in one direction or in a plane, depending upon the fabrication method.

An important consideration in fiber architecture is the extent to which a fiber can be bent without fracturing. This is a function of both the elastic modulus and the diameter. This can be quantified in terms of the minimum diameter of a mandrel over which a fiber of a given diameter can be bent without fracturing. Figure 1 [2] indicates critical (i.e. maximum) fiber diameters as a function of mandrel diameter. Most of the ceramic and carbon fibers shown are less than 20 μm in diameter, so they can be bent over a 2-mm diameter mandrel without fracturing. Based on this result, it is clear that most of the ceramic fiber tows can be arranged into relatively complex configurations without significant fiber fracture. However, the 143-μm diameter SCS-6 fibers are much more limited in this respect. Figure 2 [3] is a cross-sectional view of a composite containing both Nicalon fibers and SCS fibers, which illustrates the large difference in cross-sectional area between the two types of fibers.

II. TYPES OF FIBER ARCHITECTURE

A. Uniaxial and Cross-Plied Fiber Alignment

The simplest fiber arrangement is uniaxial alignment, also termed unidirectional alignment. Typically, this type of orientation is obtained by winding fibers or fiber tows onto a drum that is translating along its axis as well as rotating (or the fiber feed is translating). A removable binder, either alone or in combination with material that will form the matrix when consolidated, is used to hold the fibers in the proper alignment. When a fiber layer with the desired spacing has been formed, the winder is stopped and the aligned fiber structure is removed

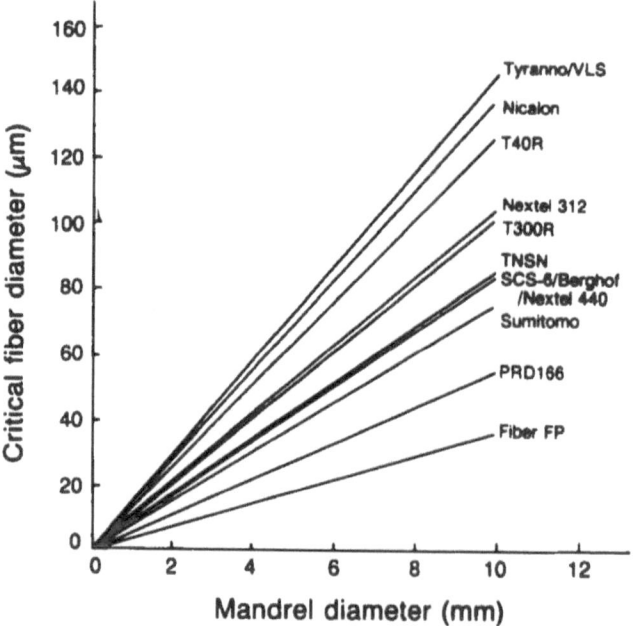

Fig. 1 Critical diameters of high modulus fibers as a function of radius of curvature. (From Ref. 2.)

by cutting parallel to the axis of the drum. This is equivalent to the "prepreg" used in polymer matrix composite fabrication. Figure 3 illustrates drum winding.

Curvature in the prepreg is produced by winding unto a drum. Since this can be objectionable if a flat structure is desired and the binder system used results in some rigidity of the prepeg, a plate winder is sometimes used instead. In this case, an aligned, straight fiber layer can be cut from each face of the plate.

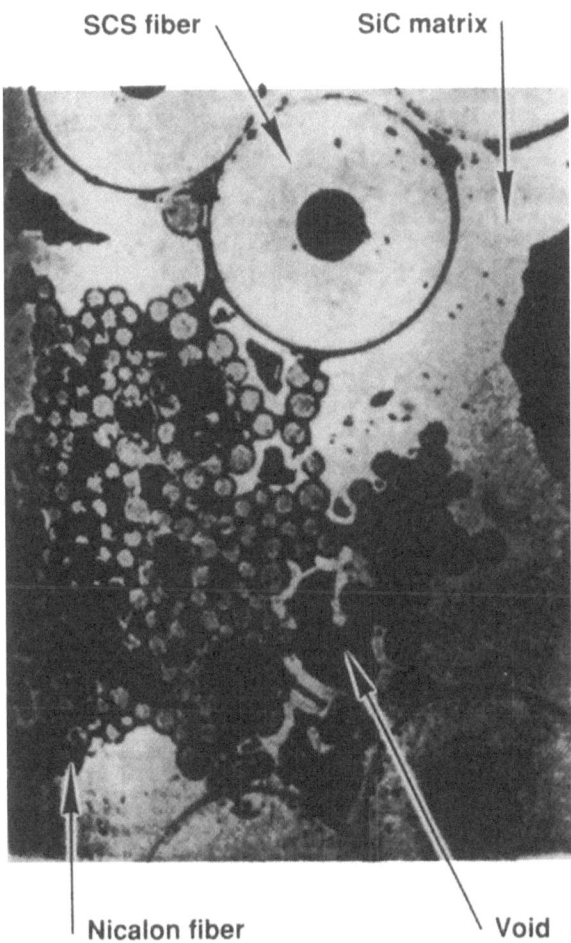

FIG. 2 Cross-sectional view of a composite containing both Nicalon™ and SCS-6 fibers. (From Ref. 3.)

A sufficient number of layers of the aligned fibers required to form a structure of a desired thickness after processing can be stacked together with the fibers aligned in the same direction. The result, after appropriate processing to form the composite, is uniaxial reinforcement. Composite mechanical properties are very anisotropic (directional). Strength and especially toughness are enhanced only in the direction of fiber alignment. The composite is susceptible to delamination between fiber layers, in both orthogonal directions to fiber alignment. That is, intralaminar and interlaminar shear strength are usually low. Such a composite is still subject to catastrophic, brittle shear failure.

The next level of architectural complexity is a cross-plied arrangement. In this case, the initial step of forming single layer structures of uniaxially aligned fibers is the same as described above, but instead of aligning the layers with the fibers in one direction, the directions are altered in successive layers. An arrangement with successive layers having the fiber axes at 90° to each other is common. As would be expected, mechanical properties in any single direction are reduced, but now they are the same in two orthogonal directions.

FIG. 3   Drum winding of ceramic fibers. (Courtesy of Avco Specialty Materials/ Textron.)

To enhance mechanical properties in other directions within a plane, a more varied stacking sequence such as 0°/+45°/90°/-45°/0°, or the sequence 0°/+30°/+60°/90°/-60°/-30°/0° can be used. With these orientations, properties become more nearly the same in any direction within the plane. However, since there is still no layer-to-layer fiber reinforcement, layers can delaminate.

B. Cloth Reinforcement

An architectural arrangement that decreases possibilities for delamination with a given volume fraction of fibers while providing multidirectional reinforcement within a plane is fiber cloth. Another advantage of cloth reinforcement is that composite fabrication is simplified. Cloth formation is faster and less labor intensive than forming prepregs on a drum or plate winder. Methods of cloth fabrication using fibers suitable for ceramic reinforcement have been adapted from textile manufacture. Fiber tows can be interlaced, intertwined, or interlooped.

Weaving is the oldest method of creating textiles.[1] The common right-angle weaving technique (such as used to make "paper baskets") has been adopted for cloth production for high performance fibers for ceramic reinforcement. This technique is also known as biaxial or flat weaving. The simplest biaxial weave pattern (plain weave) is schematically illustrated in Figure (a).[4]

In the basic biaxial weaving process, the fiber tows are brought together from two sources on the "loom" (a machine for weaving). The "warp" tows, which span the length of the fabric, are typically unwound from a rack of bob-

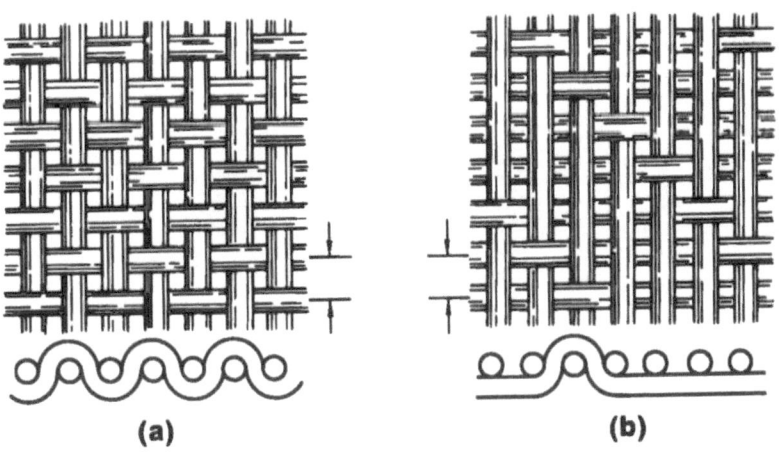

(a)   **(b)**

FIG. 4 Schematic illustrations of plain weave (a) and five-harness satin weave (b) cloths. (From Ref. 4.)

bins (cylinders on which the fiber tows are wound). These tows are kept taut and parallel to each other. These lengthwise tows are fed through guides (heddles). For a plain weave, odd numbered tows are moved up and even numbered tows down for the insertion of one crosswise, or filling, tow. The procedure is reversed for the next filling tow, and so on.

The filling or "weft" tows are inserted in lengths (called picks) to produce the crosswise tows in the fabric. The traditional insertion device, the "shuttle," carries a load of filling tows along with it. Another step is used to straighten the filling. A loop called the "selvage" forms a finished edge for the weft tows.

Other types of biaxial weaves are produced also. The type that has been most commonly used in producing ceramic composites is the "satin weave." These weaves do not have the over-and-under configuration with respect to every tow oriented at 90°, so that they have sections of unbent fiber tows. Figure 4(b) shows a "five-harness" satin weave, which means that only every fifth fiber tow is interlaced. Many composites fabricated using ceramic fibers have involved satin weave cloth. In principle, at least, satin weaves might be expected to be superior to plain weaves because of the sections of unbent fibers.

Fɪɢ. 5   Schematic illustration of a triaxial weave pattern. (From Ref. 1.)

The nature of the pore structure in a cloth reinforced ceramic can also depend on the weave pattern. Differences due to cloth weave patterns in ceramic matrix composites have not yet been well documented, but in a study on carbon/ carbon composites [5], eight-harness satin weave was reported to be preferred over plain weave. Cloths having another basic biaxial weave, twill weave, have apparently not been used to any extent for reinforcing ceramics.

Another form of weave is the triaxial weave. This type of weave results in a high level of in-plane shear resistance, and high levels of isotropy and dimensional stability can be attained with triaxial weave at a low volume fraction. Although triaxial weaves have been commonly used in basket weaving, chair caning, and the making of snowshoes [1], they are not traditional in the textile industry. The basic triaxial weave is shown in Figure 5. Other variations are possible. Triaxial weaving is mainly of future interest for ceramic matrix composites.

Other methods, such as braiding, are also available for producing cloths. In braiding, fiber tows from three or more source points are intertwined using an under-and-over sequence to form a structure based on diagonal patterns. Braided fabrics can readily be produced in tubular form, which is ideal for some ceramic matrix composite applications. An example of a tubular braided structure is the traditional "maypole." Although the outline of a tubular braided structure is three dimensional, conventional braiding produces a single layer thickness or often a series of single layers. Braided fabrics are highly conformable, shear resistant, and tolerant to impact damage.

A related process for producing tubular shapes is filament winding. However, in traditional filament winding the mandrel rather than the fiber tow is

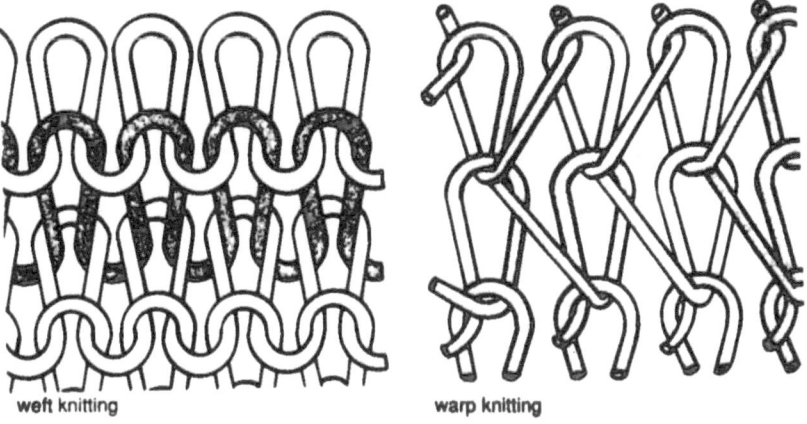

weft knitting   warp knitting

FIG. 6   Schematic illustrations of weft knitting and warp knitting. (From Ref. 1.)

rotated, and there is no interlacing of the fibers. Hence, impact damage tolerance is more limited with filament winding. However, versions are available with interlaced layers. A concave contour is difficult to produce with filament winding.

Knitted fabrics are interlooped structures in which the loops are produced by introducing the knitting yarn either along the machine direction, warp knit, or in the cross-machine direction, weft knit (Figure 6). Weft knitted fabric is very conformable, while warp knitted fabric has a greater dimensional stability. As with the other cloth producing methods described above, there are many variations on the basic knitting process. Because of the severe bending of fibers

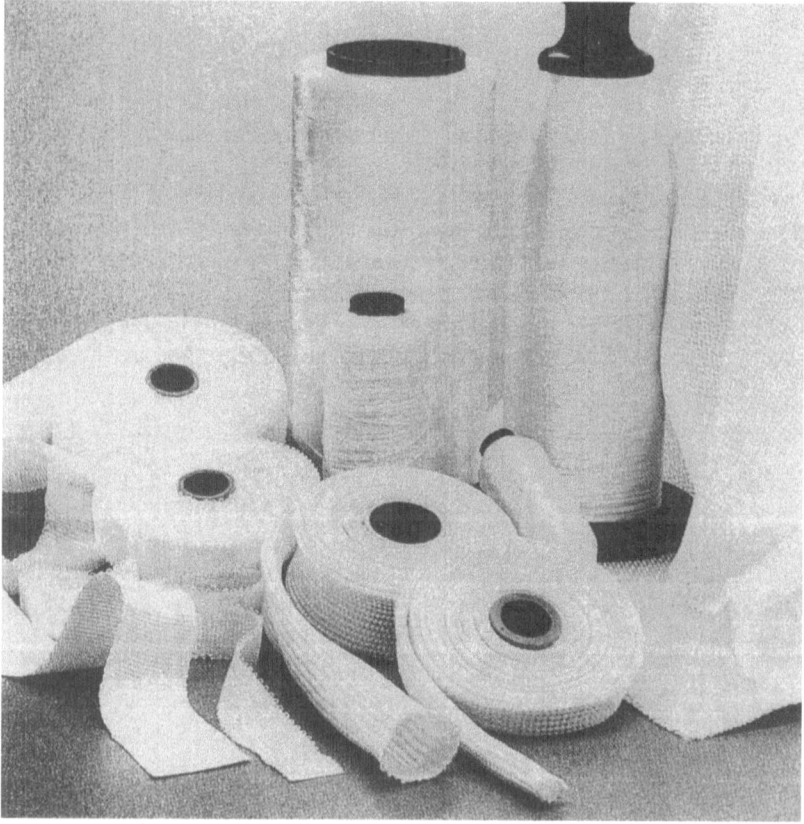

FIG. 7   Examples of ceramic fibers, cloths, and tubular materials. (Courtesy of 3M Ceramic Materials Department).

inherent with the knitting process, knitted cloths are not particularly attractive for the high modulus fibers of interest for ceramic reinforcement.

As with the uniaxial fiber prepregs, cloth layer prepregs can be stacked to provide the desired thickness of the finished composite. Although the cloth layers inherently have reinforcement in several directions, layers can be oriented in different directions to achieve desired properties. For example, with biaxial woven cloth, a 0°/45°/0° arrangement can be used. Figure 7 gives some illustrations of ceramic fibers, cloths, and tubular materials.

Even though cloths have some advantages over laying up of uniaxial plies in composite fabrication, the basic problem of low interlaminar shear strength remains. This is addressed by various methods of placing reinforcing fibers in three dimensions.

C. Three-Dimensional Fiber Architecture

To eliminate the problem of low interlaminar shear strength for continuous fiber reinforced ceramics, a number of methods for placing fibers in three dimensions

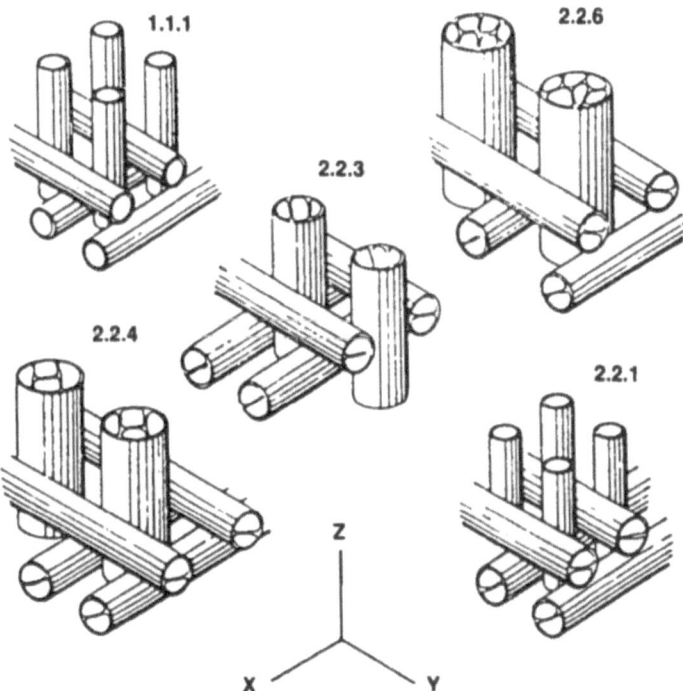

FIG. 8 Examples of selective fiber reinforcement. (From Ref. 4.)

have been developed or are under development. The term "three-dimensional" will be used here to include any arrangement in which fibers are placed in at least three orthogonal directions. Placing fibers in additional directions will not be considered to constitute "four-dimensional," etc. reinforcement, although this terminology is often used in describing fiber architecture.

Orthogonal nonwoven fabric technology has been developed by a number of aerospace companies such as General Electric and Fiber Materials Corporation in the USA, and Aerospatiale, Société Européenne de Propulsion, and Brochier in Europe.[2] In this process, selective reinforcement in one or more directions may be used, depending on application requirements. For example, Figure 8 [4] shows various types of arrangement for selective reinforcement.

In a cylindrical three-dimensional configuration, selective fiber tow additions must be made to account for the larger outside diameter than inside diameter. Either axial yarn compensation (axial yarn diameter varies) or radial yarn compensation (axial yarn diameter constant) can be used.

The weaving process can be adapted for three-dimensional structure production. Three dimensional woven fabrics have been used in industrial belting and webbing for many years.[1] These are composed of layers of warp and filling yarns, bound together by interlacing warp ends with the filling of adjacent layers (angle interlock) or with the ends interlaced between the face and back layers (warp interlock). These weaves are illustrated in Figure 9. It has been found that interlacing throughout the fabric need not be done. The addition of vertical yarns interlaced with the top and bottom horizontal yarns provides adequate reinforcement.

Multilayer structures can be given additional strength by inserting "stuffing yarns" in each layer. These remain straight and contribute their full strength to that direction. Yarns that interlace between layers as binding yarns contribute

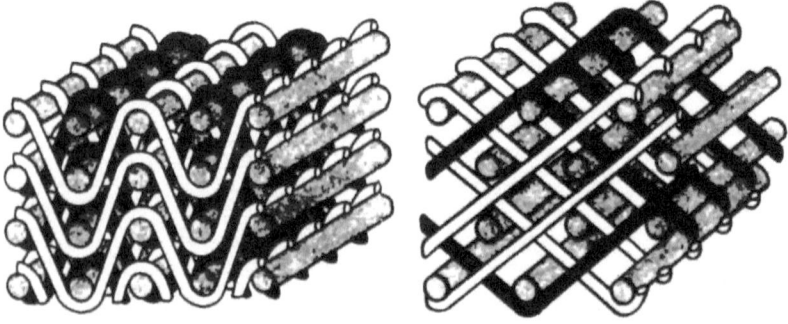

Fıɢ. 9   Schematic representations of angle interlock weaving (left) and warp interlock weaving (right). (From Ref. 1.)

only partially to the strength in their direction. Hence, there are compromises that must be made. An increase in fiber volume fraction in one direction is achieved only at the expense of one or both other directions.

Traditional weaving machines can be adapted for three-dimensional weaving. A schematic illustration of three-dimensional weaving is given in Figure 10. From the left are: a creel of bobbins to supply the lengthwise and vertical yarns; harnesses to exchange rows of vertical yarns; a reed, which moves to pack the fibers in the material; filling needles that insert crosswise yarns; selvage needles; a crossbar to hold the selvage yarn; a pair of knitting needles; and, apparatus for maintaining tension of the finished material.

Three-dimensional braiding technology is an adaptation of two-dimensional braiding. A variety of three-dimensional braided arrangements resulting in a highly damage-resistant preform can be produced.[2] The basic braiding motion includes the alternate x and y displacement of yarn carriers, followed by a compacting motion. Different shapes are formed by varying the positions of the carriers and by joining various rectangular groups by selected carrier movements. Figure 11 shows an example of three-dimensional braiding of a rocket motor exit cone preform.

Three-dimensional knitting has undergone considerable development for advanced composite manufacture although, as with the two-dimensional variation, the severe fiber bending associated with the knitting process limits the fiber types that can be used. Knitted three-dimensional structures can be produced by

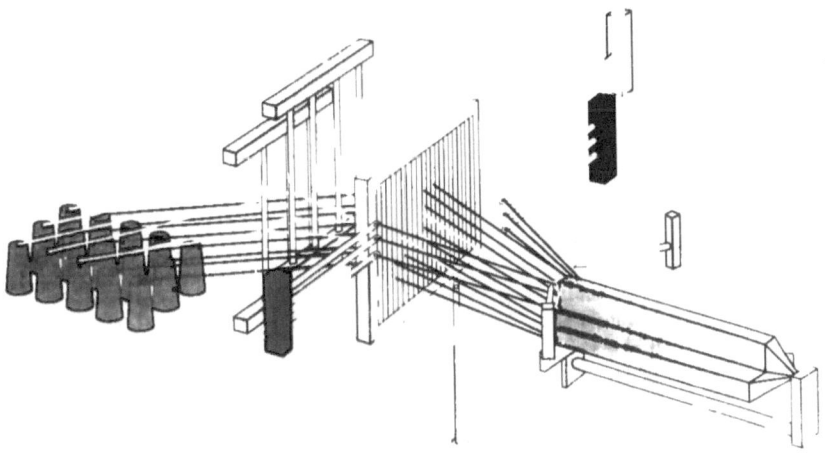

FIG. 10 Schematic illustration of three-dimensional weaving. (From Ref. 1.)

either warp knitting or weft knitting. Weft-knitted structures are highly conformable. The Pressure Foot™ weft-knit process by Courtalds has been utilized for fabricating preforms which, in collapsed form, are used for carbon/carbon aircraft brake production. An undesirable feature of weft-knitted three-dimensional structures is bulkiness, which means that volume fraction of the reinforcing fibers is low.

Another three-dimensional knitting technique is the multiaxial, warp-knit (MWK) method. There are a number of variations of this process. In the "nonimpaled process," stitches are formed without piercing through the reinforcing

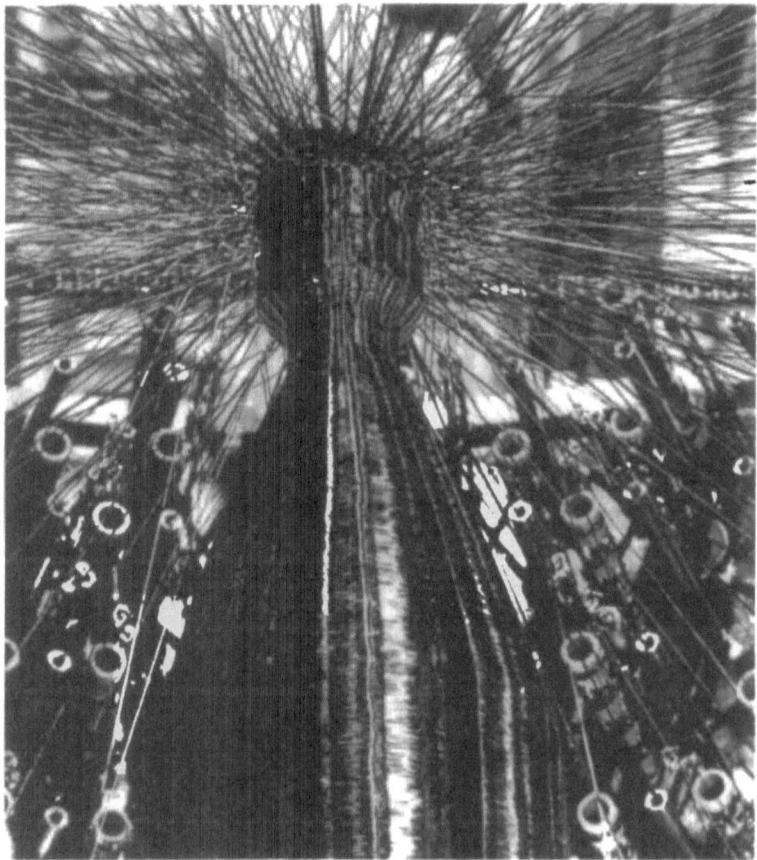

FIG. 11   Illustration of three-dimensional braiding of a rocket motor exit cone preform. (From Ref. 1.)

yarns. In the "impaled process," layers of linear yarns are assembled in various stacking sequences and stitched together by knitting needles piercing through the yarn layers. Unfortunately, the piercing action unavoidably damages the reinforcing fibers.

D. Architecture with Whisker Reinforcement

The concept of architecture is much less relevant for whiskers (or discontinuous fibers). In general, whiskers are more-or-less randomly oriented in whisker reinforced ceramics. However, there can be some partial orientation of whiskers, depending on the fabrication method. For example, uniaxial pressing of a whisker/matrix slurry to form a green ceramic, which will be completed by pressureless sintering or some other method, will tend to result in some degree of two-dimensional orientation in the plane normal to the pressing direction. In contrast, extrusion to form the green ceramic can be expected to result in some preferred orientation in the extrusion direction.

Long whiskers (such as those produced by the VLS process) could be formed into "staple" (discontinuous) yarns by appropriate selection of whisker length and twist in the yarn. Some of the techniques used in continuous fiber placement technology could then be utilized. However, this has apparently not yet been successfully accomplished.

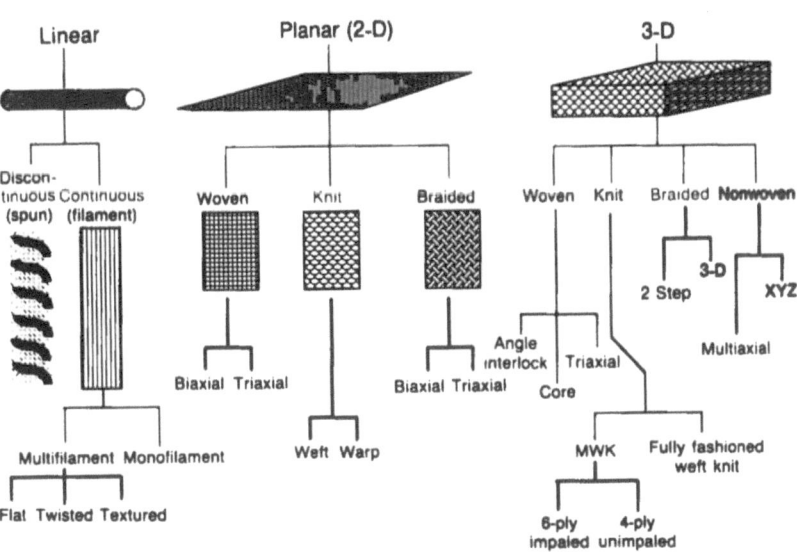

FIG. 12 Summary of various fiber architectures. (From Ref. 2.)

III. SUMMARY

In summary, there are many methods of two- and three-dimensional fiber placement being developed for advanced composite usage. These are schematically illustrated in Figure 12.[2] Many of the methods have their origins in textile technology, but the unique requirements of preforms for composite manufacture and the differences in stiffnesses between textile fibers and high performance ceramic or carbon fibers have necessitated significant alterations. There is no reason to believe that any single method of fiber placement will predominate in ceramic matrix composite fabrication. Rather, a number of methods will be used, depending on technical requirements and economics.

REFERENCES

1. Mohamed, M. H., Three-dimensional textiles, *Scientific American*, vol. 78, 530–541 (November–December 1990).
2. Ko, F. K., Preform fiber architecture for ceramic-matrix composites, *Ceramic Bulletin*, vol. 68, no. 2, 401–414 (1989).
3. Hopkins, G. R., and Chin, J., SiC matrix/SiC fiber composite: a high-heat flux, low activation, structural material, *J. Nucl. Mater.*, vol. 141–143, 148–151 (1986).
4. Lachman, W. L., Crawford, J. A., and McAllister, L. E., Multidirectionally reinforced carbon/carbon composites. In *Proc. of Internat. Conf. on Composite Materials*, (ICCM 2) B. Noton, R. Signorelli, K. Street, and L. Phillips (Ed.), TMS, Warrendale, PA, 1978, 1302–1319.
5. Manocha, L. M., and Bahl, O. P., Influence of carbon fiber type and weave pattern on the development of 2D carbon-carbon composites, *Carbon*, vol. 26, no. 1, 13–21 (1988).
6. Yang, J.-M., Lin, W., Shih, C. J., Kai, W., Jeng, S. M., and Burkland, C. V., Mechanical behaviour of chemical vapour infiltration-processed two- and three-dimensional Nicalon/SiC composites, *Journal of Materials Science*, vol. 26, 2954-2960 (1991).

7

Selection of Compatible Reinforcement and Matrix Materials

I. BACKGROUND

Continuous fibers available for relatively near-term applications are limited to those already discussed and perhaps a few others, since development of continuous fibers on a commercial scale has invariably proven to be a long process. Commercial development of a new whisker type might be a little more rapid, but would still be expected to require several years of effort. Hence, near-term ceramic matrix composite development requires the utilization of already available carbon or ceramic fibers or whiskers along with a compatible matrix material. For the longer range, compatibilities of materials that can potentially be produced as fibers or whiskers with candidate matrix materials are of interest.

Since high temperatures are generally encountered during ceramic matrix composite fabrication or application or both, fiber and matrix stability as well as interactions between fibers and matrix at high temperatures are of particular concern. Fiber coatings can also be necessary (as will be covered in the next chapter in more detail), so that fiber/coating and matrix/coating reactions at fabrication and use temperatures are important. The composite materials must not melt, vaporize, dissociate, or otherwise chemically degrade at the temperatures to be encountered during fabrication or use. The stabilities of available reinforcing fibers have been discussed in an earlier chapter, so this factor will not be emphasized here. Stabilities of potential matrix materials and of fiber/

matrix combinations (or fiber/coating or matrix/coating combinations) will be discussed.

II. MATRIX SELECTION CRITERIA

A. Melting Point

A number of materials having high melting or decomposition points as well as relatively low densities were shown in Chapter 2, Figure 6. Although this figure included many selected ceramic materials of major interest, the actual number of possible matrix materials is far greater.

Hillig [1] has listed materials with high melting or decomposition temperatures by classes, which include carbides, borides, nitrides, oxides, silicides, sulfides, selenides, and tellurides. Thirty six carbides melt or decompose at over 1750°C, several of these at over 3500°C. There are 16 nitrides, 49 borides, 30 silicides, and 42 sulfides, selenides, and tellurides that melt or decompose at over 1750°C. The largest category of materials with melting or decomposition temperatures of over 1750°C is the oxide category, with 101 oxides meeting this criterion. For purposes of comparison, 41 metals and intermetallic compounds also meet this criterion, although over half of these are at the low end of the range, 1750°–2000°C.

The number of available materials decreases as melting/decomposition range increases. Excluding the metals and intermetallic compounds, there are 93 materials melting or decomposing between 1750° and 2000°C. The number decreases to 79 for the range 2001°–2250°C, to 43 for the range 2251°–2500°C, to 30 for the range 2501°–2750°C, and to 13 for the range 2271°–3000°C. Only 16 ceramic materials melt or decompose above 3000°C. Considering only non-oxide fibers and an oxide matrix, there are over 20,000 combinations of materials having melting/decomposition points over 1750°C.

B. Volatility

Melting temperature is, of course, only one consideration for a matrix material. Another consideration is volatility. Volatile species can be formed by sublimation (e.g. MgO at high temperature), by disproportionation (e.g. reaction of silicon dioxide to form silicon monoxide and oxygen), oxidation (e.g. reaction of Cr_2O_3 with oxygen to form CrO_3), reduction (e.g. reaction of SiO_2 and SiC to form CO and SiO), and hydration (e.g. reaction of B_2O_3 with water to form HBO_2).

Figure 1 [2] shows vapor pressure as a function of temperature for selected ceramic materials of interest for composites. Hillig reported that a vapor pressure of 0.13 Pa presents no long term problem, while a pressure of 1.3 kPa is probably excessive. Hence, 1.3 Pa is suggested as the upper bound for marginal

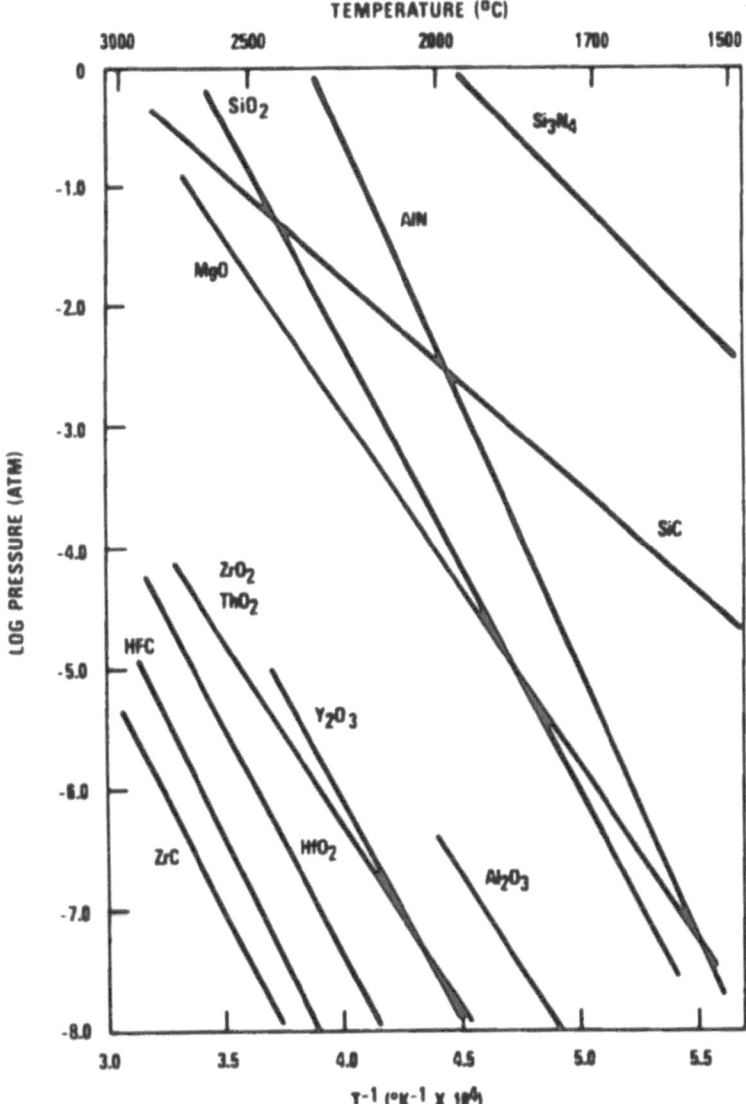

FIG. 1 Vapor pressure as a function of temperature for selected ceramic materials. (From Ref. 2.)

utility. On this basis, materials currently used for ceramic composite matrices, such as silicon carbide and silicon nitride, are not ideal for temperatures as high as 1400°C.

Among oxide materials, zirconia has the highest temperature for the 1.3 Pa maximum vapor pressure requirement, about 2500°C. However, zirconia has a high permeability with respect to oxygen transport and, hence, would not be particularly good for use with a non-oxide fiber or fiber coating that is susceptible to reaction with oxygen.

Wiedemeier and Singh [3] carried out a comprehensive computational thermodynamic analysis of the stabilities of group IV, V, and VI transition metal borides, carbides, nitrides, and oxides. Their calculations suggested that at about 1700°C, V_3B_4 is the most stable boride, followed by V_2B_3 and HfB_2. HfC was predicted to be the most stable carbide and HfN the most stable nitride. Ti_3O_5 and Ta_2O_5 were predicted to be the most stable oxides. For the materials studied, oxides, as a group, were calculated to be the most stable, followed by borides, carbides, and nitrides.

C. Oxidation

Oxidation resistance is obviously an important consideration since many applications involve high temperatures in air. Most non-oxide materials will oxidize at high temperatures. Typically, oxidation at high temperatures follows a parabolic rate law. An acceptable parabolic rate constant is considered to be about 10 $\mu m^2/h$, or 100 μm of oxidation product in 1000 h.[4] The rate is best obtained by plotting oxide thickness versus the square root of time.

Most carbides, borides, and nitrides oxidize at rates that are several orders of magnitude too high to be useful for long term structural applications at 1600°C or higher. At suitably lower temperatures, such materials can, of course, be useful. However, silicon carbide and silicon nitride can potentially be used at up to 1600°C, because one of the oxidation products in each case is silicon dioxide, which forms a coating that significantly retards further oxidation. This film is ineffective when the maximum temperature of silicon dioxide stability is exceeded. Rate constants for selected materials are shown in Figure 2.[2]

If the cation of an oxide is not in its highest possible oxidation state, further oxidation can occur, as described above for Cr_2O_3. Oxide base materials that represent the highest oxidation state of the cation are stable toward further oxidation and, hence, can be attractive matrix candidates providing that the other criteria outlined in this chapter are met.

Ternary oxides, such as $SrZrO_3$, can "demix" in an oxygen gradient due to oxygen diffusion effects, if the diffusion coefficients of each of the component binary oxides and the ternary oxide are sufficiently different.[5] For example, $SrZrO_3$ heated at 1700°C with argon on one side and 20 kPa of oxygen on the

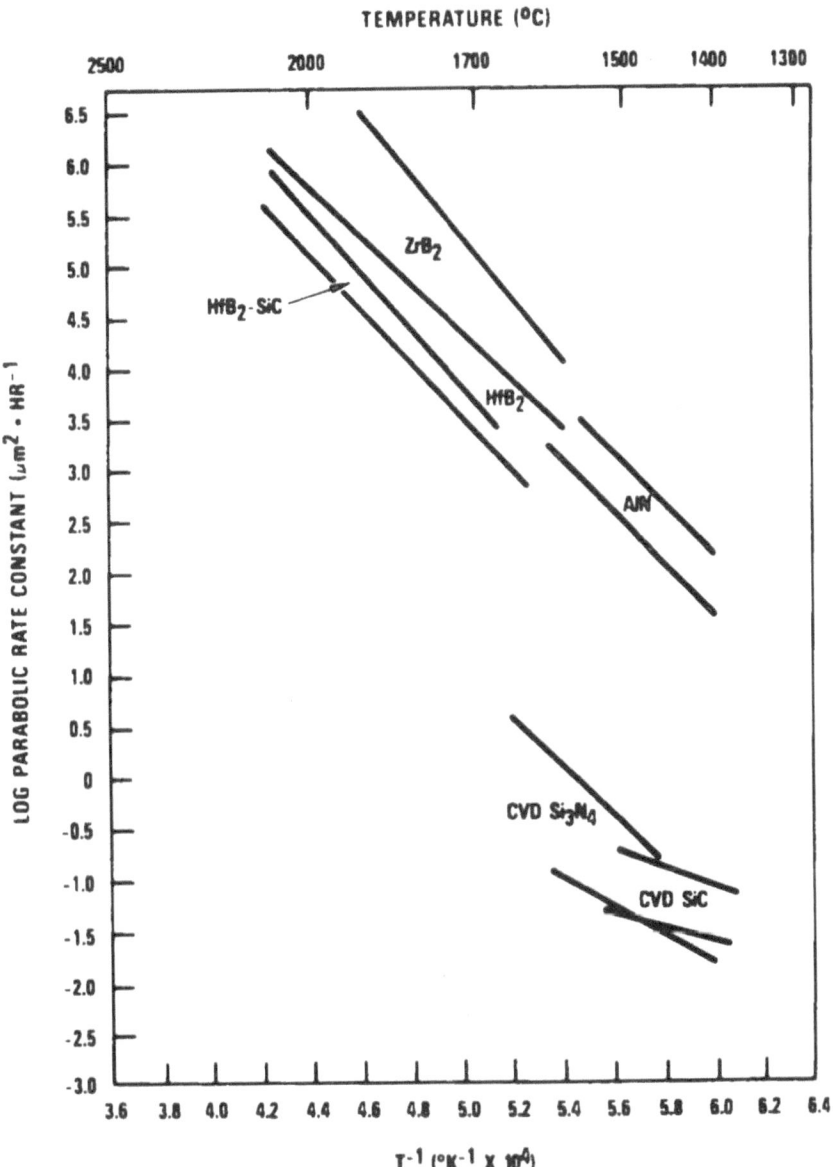

Fig. 2 Parabolic rate constant as a function of temperature for oxidation of selected ceramic materials. (From Ref. 2.)

other completely demixed, although LaHfO$_2$ was relatively stable under the same conditions. In a composite having a non-oxide reinforcement and an oxide matrix, oxidation of the non-oxide material will serve as an internal oxygen sink, and cause an oxygen gradient.

Even a binary oxide with the cation at its highest oxidation state is not necessarily an ideal matrix material. Oxygen transport through the oxide to the reinforcement phase can occur. The permeability constant for the diffusion of oxygen at the application temperature must be about 10^{-10} g O$_2$/cm·s or less to be useful.

D. Chemical Reactivity Between Reinforcement and Matrix Materials

Experimentally determined reactions occurring at high temperatures in available reinforcing fibers and between fibers and oxygen were discussed earlier. Reactions between fibers and matrix (and coating, in many cases) are also a concern. Experimentally determined reactions between fibers and matrix in existing systems will not be discussed in this section, but will be selectively described in later chapters dealing with specific composite systems. This section will briefly cover predictions based on thermodynamic considerations.

Although free energies of formation as a function of temperature are available for most of the ceramic materials of interest, permitting calculations of potential reactions, most of the ceramic fiber/matrix, or fiber/coating/matrix combinations have many components, making such calculations difficult. The complex composition of the most commonly utilized ceramic fiber, Nicalon, without even considering a matrix material, makes calculations of possible reactions a formidable challenge. However, with modern digital computers it is possible to routinely predict reactions for complex systems.

Perhaps the most widely used computer program for thermodynamic calculations is the SOLGASMIX program developed by Gunnar Eriksson of the University of Umea in Sweden.[6] Several versions and many independent modifications of this program have been made. In simple terms, this program is based on minimization of the Gibbs free energy as a function of the chemical potentials and amounts of the possible reacting species and products. External variables such as temperature and pressure are kept constant and mass balance satisfied. Very complicated systems can be examined rapidly, and changing of assumed conditions is not difficult.

There are, of course, limitations to predicting reactions by thermodynamic calculations. Obviously, the calculations are only as good as the database of thermodynamic properties for the materials used in the composite and the possible reaction products. Relatively little high quality thermodynamic data at high temperatures are available for many materials of potential interest. A particular problem occurs when a possible reaction product is not even included in the database. In this case, a prediction of stability under the conditions chosen can be wrong. On the other hand, it is possible that a system predicted to be unstable from the thermodynamic calculations could be adequate for a particular time/ temperature fabrication or application cycle due to favorably slow kinetics. However, at the high temperatures of interest for ceramic matrix composite fabrication and use, one should not be overly optimistic about kinetic favorability.

Some examples of thermodynamic calculations of potentially useful systems by using a modification of the SOLGASMIX program were reported by Greil.[7] In preliminary calculations, he predicted reactions of compositions characteristic of polymer-derived silicon carbide-rich fibers and of silicon carbide fibers deposited by CVD onto a carbon fiber core. The results of these preliminary calculations will not be mentioned here, but they are in reasonably good agreement with experimental results described earlier.

Based on further calculations, carbon is predicted to be stable with silicon carbide in an inert atmosphere over a wide temperature range. Hence, carbon can be used as a low friction coating on silicon carbide fibers or silicon carbide can serve as a protective coating toward air on carbon fibers.

Low stability is predicted for carbon in silicon nitride. At temperatures below 1000°C, carbon can react with silicon nitride to form silicon carbide and nitrogen. However, if the inert gas is replaced by nitrogen, stability increases. A nitrogen pressure of about 10 MPa is predicted to shift the equilibrium to the left. Carbon is also predicted to react with tin nitride, forming tin carbide and nitrogen. Carbon in contact with aluminum nitride is predicted to be more stable, with reaction to form aluminum carbide and nitrogen occurring appreciably at temperatures above 1600°C. With all of the nitride materials, fabrication can be carried out in a nitrogen atmosphere to suppress reactions. In addition, during high temperature use, nitrogen release might be retarded with a dense matrix, again suppressing reaction.

In carbon/oxide systems, metal carbides are expected to form. For example, carbon and titanium dioxide will form titanium carbide and carbon monoxide, with a carbon monoxide pressure of 0.1 MPa at 1300°C. Carbon with aluminum oxide is more stable, with the critical gas pressure being reached at about 1730°C.

Pure silicon carbide fibers should be stable up to very high temperatures in silicon nitride and aluminum nitride matrices. In oxide matrices, silicon carbide shows no stability with titanium dioxide, forming titanium carbide and silicon dioxide, but remains stable in silicon dioxide up to 1600°C. At that temperature, carbon and silicon monoxide are predicted to form.

Work carried out on calculated multicomponent phase diagrams of ceramic materials of potential interest for ceramic matrix composites by Kaufman of

ManLabs, Inc. under sponsorship of the Air Force Office of Scientific Research should be mentioned also. This has included binary and ternary phase diagrams of combinations of Cr_2O_3, MgO, Al_2O_3, SiO_2, CaO, Si_3N_4, AlN, BeO, Y_2O_3, Ce_2O_3, GeO_2, HfO_2, ZrO_2, and TiO_2.[8]

E. Mechanical Properties

Retention of strength and Young's modulus with temperature are important as is creep resistance at high temperatures. Strength, modulus, and, to a lesser extent, creep as functions of temperature for available fibers and whiskers were described in some detail earlier. In this section, emphasis is on a more general overview of other potential reinforcing materials, and on matrix materials, based on Hillig's analysis.[1]

Hillig considers that for ceramic composites to be used in applications currently served by superalloys, stiffness on an equal weight basis (modulus/density) should be comparable, or at least 1.7×10^6 m. On this basis, the calculated acceptable temperature maximum for a material does not directly correlate with its melting point, and can be substantially below the melting point. Approximate values at which modulus/density falls to 1.7×10^6 m for some candidate materials for ceramic matrix composites are as follows: HfC, 2900°C; MgO, 2600°C; AlN, 2500°C; BeO, 2400°C; Al_2O_3, 1900°C; and ThO_2, 1100°C.

Although strength is a very important factor, it is more difficult to calculate comparative values for materials because it is closely related to microstructure, which in turn is very fabrication sensitive. Hillig indicates that at any given temperature, the decrease in strength relative to room temperature will be much greater than the decrease in modulus. It is indicated that strength at one-half the melting point will be about one-half the room temperature strength.

Courtright [4] considers a strength of 150 MPa at the application temperature to be required for typical aerospace applications. Most oxide ceramics do not possess this strength at temperatures of 1600°C or higher, although single-crystal alumina can, if properly oriented.

Many models of steady state creep have been formulated. All involve dislocation movement and/or some mechanism for relative grain boundary motion. Hillig states that a creep rate of 10^{-7}/s is probably close to the maximum level tolerable for long term applications of structural materials where strain is at least 0.0001. For a composite, strain on the fibers can be much greater. For a variety of ceramic materials, temperature for acceptable creep rate divided by the melting temperature varies from about 0.5 to 0.8, with an average of about 0.6. Highest value for single-crystal alumina is 0.83, compared with a value of 0.63 for polycrystalline alumina. For other materials, values for single crystals are not necessarily higher than for polycrystalline materials, indicating the importance of an orientation providing minimum creep motion.

Perfect whiskers, grown with a single screw dislocation, cannot creep because they lack the necessary dislocation structure. However, it is possible that local stress-raising imperfections or, less likely, thermal activation can induce additional dislocations that enhance creep.

Cyclic fatigue is another consideration in ceramic matrix composite applicability. Fatigue behavior has not been studied for ceramic materials to the extent of that for structural metal alloys. Matrix microcracking may enhance fatigue damage (as well as increasing the possibility of atmospheric reaction with reinforcing fibers), so design stress below the matrix cracking stress is probably desirable.

F. Thermal Expansion Coefficient

Another factor in composite fiber/matrix selection is the difference in thermal expansion coefficients. Figure 3 [2] shows expansions of some ceramic and carbon materials as a function of temperature. A few materials, such as some

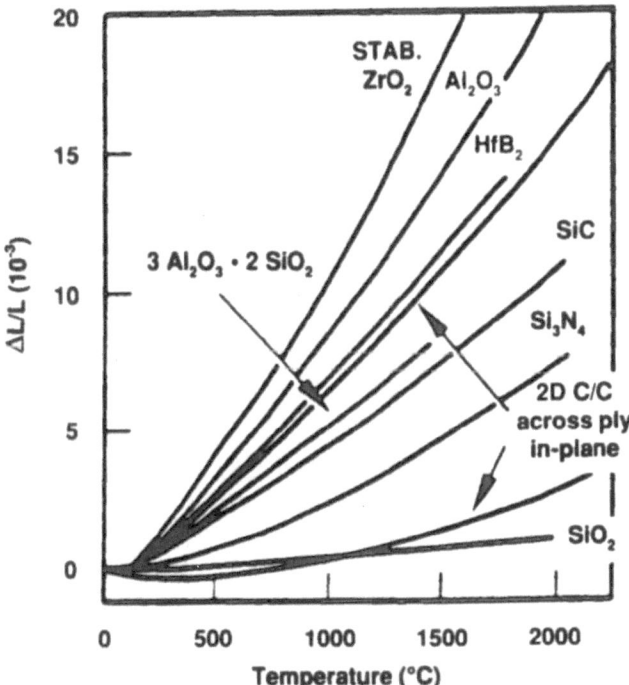

FIG. 3 Linear expansion for selected carbon and ceramic materials as a function of temperature. (From Ref. 8.)

carbon fibers along the fiber length, have little expansion or actually "negative expansion" (contraction) in some temperature regimes. Some materials expand relatively linearly with temperature, while others do not. There are large differences among materials.

There is probably a tolerable limit with respect to differences in thermal expansion coefficients between the fibers and matrix, in order to avoid excessive cracking due to thermal stresses when cooled from the fabrication temperature or heated during application. However, a general rule-of-thumb does not seem to be available at this time.

Some difference in thermal expansion coefficients between the fibers and matrix can be beneficial. For example, at an application temperature below the original fabrication temperature, a matrix having a coefficient of expansion greater than the radial thermal expansion coefficient of the fiber will result in a compressive stress (or "clamping effect") on the fibers, whereas the opposite relative thermal expansion coefficients will result in a tensile stress at the fiber/ matrix interface. Depending on the specific system, including the nature of the interface, either might be beneficial. In general, the requirement for a relatively weak interface might be favored by a tensile stress. Stress in the longitudinal direction due to differences in thermal expansion coefficients between the fibers and matrix might also influence composite performance.

Although the types of interactions described above are undoubtedly significant in affecting composite behavior and are often discussed in composite literature, it is difficult to predict general applicability. The analysis usually conducted is for a composite that has been fabricated at a high temperature (the usual case) but which is now at room temperature (not the usual case in most applications). If application temperature is about the same as the fabrication temperature, there are no internal stresses due to differences in thermal expansion coefficients. If application temperature is higher than fabrication temperature, the stress will be the opposite of that described above. Most often, application temperature is cyclic, which further complicates the situation.

III. OVERALL PREDICTED SUITABILITIES OF CERAMIC MATERIALS FOR HIGH-TEMPERATURE COMPOSITE APPLICATIONS

Some of the individual criteria given by Hillig for selecting ceramic materials for high temperature composite applications (melting point, creep, Young's modulus/density, and volatility) were combined to show potential relative maximum use temperatures of binary compounds and carbon in ceramic composites. Although this analysis does not take into account material strengths or other factors such as the moisture sensitivity of calcium oxide at low temperatures, the limited availability of hafnia, or the health hazards of materials such as beryllia, thoria, and uranium oxide, it is nonetheless informative and will be summarized here.

For binary oxide materials, relative rankings (along with the potential maximum use temperature for the highest and lowest ranked of the the ten most promising materials) are CaO (2150°C) > BeO > ThO_2 > ZrO_2 > UO_2 > HfO_2 > Al_2O_3 > Cr_2O_3 > MgO > SrO (1475°C). For non-oxide materials, the rankings are SiC (2500°C) > C > ZrC > TiC > (HfN) > NbC > HfB_2 > (TaB_2) > TaN (1575°C). HfN and TaB_2 are given in parentheses because data on Young's modulus/density were not available. Also, TaC and HfC, which were favorable from a melting point and creep basis, failed the minimum Young's modulus/density criterion.

IV. SUMMARY

A number of considerations are important in the selection of matrix materials for use with available reinforcing fibers and whiskers, as well as for selecting materials that might be useful candidates for fiber or whisker production in the future. These include melting point, volatility, oxidation, chemical compatibility between fibers and matrix, mechanical properties at elevated temperatures, and thermal expansion coefficient. Experimental data are limited with regard to many of these properties, but available data permit general calculations that provide the framework for design of ceramic composite systems.

REFERENCES

1. Hillig, W. B., Prospects for ultra-high-temperature ceramic composites, *Mater. Sci. Res.*, vol. 20, 697–712 (1986).
2. Strife, J. R., and Sheehan, J. E., Ceramic coatings for carbon-carbon composites, *Ceramic Bulletin*, vol. 67, no. 2, 369–374 (1988).
3. Wiedemeier, H., and Singh, M., Thermal stability of refractory materials for high-temperature composite applications, *Journal of Materials Science*, vol. 26, 2421–2430 (1991).
4. Courtright, E. L., Engineering property limitations of structural ceramics and ceramic composites above 1600°C. Paper presented at the 15th Conference on Composites and Advanced Ceramics, Cocoa Beach, FL, January 14–18, 1991 (Pacific Northwest Laboratory Report PNL-SA-1890).
5. Courtright, E. L., Prater, J. T., Henager, C. H., and Greenwell, E. N., Oxygen permeability for selected ceramic oxides in the range 1200°C–1700°C. Pacific Northwest Laboratory, Final report for period September 1987–May 1989 (May 1991). Wright Laboratory Air Force Systems Command Report No. WL-TR-91-4006.
6. Eriksson, G., Thermodynamic studies of high temperature equilibria, *Chem. Scripta*, vol. 8, 100–103 (1975).
7. Greil, P., Thermodynamic calculations of Si-C-O fiber stability in ceramic matrix composites, *Journal of the European Ceramic Society*, vol. 6, 53–64 (1990).

8. Kaufman, L., Application of computer methods for calculation of multi component phase diagrams of high temperature structural ceramics. ManLabs, Inc, Final report on Contract F-49620-84-C-0078, August 1, 1984–July 31, 1987. Submitted to Air Force Office of Scientific Research, Bolling Air Force Base, DC.
9. Strife, J. R., and Sheehan, J. E., Development of protective coatings for high temperature composites, *Materially Speaking*, vol. 7, no. 1 (1990).

8

Importance of the Fiber/Matrix Interface in Ceramic Matrix Composites

I. BACKGROUND

The nature of the fiber/matrix interface in a ceramic matrix composite is critical to the performance of the composite. As discussed in Chapter 3, several of the mechanisms for toughening that are operable in reinforced ceramics require debonding of the fibers. With strong interfacial bonding, a crack propagating through the matrix will continue through a fiber, if stress is sufficient. With weaker bonding, debonding can occur under stress, resulting in toughening by crack deflection, fiber bridging, and fiber pullout. This is in contrast to a typical metal matrix or polymer matrix composite, where the major goal is more often strengthening, not toughening. In this case, a stronger fiber/matrix bond results in good load transfer from the matrix to the fibers and the presence of the strong fibers results in a stronger material than the monolithic matrix.

The importance of interfacial bonding in a ceramic matrix composite has been recognized in a qualitative manner for some time. Even within a single Nicalon fiber, silicon carbide matrix composite, areas in which the bonds were relatively strong, with fracture occurring through fibers and matrix together, and areas in which there were relatively weak bonds, leading to toughening by bridging and appreciable pullout, could be identified (Figure 1).[1] Examination of the interfaces by scanning electron microscopy showed that intermediate layers were occasionally present at the interface between the fibers and the matrix. Carbon-rich interfacial layers were postulated to be responsible for the desirable, weak bonding areas.

In addition to interfacial bond strength, frictional resistance to sliding of broken fibers through the matrix influences toughening. Hence, the measurement of interfacial friction after debonding is also important.

It might be worthwhile at this point to comment on terminology used in this area. A number of terms including "interfacial shear stress" and "debond fracture energy" are used to describe the strength of the bonds between fibers and matrix (or between fibers, intermediate layers, and matrix). In general, these terms have the same meaning. In the same manner, a number of terms are used to described the extent of friction between fibers and matrix. The terms, "friction(al) stress" and "friction(al) shear strength" are commonly used.

II. MEASUREMENT OF INTERFACIAL AND FRICTIONAL STRESSES

Several types of tests have been developed to determine interface shear strength and frictional stress in ceramic composites. In many cases, the tests were developed for studying other types of composites, but have been adapted for use with ceramic matrix composites. This section outlines the kinds of experimental techniques that have been used for determining interfacial properties of ceramic

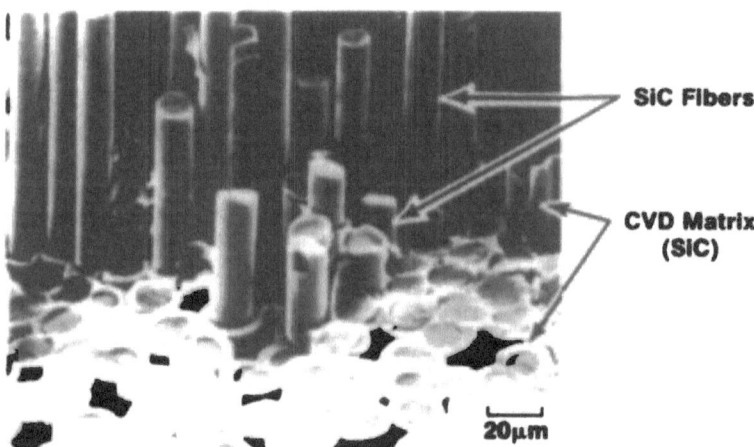

FIG. 1   Scanning electron micrograph of a ceramic matrix composite surface showing areas having strong interfacial bonds and areas having weak interfacial bonds. (From Ref. 1.)

matrix composites.[2] The most widely used and useful type of test will be covered in more detail in the next section.

One method proposed for determining interfacial shear strength is matrix crack spacing after composite tensile testing. A model by Aveston et al [3] proposes a uniform crack spacing in the matrix of a unidirectional composite, with the spacing dependent upon the interfacial bond strength. Multiple matrix cracking has been observed in non-ceramic systems, and reasonable values for interfacial shear strength calculated. More recently, matrix crack spacing has been used in analysis of several glass and ceramic matrix systems. Values for interfacial shear strength were in the range 1-50 MPa. However, it must be noted that pre-existing flaws in a typical ceramic matrix material are likely to affect the location of matrix cracks to a considerable extent, so that the method may not be very applicable to most systems.

A fiber pullout test was originally designed for polymer matrix composites, and has been adapted for investigating interfacial bonding in ceramic matrix composites. In this test, a model composite containing a single embedded fiber with part of it extending from the matrix is fabricated. The test is difficult to perform with small diameter fibers, so large diameter fibers such as SCS-6 have generally been used. In another variation of the test, the two ends of a single fiber are embedded in the same matrix material, with the middle part of the fiber exposed. Figure 2 schematically shows these variations. In both cases, the fiber is pulled out of the matrix using a tensile-testing machine, and load and displacement are measured.

Typically, a peak load is reached and then load drops off abruptly. The peak load is believed to occur when the fiber debonds. A further decrease in load with displacement is presumed to be due to fiber pullout from the matrix. This latter part of the stress-strain curve is thus considered to be indicative of the frictional stress. Unloading and reloading before complete pullout makes it possible to measure both static and dynamic frictional stresses.

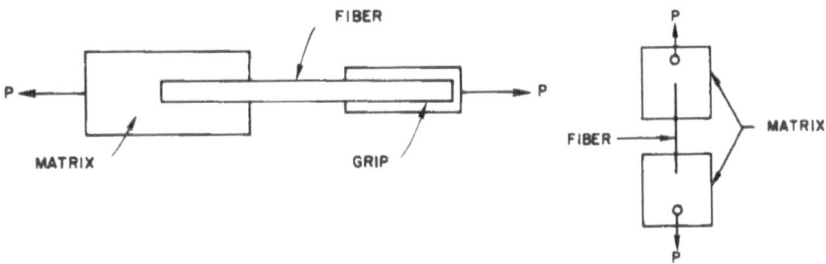

FIG. 2 Schematic illustrations of single fiber pullout tests for determining interfacial bond strengths. (From Ref. 2.)

Analysis of the results of the pullout test must take into account a number of factors. The tensile stress produces a Poisson's contraction of the fiber, which produces a radial tensile stress at the interface, facilitating debonding and pull-out. In addition, the imposed shear stress is not constant along the length of the embedded fiber, but decays with distance from the surface. Taking these factors into account and incorporating fiber radius, embedded length, and elastic constants of fibers and matrix, the peak shear strength, friction shear strength, and coefficient of friction can be calculated.

In a typical study [4], SCS-2 fibers, SCS-6 fibers, uncoated SCS fibers, and experimentally carbon coated SCS fibers were investigated. Hence, the effect of coating on interfacial bond strength could be determined. Matrix material was soda-borosilicate glass. Matrix compositions were varied to provide different coefficients of thermal expansion. In all cases, glass CTE was greater than fiber CTE, so that effects of clamping (compressive) stresses on the fibers from the different glass compositions could be examined. Effects of fiber coatings were apparent. In addition, interfacial shear stress increased with increasing matrix CTE to a maximum value, then decreased. The decrease was shown by SEM measurements to be due to radial matrix cracks caused by an excessive thermal expansion mismatch. Other very extensive work, which will not be described here, on effects of glass matrix CTE on interfacial strength was reported by Hegeler and Brückner.[5]

Various other workers using this technique on a number of SCS-6 fiber, glass matrix specimens have reported peak shear strengths ranging from about 5–17 MPa.[2] Friction shear strengths ranged from about 4–14 MPa. Friction coefficients were calculated to be 0.05 and 0.2 on two of the systems.

It appears that fiber pullout testing is a reasonable method for determining interfacial properties of ceramic matrix composites. However, sample preparation can be tedious and the test is difficult to run with the small diameter fibers of most interest for ceramic reinforcement.

Most commonly used for measuring interfacial shear and frictional properties in ceramic matrix composites is the indentation technique. Marshall [6] first reported on the use of a sharp indenter (the Vickers Pyramid indenter, designed for determining hardness) to push on a fiber perpendicular to the polished surface of a ceramic composite specimen to measure interfacial properties.

Several versions of this method have since been developed. Both thick specimens, in which only part of the fiber debonds, and thin slices, in which the total length of the fiber section is debonded, have been used. The latter is commonly called a pushout test. Calculation of frictional sliding stress is difficult with thick specimens because the length of fiber that has debonded is not known, although approximate calculations can be made using appropriate assumptions. Interpretation with thin slices is more straightforward, but thin sections of composites with small diameter fibers are more difficult to prepare.

Because indentation testing has been the major method of studying interfacial properties in continuous fiber reinforced ceramics, the indentation technique (including the pushout modification) will be described in more detail in the next section.

A method of studying interfacial shear strengths between potential fiber and matrix materials has been proposed for cases where fibers of a material of interest have not yet been produced.[7] The method involves production of a sandwich structure of the two kinds of materials. A model system using a silicon carbide layer sandwiched between Pyrex glass layers was tested. The experimental interfacial shear stress values were close enough to those determined in composites with silicon carbide fibers in Pyrex to offer support for the method, but more development will be required for verification of its usefulness.

III. THE INDENTATION TEST FOR STUDYING FIBER/ MATRIX INTERFACES

In Marshall's original version of an indentation test, the end of an individual fiber normal to the direction of the surface was pushed upon by a pyramidal indenter, and the resulting displacement of the fiber below the surface of the matrix due to sliding was determined. The frictional stress was calculated from the force-displacement relation. The force and displacement were measured only at the peak load. A constant shear resistance at the interface was assumed. It was assumed also that the length over which sliding occurred was large compared to the fiber diameter. The technique provided values for average frictional stresses at individual fibers.

Marshall and Oliver [8] subsequently reported on an improved test version. They used an ultra-low-load indentation instrument with which forces could be measured with a resolution of 0.3 μN and displacement measured with a resolution of 0.16 nm. The indenter was a triangular pyramid. A coil and magnet arrangement was used to move the indenter toward the surface and to apply the load. The position of the indenter was measured using a capacitive displacement gage. Loads up to 0.12 N were applied at a constant velocity of 10 nm/s, resulting in a typical loading time of 100 s. Figure 3 is a schematic representation of the indentation process. Figure 4 is a scanning electron micrograph of a fiber and surrounding matrix after indentation.

In these experiments, force was increased monotonically to 0.1 N, decreased to 0.002 N, and cycled between the two values five times. A typical force/displacement curve for a Nicalon fiber in a lithium aluminosilicate matrix was roughly linear to a displacement of about 1.5 μm. Unloading resulted in a permanent displacement of about 0.5 μm. Reloading produced about the same 1.5 μm displacement as found with the original loading. However, as shown in Figure 4, a correction must be made for the permanent indentation of the fiber

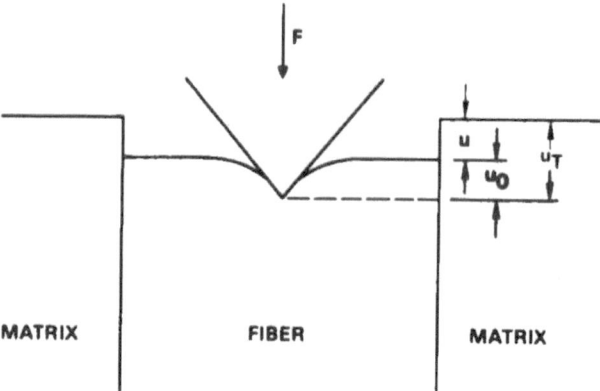

FIG. 3   Schematic representation of the indentation process for determining fiber/matrix interfacial strength. (From Ref. 6.)

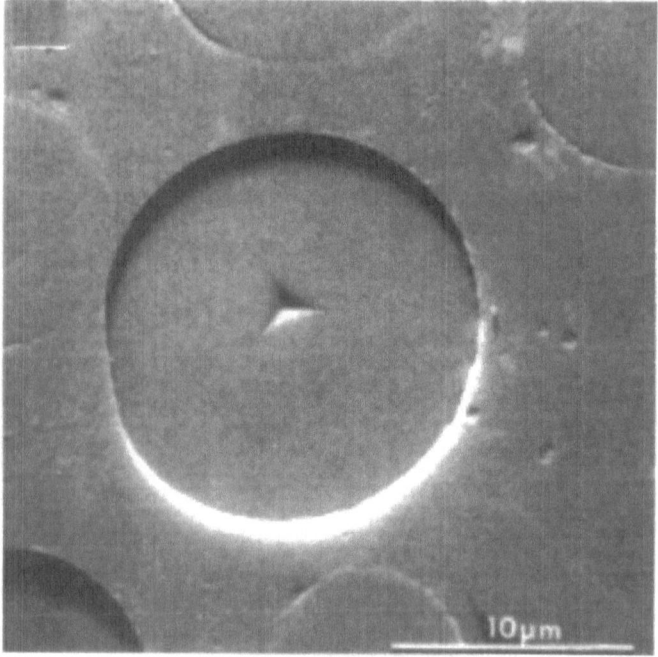

FIG. 4   Scanning electron micrograph of a fiber and surrounding matrix after an indentation test for determining interfacial bond strength. (From Ref. 6.)

surface after removal of the indenter. This correction, u_o in Figure 3, was subtracted from the measurements to indicate the true fiber indentation behavior. Frictional stress was calculated to be 3.6 MPa.

Marshall and Oliver presented mathematical models for the interface debonding and frictional sliding. From their analysis, they concluded that the fracture (or debond) energy was small relative to the frictional stress. They concluded also that frictional stress was very uniform along a fiber and not affected by irregularities of fiber shape or thickness that occur over lengths equal to or larger than the fiber radius.

For a typical fiber in a composite specimen that had been heat-treated for 10 min at 1000°C in air, loading and unloading produced little displacement. Behavior was similar to that when an indenter is used on a monolithic material. Thus, heat treatment resulted in much stronger interfacial bonding.

Another version of an indenter test was reported by Mandell et al.[9] Originally, the test was used for studying carbon and glass fibers in organic matrix composites. In this test version, the compressive load was applied in a series of steps of increasing force, with the specimen inspected for the onset of debonding after each step. The force necessary to produce debonding was recorded, and a micrograph was taken of the fiber and the surrounding region. A rounded indenter was used.

A micrometer base allowed the initial location of the indenter to be loaded to within 1 μm. A turntable with magnetic stops permitted repeated rotation from the loading to the viewing position, also with an accuracy of better than 1 μm. The indenter, with a 5 to 15-μm radius ground diamond tip allowed soft loading contact through an extensometer, with a load measurement to within 0.1 g and frictional sticking of less than 0.1 g. A research quality light microscope capable of detecting debonding was used. The apparatus was mounted on a vibration isolation table.

Interfacial shear strength was calculated from the debonding force using the results of a finite element calculation incorporating fiber diameter, spacing to the nearest adjacent fiber, and elastic properties of the fibers, matrix, and "far field composite." For the frictional stress calculation, they used the method of Marshall.

Mandell et al [10] reported on the use of the apparatus described above for determining interfacial properties of glass and glass-ceramic matrix composites with Nicalon and carbon fibers. Debond strengths were calculated to be about 10 MPa for high modulus carbon fibers in borosilicate glass, 54 MPa for Nicalon fibers in lithium aluminosilicate glass-ceramic, 60 MPa for Nicalon fibers in barium magnesium aluminosilicate glass-ceramic, 239 MPa for Nicalon fibers in Corning Code 1723 glass, and 249 MPa for Nicalon fibers in calcium aluminosilicate glass-ceramic. Frictional stress for the Nicalon/lithium aluminosilicate system was calculated to be 3.2 MPa, and that for the Nicalon/barium magnesium aluminosilicate system was 5.5 MPa. Although they calculated a frictional stress of 143 MPa for the Nicalon/Code 1723 glass system, they concluded that a frictional stress value cannot be validly calculated for a thick specimen having a high debond strength.

Bright et al [11] gave details of a thin section fiber pushout test, a method that had been used earlier by Laughner et al [12] and by Brun and Singh.[13] A schematic diagram of the apparatus used by Bright et al is given in Figure 5. A special tungsten carbide pushout tool with a conical tip was used. A typical experimental sequence involved making an indent at an arbitrary location on the pushout specimen, using a low (10-N) load. The micropositioner and specimen assembly were then moved to a stereomicroscope and the positioning indent was centered under the cross wires of the eyepiece by moving the entire assembly. The coordinates of the centers of a number of fibers were then selected. The

CROSSHEAD

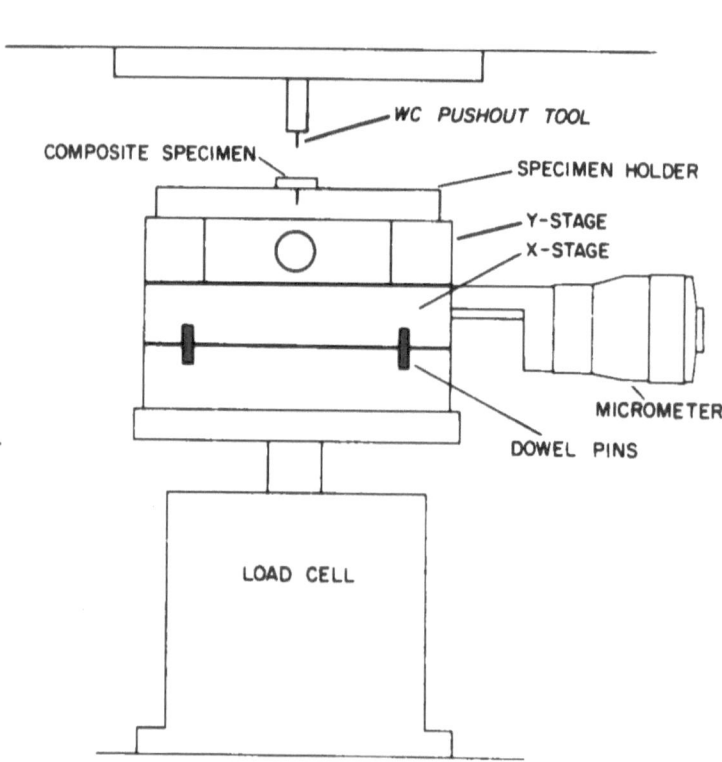

FIG. 5   Schematic illustration of a thin section fiber pushout test. (From Ref. 11.)

micropositioner and specimen assembly were transferred back to the universal testing machine and the selected fibers were, in turn, identified by using the coordinates. The pushouts were conducted at a constant displacement rate of 50 μm/min.

A typical load-time curve on an SCS-6, Code 7740 borosilicate glass specimen is shown in Figure 6. The load drop P_D corresponds to partial debonding of the fiber. Following debonding, the load increased slightly and then gradually decreased. P_{max} corresponds to the point at which the entire length of the fiber is debonded and the fiber starts to exit from the opposite face of the specimen. Several mathematical models were used to evaluate the interfacial properties.

Interfacial bond strengths for a unidirectional SCS-6 fiber, reaction bonded silicon nitride matrix composite, a unidirectional SCS-6 fiber, Code 7740 glass matrix composite, and a 0/90 cross-ply SCS-6 fiber, Code 7050 glass matrix composite were similar, ranging from 17.5–19.1 MPa. However, sliding friction parameters were quite different for the third material. For example, coefficients of friction values were calculated to be 0.27 and 0.29 for the first two materials, but 0.94 for the third material. It was postulated that this was due to curvatures

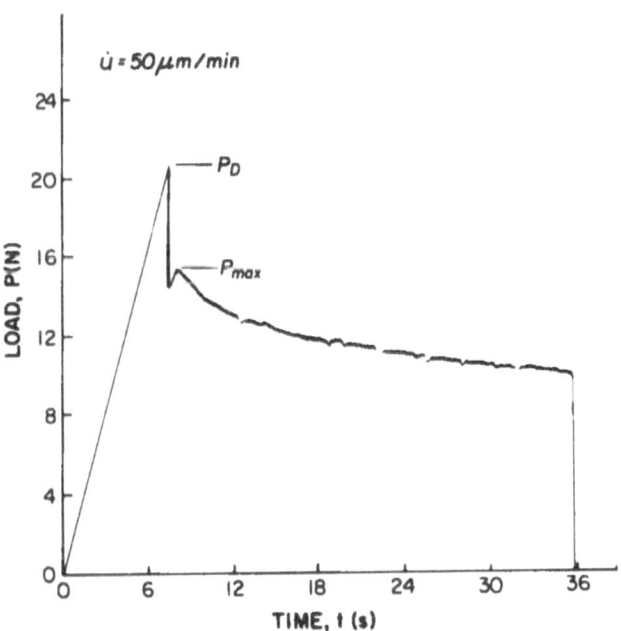

FIG. 6 Typical load-time curve for an SCS-6 fiber, Code 7740 borosilicate glass matrix composite using a thin section fiber pushout test. (From Ref. 11.)

developed in the fibers in the cross-ply composite during hot pressing. A fiber with a periodically varying curvature would be expected to result in greater frictional resistance to sliding than a straight fiber.

IV. INTERFACIAL STRESSES FOR WHISKER REINFORCED CERAMICS

It can be assumed that whisker debond strengths and interfacial frictional properties of whisker reinforced ceramics are also important. However, direct measurement would be difficult. Typical problems are random orientation, small diameters, and variable lengths of whiskers. Effects of interfacial properties have been inferred for some whisker systems by comparisons of whisker surface chemistry with composite mechanical properties. Studies of this type will be mentioned later in conjunction with specific whisker reinforced materials.

V. SUMMARY

A number of methods have been developed for investigation of fiber/matrix interfacial debond stress and frictional stress after debonding. An indentation test with either relatively thick specimens, for which only partial fiber debonding occurs, or with thin specimens, for which a fiber can be totally debonded and pushed through the matrix, has been used most often. The thick specimen test is easier to conduct experimentally, but more difficult to interpret because the debonded length cannot be measured directly. Although results from the thin section test are easier to interpret, experimental difficulties with small diameter fibers have generally limited this test to composites containing large diameter fibers such as SCS-6 fibers.

Although values for debond stress and frictional stress measured on similar specimens by different researchers are not always in good quantitative agreement, similar qualitative trends are invariably found. For a composite having significantly improved toughness over the monolithic matrix material, values for both debond stress and frictional stress are relatively low, usually less than 50 MPa, and often lower than 10 MPa. Higher values, usually over 100 MPa, are generally found for poorly performing composites.

REFERENCES

1. Stinton, D. P., Caputo, A. J., Lowden, R. A., and Besmann, T. M., Improved fiber-reinforced SiC composites fabricated by chemical vapor infiltration, *Ceram. Eng. Sci. Proc.*, vol. 5, no. 7–8, 983–989 (1986).
2. Kerans, R, J., Hay, R. S., Pagano, N. J., and Parthasarathy, T. A., The role of the fiber-matrix interface in ceramic composites, *Ceramic Bulletin*, vol. 68, no. 2, 429–442 (1989).

3. Aveston, J., Cooper, G. A., and Kelly, A., Single and multiple fracture. In *The Properties of Fiber Composites, Conference Proceedings*, National Physical Laboratory, Guildford IPC Science and Technology Press, Surrey, England, 1971, 15–26.

4. Goettler, R. W., and Faber, K. T., Interfacial stresses in fiber-reinforced glasses, *Composites Science and Technology*, vol. 37, 129–147 (1989).

5. Hegeler, H., and Brückner, R., Fibre-reinforced glasses: influence of thermal expansion of the glass matrix on strength and fracture toughness of the composites, *Journal of Materials Science*, vol. 25, 4836–4846 (1990).

6. Marshall, D. B., *J. Am. Ceram. Soc.*, vol. 67, no. 12, 259–260 (1984).

7. Parthasarthy, T. A., Pagano, N. J., and Kerans, R. J., A technique to measure the interfacial bond strength, *Ceram. Eng. Sci. Proc.*, vol. 10, no. 7–8, 872–881 (1989).

8. Marshall, D. B., and Oliver, W. C., Measurement of interfacial mechanical properties in fiber-reinforced ceramic composites, *J. Am. Ceram. Soc.*, vol. 70, no. 8, 542–548 (1987).

9. Mandell, J. F., Grande, D. H., Tsiang, T.-H., and McGarry, F. J., Modified microdebonding test for direct *in situ* fiber/matrix bond strength determination in fiber composites. In *Composite Materials–Testing and Design* (7th Conference), J. M. Whitney (Ed.), American Society for Testing and Materials, Philadelphia, PA, 1986, 87–108.

10. Mandell, J. F., Hong, K. C. C., and Grande, D. H., Interfacial shear strength and sliding resistance in metal and glass-ceramic matrix composites, *Ceram. Eng. Sci. Proc.*, vol. 8, no. 7–8, 937–940 (1987).

11. Bright, J. D., Shetty, D. K., Griffin, C. W., and Limaye, S. Y., Interfacial bonding and friction in silicon carbide (filament)-reinforced ceramic- and glass-matrix composites, *J. Am. Ceram. Soc.*, vol. 72, no. 10, 1891–1898 (1989).

12. Laughner, J. W., Shaw, N. J., Bhatt, R. T., and DiCarlo, J. A., Simple indentation method for measurement of interfacial shear strength in SiC/Si_3N_4 composites. Presented at the 10th Annual Conference on Composites and Advanced Ceramic Materials, Cocoa Beach, FL, January 1986.

13. Brun, M. K., and Singh, R. N., Effect of thermal expansion mismatch and fiber coating on the fiber/matrix interfacial shear stress in ceramic matrix composites, *J. Am. Ceram. Soc.*, vol. 3, no. 5, 506–509 (1988).

9

Fiber Coatings

I. BACKGROUND

Fiber coatings can have an important influence on the debond strength and frictional stress in a ceramic matrix composite. A coating providing an appropriately weak interface results in a tough composite, with fiber pullout evident at fracture surfaces (Figure 1). An early indication of the presence of a variable carbon-rich layer in Nicalon fiber reinforced silicon carbide was outlined earlier, and many more investigations into fiber coatings have been conducted. These have involved both coatings formed *in situ* and coatings deposited on fibers prior to composite fabrication. In addition to influencing interfacial properties, coatings have the possibility of retarding fiber-matrix reactions and retarding fiber oxidation.

II. FIBER COATINGS FORMED *IN SITU*

A well documented example of the presence of a beneficial carbon-rich coating formed *in situ* has been found for Nicalon fiber/glass matrix composites such as Nicalon/lithium aluminosilicate (LAS). Development of tough composites of this type at United Technologies Research Center and elsewhere will be discussed in more detail in a later chapter, but some coverage will be given here because of the relevance to the present subject. In conducting investigations in this area, it was found that certain fabrication conditions or certain matrix compositions resulted in brittle composites different from those usually observed in the developmental work. Fracture surfaces of the brittle composites were indicative of strong interfacial bonding, since there was little fiber pullout. The fracture surfaces of the tough composites indicated relatively weak interfacial bonding, since there was extensive fiber pullout. Little else differed between the tough composites and the brittle composites.

In a key investigation in this area, Brennan [1] studied composites of Nicalon fibers in three LAS compositions. These included LAS-I, which is similar in composition to Corning 9608, but the titanium oxide (~3 wt%) used as a nucleating agent is replaced by the same amount of zirconium oxide. LAS-II is identical to LAS-I, except that it contains ~5 wt% niobium oxide to form a niobium carbide barrier between the fibers and matrix. LAS-III also contains the niobium carbide addition, but differs from the other materials in that it is formulated to be more refractory (have a higher softening point).

Fabrication details will given in a later chapter. However, at this point it will be mentioned that the composites were hot pressed at 1000°–1350°C. Some materials were also ceramed (post heat treated to convert the matrix from a glassy state to a partially crystallized state) at 900°–1350°C. The major methods of interface investigation were the scanning transmission electron microscope

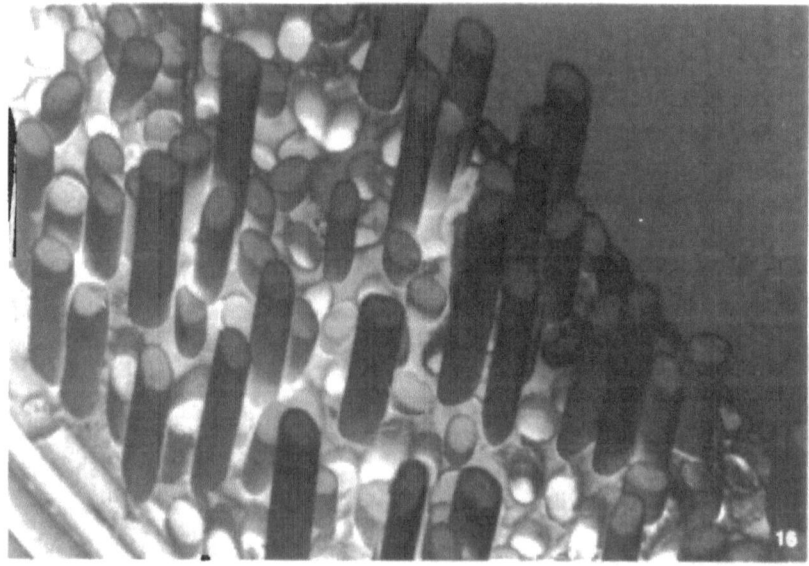

Fɪɢ. 1   Fracture surface of a fiber reinforced ceramic exhibiting fiber pullout. (Courtesy of 3M Ceramic Materials Department.)

(STEM) and the scanning Auger electron microprobe (SAM), but other techniques were used as well.

Expected matrix phases were observed with these composites. With matrices containing the niobium oxide addition, a reaction layer of niobium carbide could be detected near each fiber. The niobium oxide added to the LAS composition apparently reacts with the fibers during hot pressing to form this layer. In addition, in many cases a second thin ring was detected around the fibers. This ring was not present on the fibers initially, and was never observed when Nicalon fibers were incorporated into a resin matrix composite. From TEM investigation of fractured composite surfaces, it was found that a very thin film of material often adhered to the fiber surfaces. The diffraction pattern of this material correlated quite well with that of carbon. Some of the films also contained small amounts of silicon and/or niobium carbide particles.

It was deduced that this film was at least part of the ring observed between the niobium carbide layer and the Nicalon fiber, and that it was weakly bonded to both the matrix and fiber. It was postulated that it was the main factor leading to low debond strength and, hence, high toughness. This layer was not detected or was very much reduced in carbon content in composite specimens exhibiting weak, brittle behavior.

It was found from SAM depth profiling that most fiber surfaces (from which the vinyl acetate size was removed) were oxygen rich and low in carbon. However, in the composites the surface composition often changed from oxygen rich to extremely carbon rich. In a composite having an LAS-I matrix, the carbon layer appeared to be of the order of 100 Å of pure carbon plus another 200 Å of carbon that graded into the usual fiber composition. On a specimen with an LAS-III matrix, a very carbon-rich layer of 300–500 Å was identified. Carbon-rich layers on matrix "troughs" on fracture surfaces could be detected also.

The detailed mechanism resulting in the carbon-rich layers in these composites is not known, but it is obvious from these studies that the chemistry and/ or microstructure of the Nicalon fibers is involved with the type of interface formed. Cooper and Chyung [2] conducted additional investigations on composites of the general type described above. They included composites having matrices of Code 1723 glass and calcium aluminosilicate (CAS) and LAS glass-ceramic matrices. One of the composites having the Code 1723 glass matrix had a low ultimate strength of 200 MPa and a brittle fracture morphology whereas the other had an ultimate strength of 870 MPa and a fibrous morphology. The only difference between the two composites appeared to be the thermal history.

Although both of the CAS matrix composites had a fibrous fracture morphology, one had an ultimate strength of 325 MPa and the other had a strength of 995 MPa. The only difference between these composites was the addition of 1 wt% arsenic oxide as a "fining agent" in one case. (In the manufacture of glass, a fining agent is a material added to molten glass to make the melt blister-free by either resorption or expulsion of gas bubbles.) The single LAS matrix specimen had a strength of 925 MPa and fibrous fracture morphology. Methods of analysis included high-resolution transmission electron microscopy (HRTEM), and analytical, scanning transmission electron microscopy (AEM, STEM).

For each composite having a tough, fibrous fracture morphology and high strength, a reaction-layer type fiber-matrix interface could be identified by TEM. This layer ranged from 0.02 to 0.2 μm in thickness. In the tough composite with the Code 1723 matrix, a 0.08-μm thick reaction-layer interface was found. With the weak, brittle composite having the same matrix, no layer could be detected. Differences were also found between the two CAS matrix composites. In one case, the layer was about 0.02 μm thick, while in the other thickness was about 0.2 μm. As found earlier by Brennan, the films were carbon rich, and this work indicated that the layer around a fiber contained very fine crystallites.

Comparisons among specimens indicated that the interfacial layer is enhanced by an increase in processing temperature for a constant time, an increase in time at a relatively high constant temperature, and the addition of a fining agent to the matrix. The Code 1723 matrix composite with no detectable interfacial layer was processed at 950°C, while the other Code 1723 matrix composite was processed for an identical time at 1350°C. The CAS matrix composite developed a thicker interfacial layer (from 0.02 to 0.06 μm) when given an additional 2-h anneal at 1200°C. The fining agent added to a CAS matrix increased the thickness of the reaction layer from 0.02 to 0.1 μm. These observations indicate that the reaction layer formation is a dynamic diffusional process controllable with matrix composition and/or processing variables.

CAS matrix specimens were exposed to 600° and 900°C in air for 20 min. After these treatments, the carbon-rich interfacial layers were either appreciably reduced in thickness or removed, since they could not be identified by the methods used on the original samples. At some interfaces, there was evidence that the carbon-rich layer had been replaced by an amorphous silica-rich layer. Evidence for a silica containing layer on Nicalon fibers heat treated in air was indicated in an earlier chapter.

III.   FIBER COATINGS APPLIED PRIOR TO COMPOSITE
FABRICATION

A.   General

Since it had been demonstrated in studies such as the ones described above that a carbon layer generated by *in situ* reactions can provide an appropriately weak interfacial bond strength, a natural outgrowth has been to deposit thin carbon coatings onto fibers prior to composite fabrication. Indeed, since the identification of the *in situ* coatings, much of the work conducted to date on ceramic matrix composites has utilized carbon-coated fibers.

However, carbon is not an ideal coating for composites intended for use in an oxidizing atmosphere at a temperature exceeding about 400°C, unless the matrix (or an external composite coating) is impervious to oxygen and is not microcracked during use. Even with an impervious matrix, the structure must not have fiber ends terminating at the surface, since the phenomenon of ''pipeline oxidation'' down the length of the coating is possible. Hence, improved coating materials have been sought. Only boron nitride has been well documented to give good interfacial properties in some composite systems. The application of other coatings, including duplex coatings, may be beneficial, but there is, as yet, little published documentation.

B. Carbon Coatings

1. CARBON COATINGS BY CHEMICAL VAPOR DEPOSITION

Although diamond and diamond-like carbon coatings have been developed recently and have many potential applications, it is the graphite structure (Chapter 5, Figure 1) that is of value for fiber coatings for ceramic matrix composites. The weak bonds present in the c-direction facilitate cleavage, and help produce the desirably weak interface useful for these composites.

Only when the graphite-type carbon material has been heat treated to close to 3000°C does the structure have the regularity shown in the figure cited above, but the basic structure is set at much lower temperatures, such as during deposition from gaseous compounds by chemical vapor deposition (CVD). This technique was described briefly in an earlier chapter on ceramic fiber production and will be described in more detail in a later chapter on composite fabrication. CVD involves deposition of a solid material onto a heated substrate (a fiber, in this case) from gaseous precursors.

The deposition of graded coatings including carbon-rich layers during manufacture of SCS silicon carbide fibers has been described. Although carbon coatings on other fibers such as Nicalon have apparently been used in a majority of successful continuous fiber reinforced ceramic matrix composites discussed in published literature, relatively few references have given the conditions used. Often, even the fact that a carbon coating was used is not mentioned. Some conditions that have been reported are given below.

Lowden et al [3] reported preliminary removal of the size from a Nicalon cloth preform by multiple washings with acetone. Carbon was then deposited isothermally by the decomposition of propylene at 1000°C in argon under 5 kPa pressure. They also reported a secondary motivation for the carbon coating, protection of the fibers from reactants and products containing chlorine during their matrix deposition process. Lowden and More [4] described slightly different conditions, argon/propylene mixtures at 1100°C under 3 kPa for 2 h. Coating thicknesses were varied by changing the concentration of the propylene.

Studies by Fischbach and Lemoine [5] involved heating at 650°C in air to burn off the polyethylene oxide size on Nicalon fibers, heating at 980°C in nitrogen, and then depositing carbon at 1100°C from a mixture of helium, hydrogen, and ethane for less than 0.5 h. They concluded that the coating process caused some decrease in fiber strength.

Brennan [6] conducted a number of experiments in which carbon coatings were produced on fibers during deposition of a silicon carbide matrix by CVD. Carbon rather than silicon carbide was initially deposited when argon was used as a flushing gas prior to exposure to methyldichlorosilane and hydrogen or to methyltrichlorosilane and hydrogen, but not when hydrogen was used as the flushing gas. Results were similar with both Nicalon fibers and Nextel 440 fibers. Other aspects of this work will be discussed in a later chapter.

CVD-deposited carbon coatings have provided excellent room temperature properties for continuous fiber reinforced ceramics. For example, in the work of Lowden and More, Nicalon fiber/silicon carbide matrix composites fabricated using chemical vapor infiltration were weak and brittle with uncoated fibers. Flexural strength averaged 82 MPa.

The carbon coatings reduced frictional stress at the fiber/matrix interface to between 1/5 and 1/100 of that with uncoated fibers. With a 0.07-μm thick carbon coating, strength increased dramatically to 344 MPa and fiber pullout was observed. Maximum strength, 420 MPa, and increased pullout were obtained with a coating thickness of 0.17 μm. When thickness was increased to 0.28 μm, flexural strength was 390 MPa, and with a thickness greater than 0.5 μm, strength was 352 MPa. The interfaces were postulated to be overly weak as coating thickness exceeded the value giving maximum composite strength, resulting in less effective load transfer from the matrix to the fibers.

As another example of the benefit of CVD carbon coated fibers, Doughan et al [7] studied alumina fiber/borosilicate glass matrix composites fabricated by slurry infiltration and hot pressing. The fibers were Fiber FP. Coating thickness was about 1 μm. Coating conditions (by Pfizer Corporation) were not reported. In four-point flexural testing, composites with 20 vol% uncoated fibers averaged 118 MPa, while composites containing the same vol% coated fibers averaged 263 MPa. Relative results were similar with 45 vol% fibers.

2. CARBON COATINGS BY POLYMER DECOMPOSITION

Although not yet as widely used, polymer decomposition is an alternative to chemical vapor deposition for producing a carbon coating. Fiber tows can be immersed in a solution of a polymer, such as a phenolic resin. After drying to remove the solvent, the coating can be heat treated in an inert atmosphere to produce carbon. Coating thickness can be altered by varying the concentration of the polymer in the solvent. It has even been indicated that the polymeric size material applied to Nicalon fibers to improve handling can decompose to form a thin carbon coating which improved properties of Nicalon/silicon carbide matrix composites.[8]

3. DEGRADATION OF CARBON COATINGS AT ELEVATED TEMPERATURES

Unfortunately, the benefits of carbon coatings can be largely negated at elevated temperatures, or after heat treatment at elevated temperatures, in an oxidizing atmosphere. The carbon coating is normally replaced by a reaction layer, which results in strong bonding, and properties are similar to those obtained at room temperature without the carbon coating. Even if not replaced by a reaction layer, the gap left by the oxidized carbon would result in poor composite properties due to poor load transfer.

Only if the matrix is highly dense and no fiber ends are exposed, or if a dense external coating is applied, can oxidation of the carbon coatings be delayed considerably and satisfactory high temperature composite properties attained.

C. Boron Nitride Coatings

1. BORON NITRIDE COATINGS BY CHEMICAL VAPOR DEPOSITION

Boron nitride exists in two forms, cubic and hexagonal. The cubic form has a structure analogous to diamond, while the hexagonal form has a structure analogous to graphite (Figure 2).[9] The laminar-type structure of boron nitride might be expected to produce the same type of relatively weak interfacial bonding as carbon. However, boron nitride has somewhat more stability toward oxidation than carbon. One product from the oxidation of boron nitride is boron oxide which, albeit low melting, is somewhat protective toward further oxidation. With carbon, only gaseous species form.

The use of boron nitride as a fiber coating for ceramic matrix composites was patented by Rice in 1987 [10], and considerable work by others has followed. Rice demonstrated significantly improved composite properties with boron nitride coatings on fibers.

The recommended boron nitride precursor for CVD was borazine, with decomposition at a temperature of about 900°C. Coating thicknesses of 0.05 μm to several tenths of a micron were reported to be effective. Nicalon silicon carbide base fibers, Sumitomo alumina base fibers, and carbon fibers were coated. The coatings were found to be useful with a number of matrix compositions. Several matrix formation techniques were used. Consolidation of the matrix was by hot pressing at 1350° to 1500°C. Strength tests on Nicalon fibers before and after coating with boron nitride indicated that there might have been some degradation during coating but this was statistically uncertain.

A dramatic effect of the boron nitride coating is shown in Figure 3 for a Nicalon fiber/silica composite. With uncoated fibers, strength was low and the failure was catastrophic. With the fiber coating there was a fourfold increase in strength and a large increase in toughness, as indicated by the much larger area under the stress-strain curve.

Not all composite systems benefited from the boron nitride fiber coatings. For example, use with a zirconia matrix resulted in little or no improvement in strength or toughness, relative to a composite with uncoated fibers. The same was true with cordierite matrix composites. The reason for the failure of boron nitride coatings to improve these composites was not given.

Use of boron nitride coatings on Nicalon fibers with polymer-derived silicon carbide base matrix composites gave variable results. The best composites containing uncoated fibers had strengths and toughnesses as high or higher than those containing boron nitride coated fibers. However, composites made with boron nitride coated fibers were generally tougher, as indicated by the area under the stress-strain curves, due to a less rapid decrease in load-bearing capability beyond the point of maximum load-bearing capability. In addition, use of coated fibers seemed to give high strengths more frequently.

Some composites made using Sumitomo alumina base fibers also showed a clear advantage with boron nitride coatings, relative to otherwise identical composites without the fiber coatings. Uncoated fibers with a silicon carbide base matrix resulted in moderate strengths and toughnesses. Use of 0.2-μm thick boron nitride coatings on the fibers generally increased strength and tough-

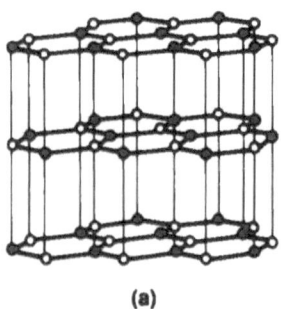   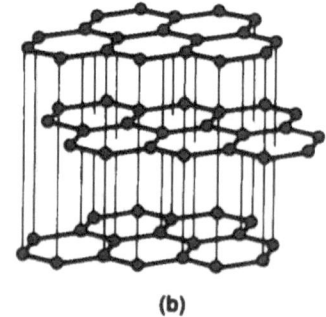

(a)   (b)

FIG. 2   Schematic comparison of the structure of hexagonal boron nitride (a) with graphite (b). (From Ref. 9.)

ness. Typical curves are shown in Figure 4. Strength with coated fibers was about double that with uncoated fibers, and area under the stress-strain curve increased considerably.

Boron nitride coated carbon fibers with a silicon carbide base matrix also resulted in improved properties, relative to those with uncoated fibers. Strength increased by about 50% and toughness about doubled.

Despite the expected better oxidation resistance of boron nitride coatings, compared with carbon coatings, reduction of composite properties still occurred with heat treatment. For example, Figure 5 shows stress-strain curves for composites having a silicon carbide base matrix and boron nitride coated Sumitomo fibers, as-fabricated and after heat treatment at 1000°C in air. Stress-strain behavior after heat treatment was similar to an as-fabricated composite with the fibers uncoated.

Other workers have given more details on boron nitride deposition conditions. For example, Gulden et al [11] described deposition of boron nitride at 700°C, a diborane (B_2H_6) flow rate (6% in a hydrogen carrier gas) of 50 cm³/min, an ammonia to diborane ratio of 10, and a pressure of about 4.7 kPa. Growth rate was 3–4 µm/h, independent of temperature over the range 500° to 900°C. Surfaces of the fibers were smooth with no discernible difference between coatings produced at different temperatures. Auger analysis indicated that nearly stoichiometric coatings were produced when excess ammonia was used.

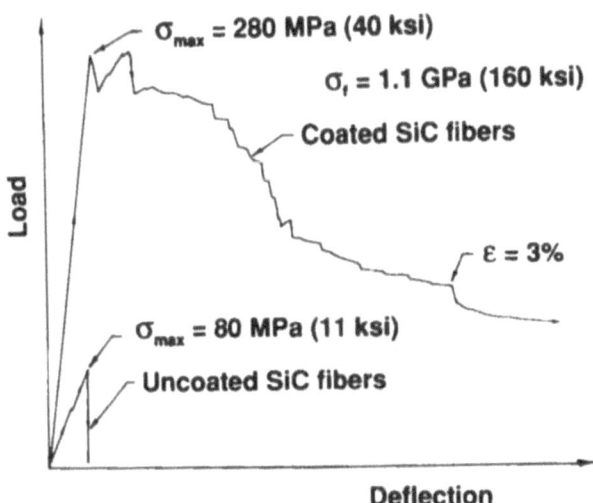

FIG. 3 Flexural stress-strain behavior for Nicalon™ fiber, silica matrix composites with and without a boron nitride fiber coating. (From Ref. 10.)

Gulden et al also reported on deposition of boron nitride using boron trichloride as the precursor. Conditions were a temperature of about 1000°C, a boron trichloride to ammonia ratio of 0.75, and a pressure of 0.03 kPa.

Since earlier studies had indicated that heat treating of silicon carbide base fibers at elevated temperatures in carbon monoxide was beneficial to subsequent property retention at elevated temperatures (as described in Chapter 5), Bender et al [12] investigated hot pressing of composites in carbon monoxide, with both uncoated and boron nitride coated fibers. (The major effect of the use of a carbon monoxide atmosphere is believed to be supression of deleterious reactions in the fiber that yield carbon monoxide and silicon monoxide.)

Nicalon fibers were first desized in air by appropriate heat treatments. Boron nitride coatings were produced by reacting boron trichloride and ammonia in a commercial reactor for 1 h at 950°C. A zirconium titanate matrix was produced by organometallic processing. Hot pressing was carried out in 110 kPa of carbon monoxide or argon at a pressure of 17.25 MPa or 34.5 MPa.

With uncoated Nicalon fibers, hot pressing in carbon monoxide did not improve mechanical properties of the composite appreciably, relative to that of a composite hot pressed in argon. Average flexural strengths were about equal, 192 MPa for hot pressing in argon and 193 MPa for hot pressing in carbon monoxide. Fracture toughness, measured by the single edge notched beam test, was 2.9 MPa·m$^{1/2}$ for hot pressing in argon and 4.4 MPa·m$^{1/2}$ for hot pressing in carbon monoxide.

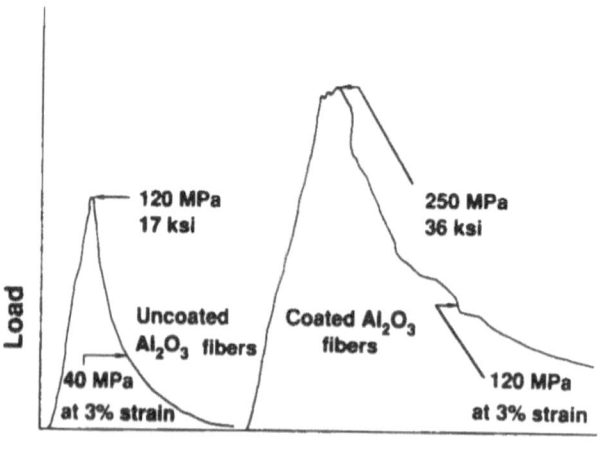

FIG. 4   Flexural stress-strain curves for Sumitomo fiber, silicon carbide base matrix composites with and without a boron nitride fiber coating. (From Ref. 10.)

Composite properties were improved considerably by the boron nitride coatings. Hot pressing in argon resulted in an average flexural strength of 684 MPa and average fracture toughness of 16.5 MPa·m$^{1/2}$. Hot pressing in carbon monoxide resulted in extremely good mechanical properties. Average flexural strength was 960 MPa and average fracture toughness was 22.4 MPa·m$^{1/2}$. The fracture toughness value is close to that of some structural metal alloys.

Other researchers have also found boron nitride to be a useful fiber coating. For example, Singh and Gaddipate [13] studied composites of SCS-6 fibers and a zirconium silicate matrix. The CVD-produced boron nitride coating was about 1-μm thick. Composites with uncoated and coated fibers were produced by slurry infiltration and hot pressing between 1500° and 1650°C.

With the as-produced SCS-6 fibers (which have a carbon coating), the stress-strain curves showed a significant drop in load after the point of maximum load. Composites with boron nitride coated fibers showed a more gradual decrease in load after the maximum load point. These results were consistent with lower interfacial shear stresses measured on the composites containing coated fibers.

2. BORON NITRIDE COATINGS BY NITRIDATION OF BORON CONTAINING FIBERS

Simpson and Verzemnieks [14] patented the production of boron nitride coatings on fibers containing boric oxide by heating at about 1300°C to cause boric oxide to diffuse to the surface, and nitriding. Partlow [15] used this technique on

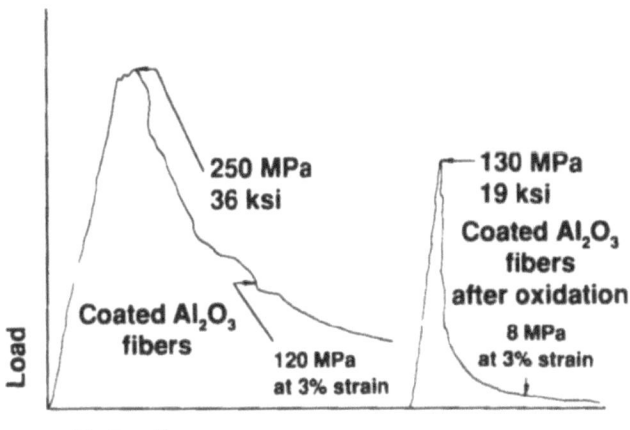

FIG. 5 Flexural stress-strain behavior for Sumitono fiber, silicon carbide base matrix composites with a boron nitride fiber coating, as-fabricated and after treatment at 1000°C in air. (From Ref. 10.)

Nextel 312 fibers, and produced composites having a matrix of calcium doped cordierite (10 MgO·10 Al_2O_3·50 SiO_2·CaO). The matrix was produced by a sol-gel technique, with consolidation by hot pressing at about 1100°C and 5 MPa.

The Nextel 312 fibers in the form of plain weave cloth were placed in an alumina boat and heated in air at 600°C for 1 h to drive off the size. The boat was then transferred to a sealed tube furnace and heated under flowing nitrogen at 800°C. At this temperature, ammonia and forming gas (90% nitrogen, 10% hydrogen) were introduced into the furnace and the temperature was raised to 1300°C for 36 min. Cooling was done under controlled conditions.

The effectiveness of the process was demonstrated by immersing the fibers in hydrofluoric acid solution for 20 min. The fibers remained intact. When fibers that only had size removal at 600°C were subjected to the same treatment, the fibers completely dissolved. This method of *in situ* formation of a boron nitride coating was not applicable to fibers containing lower amounts of boric oxide (presumably Nextel 440 and/or Nextel 480).

Comparisons were made with Nextel 312 fibers coated with boron nitride or carbon by CVD. The boron nitride coatings by CVD were not successful. Interiors of tows were not penetrated. It was also concluded that fibers were degraded by the process, even when diborane and ammonia were reacted at a temperature as low as 400°C. This indicated that the attack was chemical rather than thermal.

In comparing composites having the *in situ* boron nitride fiber coatings with those having carbon coatings produced by CVD, properties were better with the boron nitride coating. It was postulated that the carbon coating was not formed on the interiors of fiber tows, so that not all fibers were protected from chemical attack by the matrix at elevated hot pressing temperatures. Despite the fact that the *in situ* boron nitride coating performed best, flexural strength was relatively low, about 80 MPa.

C. Other Fiber Coatings

Many other fiber coatings have also been proposed and tried, but documented benefits comparable to those of carbon and boron nitride coatings are not generally available. Included are duplex coatings, often with carbon or boron nitride as one layer, but with another layer to attempt to provide oxidation protection. Because of the limited success with other coatings, this subject will not be covered in detail. However, several published works that deal with a multiplicity of coatings will be cited.

Gulden et al [11] produced a number of coatings by CVD, sol-gel processing, and polymeric precursor processing. Nicalon, Nextel 480, Fiber FP, PRD-166, and HPZ fibers were coated. Coatings included boron nitride, silicon carbide, silicon nitride, alumina, and yttria. The coated fibers were not tested in composites.

A number of broad conclusions were drawn. Stoichiometry and adherence of coatings produced by CVD were dependent upon conditions. For coatings produced by sol-gel, critical parameters included the sol production conditions, the degree to which the fiber tows wicked together, the curing environment and time, and the drying rate and time. (Sol-gel processing will be described in more detail in a later chapter.) The sol concentration and particle size and the use of fiber spreading techniques affected the degree of bridging. Coating adherence was affected by the sol type and the curing environment. Uniformity of infiltration was affected by the drying rate and the gel point of the sol.

Fischer et al [16] reviewed a number of systems having potential for fiber coating, but did not conduct any experimental coating work. Emphasis of the review was on "nontraditional" precursors as sources of metal and ceramic coatings.

A general concept of coating precursors was given. An ideal precursor is designed to contain the essential elements of the desired coating in the appropriate stoichiometric ratio (the "compositional core") and surrounded by substituents that impart desired physical properties, such as volatility, to the precursor (the "extraneous constituents"). In principle, the bonds between the elements of the compositional core are stronger than the bonds between the elements in the compositional core and the extraneous constituents. In this case, the substituents should be easily removable, preferably as relatively stable gaseous molecular by-products. Ideally, the by-products should be volatile at temperatures significantly below the decomposition temperature of the precursor. Deposition of metals, oxides, nitrides, borides, carbides and mixed-phase ceramic systems were reviewed.

A material that has recently been proposed for preventing interfacial reaction in fiber/matrix systems such as alumina/glass is tin dioxide.[17] The alumina/tin dioxide phase diagram predicts no mutual solid solubility at temperatures as high as 1620°C. Solid solubility of tin dioxide in some glass compositions also appears to be low. In the referenced study, "sandwiches" of α-alumina, tin dioxide, and a borosilicate glass were prepared by CVD of tin dioxide onto an alumina plate, and bonding of the tin dioxide to a glass plate at 900°C for 1 h. Based on flexural strength tests, it was concluded that both the alumina/tin dioxide interface and the glass/tin dioxide interface were weak, so that a tin dioxide fiber coating could be useful for alumina fiber/glass matrix composites.

Some of the work of Brennan [6] that was cited earlier was aimed at forming a carbon-rich silicon carbide coating (actually a mixture of carbon and silicon carbide) to provide the relatively weak interface of a carbon coating, but with improved oxidation resistance. Several precursors (methyldichlorosilane, methyltrichlorosilane, and dimethyldichlorosilane), flushing gases (argon and hydrogen), and carrier gases (argon, hydrogen, and methane) were tried, and ratios and other conditions were varied. The goal of about a 60 at% carbon content was not attained, but the results were promising.

IV. SUMMARY

Most, if not all, of the ceramic composites that have had significant toughening over the matrix material have had fiber coatings, either formed *in situ* during composite fabrication or specially applied, most often by CVD. Application of a carbon coating has almost invariably resulted in improved composite properties, compared with the identical system without the coating. However, because carbon oxidizes readily to gaseous products at temperatures as low as 400°C, carbon coatings have generally been destroyed during treatment at elevated temperatures. Since most of the applications for ceramic matrix composites entail high temperatures in air, carbon is far from an ideal coating. The coating can be shielded from oxidation by an impervious matrix or external composite coating. However, microcracking of the matrix and/or external coating would be expected to occur during use, and the carbon would then be subject to oxidation.

Boron nitride has been shown to result in composite performance as good as or in some cases better than that with carbon coatings, with improved oxidation resistance. However, boron nitride is not an ideal coating either. Fiber degradation appears to be a problem in some cases, and composite performance has not been improved with boron nitride coatings for some systems. Although oxidation resistance of boron nitride is better than that of carbon, oxidation is still a problem with boron nitride.

Because carbon and boron nitride are both unsatisfactory from a number of standpoints, the search continues for improved coatings. This includes other materials as well as mixtures of materials and layers of materials. However, there is little, if any, documented evidence yet that any of these new coatings will offer significantly enhanced performance.

REFERENCES

1. Brennan, J. J., Interfacial characterization of glass and glass-ceramic matrix/Nicalon SiC fiber composites. In *Tailoring Multiphase and Composite Ceramics*, (Proceedings of the Twenty-First University Conference on Ceramic Science, University Park, Pa, July 17–19, 1985), Plenum Press, New York, NY, 1986, 549–560.
2. Cooper, R. F., and Chyung, K., Structure and chemistry of fibre-matrix interfaces in silicon carbide fibre-reinforced glass-ceramic composites: an electron microscopy study, *Journal of Materials Science*, vol. 22, 3148–3160 (1987).
3. Lowden, R. A., Caputo, A. J., Stinton, D. P., Besmann, T. M., and Morris, M. D., Processing and properties of SiC/Nicalon composites. CONF-9705103-11, December 1987, Oak Ridge National Laboratory, Oak Ridge,TN.

4. Lowden, R. A., and More, K. L., The effect of fiber coatings on interfacial shear strength and the mechanical behavior of ceramic composites. CONF-891119-100, December 1990, Oak Ridge National Laboratory, Oak Ridge, TN.
5. Fischbach, D. B., and Lemoine, P. M., Influence of a CVD carbon coating on the mechanical property stability of Nicalon SiC fiber, *Composites Science and Technology*, vol. 37, 55–61 (1990).
6. Brennan, J. J., Interfacial studies of chemical-vapor-infiltrated ceramic matrix composites, *Materials Science and Engineering*, vol. A126, 203–223 (1990).
7. Doughan, C. A., Lehman, R. L., and Greenhut, V. A., Interfacial properties of C-coated alumina fiber/glass matrix fiber composites, *Ceram. Eng. Sci. Proc.*, vol. 10, no. 7–8, 912–924 (1989).
8. Jones, R. H., Henafer, C. H., Jr., Schilling, C. H., Shoenlein, L. H., Weber, W. J., and Gac, F., Interfacial chemistry-structure and fracture of ceramic composites. PNL-SA-15269, December 1988, Pacific Northwest Laboratory, Richland, WA.
9. Gardinier, C. F., Physical properties of superabrasives, *Ceramic Bulletin*, vol. 67, no. 6, 1006–1009 (1988).
10. Rice, R. W., BN coating of ceramic fibers for ceramic fiber composites, US Patent 4,642,271, February 10, 1987.
11. Gulden, T. D., Hazlebeck, D. A., Norton, K. P., and Streckert, H. H., Ceramic fiber coating by gas-phase and liquid-phase processes, *Ceram. Eng. Sci. Proc.*, vol. 11, no. 9–10, 1539–1553 (1990).
12. Bender, B. A., Jensen, T. L., and Lewis, D., III, Interfacial microstructure and mechanical properties of SiC/ZrTiO$_4$ composites hot-pressed in CO, *Ceram. Eng. Sci. Proc.*, vol. 11, 964–973 (1990).
13. Singh, R. N., and Gaddipati, A. R., A uniaxially reinforced zircon-silicon carbide composite, *Journal of Materials Science*, vol. 26, 957–963 (1991).
14. Simpson, F. H., and Verzemnieks, J., Barrier-coated ceramic fiber and coating method, US Patent 4,605,588, August 12, 1986.
15. Partlow, D. P., A ceramic-ceramic composite with low dielectric constant and non-brittle failure, *Advanced Ceramic Materials*, vol. 3, no. 6, 553–556 (1988).
16. Fischer, H. E., Larkin, D. J., and Interrante, L. V., Fiber coatings derived from molecular precursors, *MRS Bulletin*, vol. XVI, no. 4, 59–65 (April 1991).
17. Siadati, M. H., Chawla, K. K., and Ferber, M., The role of the SnO$_2$ interphase in an alumina/glass composite: a fractographic study, *Journal of Materials Science*, vol. 26, 2743–2749 (1991).

10

Slurry Processing of Whisker Reinforced Ceramics

I. BACKGROUND

The next several chapters of this book deal with specific fabrication methods for ceramic matrix composites, including those that are being used commercially, some that are not commercial but have been used to produce prototype components, and others that have been used only on a research basis.

A majority of monolithic ceramic processes involve slurry processing. This includes traditional ceramics as well as high performance ceramics such as alumina, silicon carbide, and silicon nitride. A number of ceramic matrix composite systems also utilize slurry processing. The subject of slurry processing of monolithic ceramics will not be covered in detail. However, some background information will be presented in this chapter to set the stage for descriptions of use of this method in fabricating ceramic composites. This chapter primarily covers the use of slurry processing methods to fabricate whisker reinforced ceramics.

II. GENERAL CONSIDERATIONS IN SLURRY PROCESSING

Ceramic slurries contain fine ceramic particles, a suspension liquid, and, in most cases, additives. The additives can serve a variety of functions, including dispersion aids, wetting agents, forming lubricants, and binders. The effect of the additives on the viscosity of the slurry is an important consideration also. The optimum viscosity depends upon the forming technique to be employed.

In general, slurries used in ceramic processing contain ceramic particles having average diameters of less than 1 μm and maximum particle diameters of only a few microns. For monolithic ceramics, use of fine particles results in reduced sintering time and temperatures and a reduced probability for deleterious large flaws. Since pressureless sintering is not generally applicable for ceramic matrix composites and flaws that would critically affect performance of a monolithic ceramic are not as important in composites, the reasons for preferring fine particles for slurries for producing composites are somewhat different, but nonetheless important. For whisker reinforced ceramics, the use of fine particles helps ensure that the whiskers are uniformly dispersed in the matrix. For reinforcement with tows of continuous fibers, use of fine particles facilitates the placement of matrix material around the individual fibers in a tow.

The suspension liquid is often water, but organic liquids are used also. Alcohols are perhaps the most commonly used organic liquids. Less often, nonpolar organic liquids may be used.

Many slurry additives are organic (carbon base) polymers, often with nitrogen and oxygen linkages.[1] Additives may either be soluble or dispersible (as emulsions). An example of an organic additive is polyvinyl alcohol. In order to make the polymer soluble in cold water, polar side groups are essential. Replacement of approximately 12–20% of the normal −OH side groups by acetate groups makes polyvinyl alcohol water soluble. Not only are the polar side groups attracted to water molecules (hydrophilic), but their bulk separates the main polymer chains sufficiently to permit penetration of water molecules. Cellulose is modified to a water soluble form by substituting hydroxyethyl, methyl, carboxymethyl, or hydroxypropyl side groups. Wax emulsions are examples of dispersible additives. These include synthetic waxes such as stearic acid derivatives.

Additives used as dispersion aids are generally ionic materials, since dispersion is often achieved by modifying the surface charges on the ceramic particles. Examples are sodium and ammonium polyelectrolytes. More information on additives for ceramic slurry processing is provided in reviews by Sheppard [2] and by Chan and Shanefield.[3]

A more recent type of slurry developed at Ceramics Process Systems (CPS) in Cambridge, MA [4] involves *in situ* polymerization of the carrier to produce the binder. Slurries using this system can have very high solids loading (85 wt%), while being pourable for low pressure forming. High green strengths are obtained. An example of a composition for forming alumina bodies is 85 wt% alumina, 8.5% monomers containing a polymerization initiator (90/5/5/ methylmethacrylate/methacrylic acid/diethylene glycol dimethacrylate, with 0.45 wt% benzoyl peroxide initiator), 3.5% dibutylphthalate (as a plasticizer), 1.5% decalin (as a diluent), and 1.5% of a commercial product with composition not listed (as a dispersant). Articles formed from a composition of this type exhibited

a smooth surface and high green strength. Particle packing was over 55 vol%.
The articles did not distort or crack during sintering.

From ceramic slurries, components can be formed by a variety of techniques. These include slip casting, perhaps the simplest method of consolidation.
This entails pouring the slurry into a porous mold. Much of the suspension liquid
is wicked away by the mold, leaving a consolidated body. In some cases, pressure
is applied to the liquid surface or vacuum is applied to the outside surface of the
mold. Centrifugal casting, which settles slurry particles from the bulk of the suspension liquid, is another means of consolidation. Slurry processed ceramic materials can be consolidated by uniaxial or isostatic pressing. In this case, much of
the suspension liquid is evaporated from the slurry prior to forming.

"Dry" processing, (usually with a minimal amount of additives to provide
particle lubrication in the mold during pressing), rather than slurry processing,
is also often practiced for monolithic ceramics. However, this is not generally
applicable for reinforced ceramics. In the case of whisker reinforced ceramics,
dispersion of whiskers among the matrix particles would probably be inadequate
with this processing method, and whisker attrition during blending would likely
be a problem. In the case of reinforcement with continuous, multi-fiber tows,
penetration of matrix particles into the tows would not be expected in the absence of an appreciable liquid phase.

Although the techniques described above are, of course, most well developed for monolithic ceramics, they have been used for reinforced ceramics as
well. In addition to the traditional ceramic forming processes mentioned above,
processes used in polymer forming can conceivably be used also, although their
use is not yet well developed for ceramic slurry systems. Examples of some of
these techniques are given in Figures 1–4 (Reference 5).

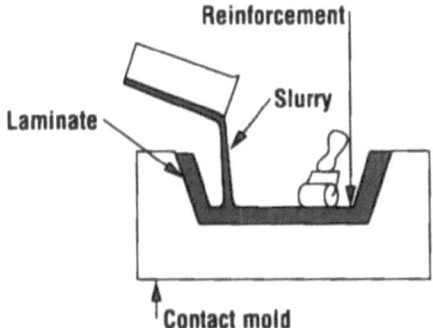

FIG. 1   Schematic illustration of hand lay-up for slurry processed ceramic matrix
composites. (Adapted from Ref. 5.)

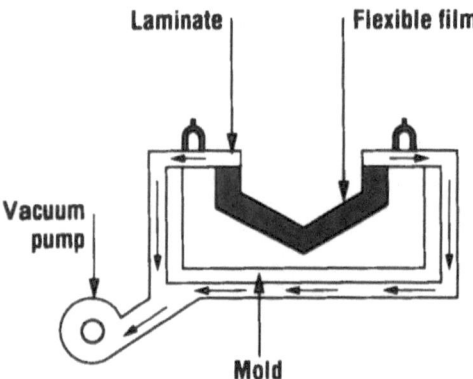

FIG. 2 Schematic representation of vacuum bagging for slurry processed ceramic composites. (From Ref. 5)

Hand lay-up (Figure 1) is probably most applicable for continuous fiber reinforced ceramics, with reinforcing fiber cloths placed in the mold, and the matrix slurry poured onto the cloth and worked into the cloth by roller, squeegee, or other means. Multiple layers can be built up in this manner. It is possible also to place compacted whisker mats into the mold. Although hand lay-up offers great design flexibility and minimal capital investment, it is labor intensive and would not be an ideal method for mass production.

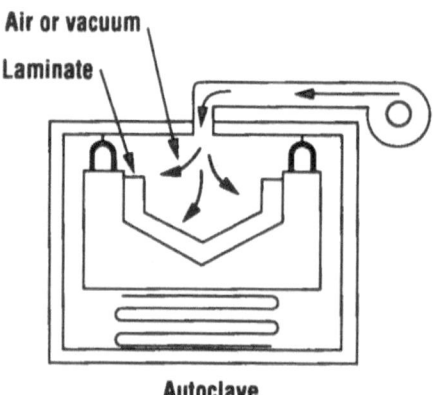

FIG. 3 Schematic drawing of an autoclave method for slurry processed ceramic composites. (From Ref. 5.)

Vacuum bagging (Figure 2) is an extension of hand lay-up. It involves the placement of a flexible film over the slurry and fibers in the mold. The joints are sealed and a vacuum pulled. The resulting atmospheric pressure on the lay-up in the mold can help to reduce voids, trapped air, and excess slurry, and improve adhesion between layers.

Additional heat and pressure over the vacuum-bag method can be applied with autoclave equipment (Figure 3). Pressures of the order of 0.7 MPa can be obtained. Voids can be further reduced over vacuum bagging and reinforcement loadings increased by the additional pressure. Undercuts are possible and cores and insets can be used. However, autoclave size limits part size, and the equipment tends to be expensive.

The fastest growing method of forming polymer matrix composites and one that is of considerable interest for forming whisker reinforced ceramics is injection molding (Figure 4). This method can produce complex, greatly detailed parts ranging from small to quite large (such as automobile bumpers, with polymer matrix composites). Overall cycle times of less than a minute are common with polymer matrix composites. As practiced for whisker reinforced ceramics, a slurry containing whiskers and matrix particles is injected under pressure into the closed mold. Slurry temperature can be elevated to improve flowability, and the mold cooled to reduce the temperature of the formed part and improve green strength. Alternately, with a slurry such as the Ceramic Process Systems type described previously, the slurry can be cool and the mold temperature elevated to induce polymerization of the binder.

Whatever the forming method, the suspension liquid and additives remaining after the forming process are ordinarily removed by heating. The amount of liquid that must be removed varies with the slurry solids loading,

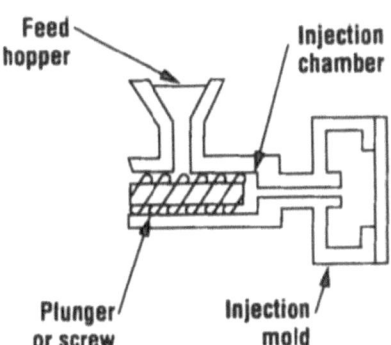

FIG. 4 Schematic depiction of injection molding of slurry processed ceramics. (From Ref. 5.)

which is tailored for the forming method used. Then, the ceramic particles must be fused together by some method. Pressureless sintering is the method of consolidation that is most often used for monolithic ceramics. In this process, the formed components are heated at ambient pressure to a significant fraction of the melting or decomposition temperature of the material, either in air or some other atmosphere. As the particles consolidate through diffusion, interparticle porosity is largely eliminated and a high fraction of the theoretically possible density for the material is obtained.

Unfortunately, pressureless sintering is not ideal for reinforced ceramics. Sintering of a monolithic ceramic, starting with a green density of about 60% of theoretical, can result in a specimen over 99% dense, but a great deal of shrinkage occurs in the process. The presence of a significant fraction of a non-

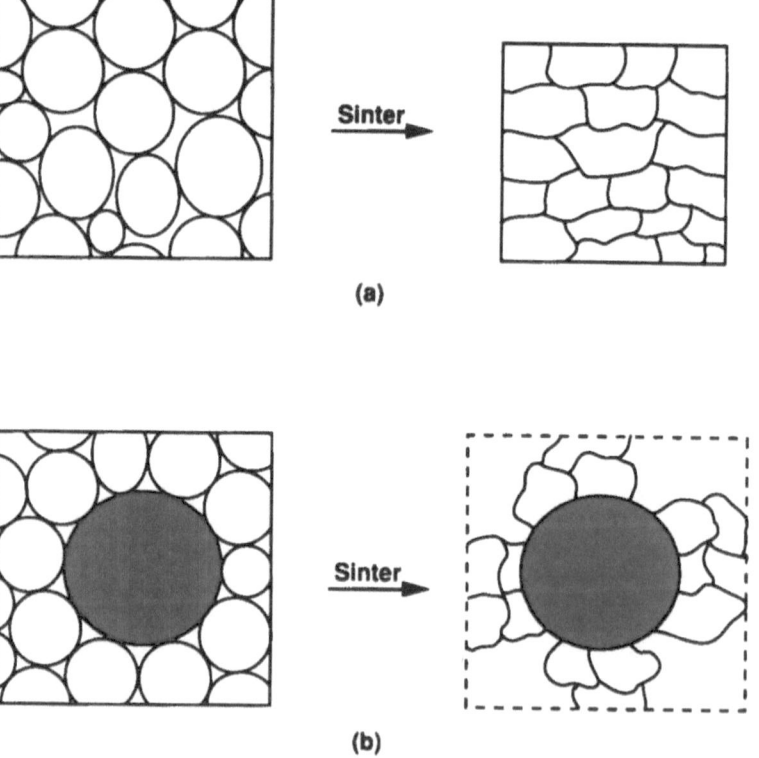

(a)

(b)

FIG. 5 Schematic comparison of sintering of a section of a monolithic ceramic (a) with a fiber reinforced ceramic (b). Shrinkage is constrained by the fiber, resulting in matrix cracking.

shrinking reinforcing phase restricts the shrinkage of the matrix (Figure 5). The external dimensions of the component cannot shrink to the extent possible in the absence of the reinforcing material and cracking of the matrix results. Another problem with sintering reinforced ceramics is that sintering temperatures are generally quite high, and the potential for reactions between the matrix and reinforcing fibers or whiskers is great.

Another common method of consolidating and densifying ceramics is hot pressing, most often using a graphite die. This technique has the disadvantages of being expensive (largely because of the low productivity of a hot press) and of allowing limited component shapes, but it often produces superior ceramics. Temperature can be several hundred degrees Celsius lower than for pressureless sintering of the same material.

An option for obtaining properties approaching those of hot pressed material is sintering, followed by hot isostatic pressing (HIPing). Hot isostatic pressing utilizes gas pressure applied directly to the component, so only closed porosity must be present after sintering (or the specimen must be encapsulated in an impervious container or coating, which increases expense). In general, hot isostatic pressing of components with closed porosity is less expensive than hot pressing because many components can be densified in the HIPing furnace at the same time.

Because of temperature limitations of state-of-the-art continuous ceramic fibers and because of the restriction to shrinkage with a high volume fraction of continuous fibers, neither pressureless sintering (with or without subsequent HIPing) nor hot pressing is ideal for continuous fiber reinforced ceramic production. However, since whiskers can often be subjected to higher temperatures without deterioration and since adequate toughness increases for many applications can be obtained with relatively low whisker contents, these methods can be used for whisker reinforced ceramics. The next section deals with processing of hot pressed and, to a lesser extent, sintered whisker reinforced ceramics. Other methods of joining the matrix particles and densifying the matrix material will be discussed as appropriate in later chapters.

III. FABRICATION OF WHISKER REINFORCED CERAMICS

A. General Aspects of Processing Whisker Containing Slurries

Incorporation of the reinforcing whiskers into a ceramic slurry is the most common method of processing, although it is possible to form a whisker preform through which the slurry is infiltrated. Addition of whiskers to a slurry complicates processing, relative to monolithic ceramic processing. One consideration is that an additive that aids in dispersing the matrix material particles may not be effective for dispersing whiskers and vice versa. Another consideration is that in the forming operation, bridging or agglomeration of whiskers can be a problem, especially if the volume fraction is high.

A method of avoiding particle agglomeration (which prevents good dispersion) is to impart an electrical charge to the particle surfaces. Electrostatic repulsion due to surface charges is commonly used for stabilization against flocculation. Surface charge is often found to be a function of slurry pH (a measure of acidity). Commonly, pH of a slurry is varied by acid or base additions and the "zeta potential" (an indication of surface charge) is measured. The slurry composition resulting in a high zeta potential is then used in processing.

As an example of work on zeta potential of whisker containing slurries, Hirata et al [6] investigated dispersion and consolidation of silicon nitride whiskers in aqueous slurries. Zeta potentials of dilute aqueous suspensions of whiskers were determined. To better understand the oxidation state of the whisker surfaces, zeta potentials of suspensions of high purity, fine silica powders were determined also. Dilute hydrochloric acid and ammonium hydroxide were used for pH adjustment. Sedimentation tests were performed on suspensions having 2 vol% whiskers in the pH range 2–11 to evaluate the dispersion state. Compacted specimens were formed from other suspensions by filtration through gypsum molds.

Zeta potentials for silicon nitride whisker and silica powder suspensions as a function of pH are shown in Figure 6. The pH dependence of both suspensions resembled each other, suggesting that silica films coat the surfaces of as-received whiskers. Figure 7 shows green densities of compacts formed by sedimentation and by filtration as functions of pH. A higher dispersion state of whiskers at higher pH values led to a denser compact with either consolidation process. At lower pH values, sedimentation formed a looser network structure of whiskers. The higher density at the lower pH values using filtration suggests that the looser network structure of whiskers was destroyed by the flux of water in the gypsum mold.

Effect of whisker content in the suspension on the green densities of compacts consolidated by filtration was studied also. In suspensions at pH 3 and 7, it was difficult to prepare a fluid suspension with a whisker content above 12 vol% due to a low zeta potential. Maximum green density was about 22% of theoretical. At a pH value of 10.5, compacts with up to 30 vol% whiskers were produced, although green density maximized at about 37% of theoretical at 10 vol%.

Microstructural analyses indicated that whiskers in suspensions resulting in higher green densities were preferentially oriented with the long axis perpendicular to the direction of filtration, whereas in suspensions resulting in lower green densities, whiskers were packed randomly during filtration. However, differences were largely eliminated by isostatic pressing at 392 MPa. Green densities ranging from 38 to 41% of theoretical after isostatic pressing were obtained for four compacts having densities ranging from 23 to 37% of theoretical prior to isostatic pressing.

The remainder of this chapter will deal with specific fabrication techniques that have been used for whisker reinforced ceramics. Although processes developed industrially (notably by Advanced Composite Materials Corporation) that are utilized in commercial products are obviously of great interest, relatively few details are available because of proprietary considerations. On the other hand, details of the work from universities and other nonprofit institutions are more available, and the literature in this area is becoming extensive.

It is difficult to make an overall assessment of various processing variations. There has been little, if any, cooperative testing of composites from dif-

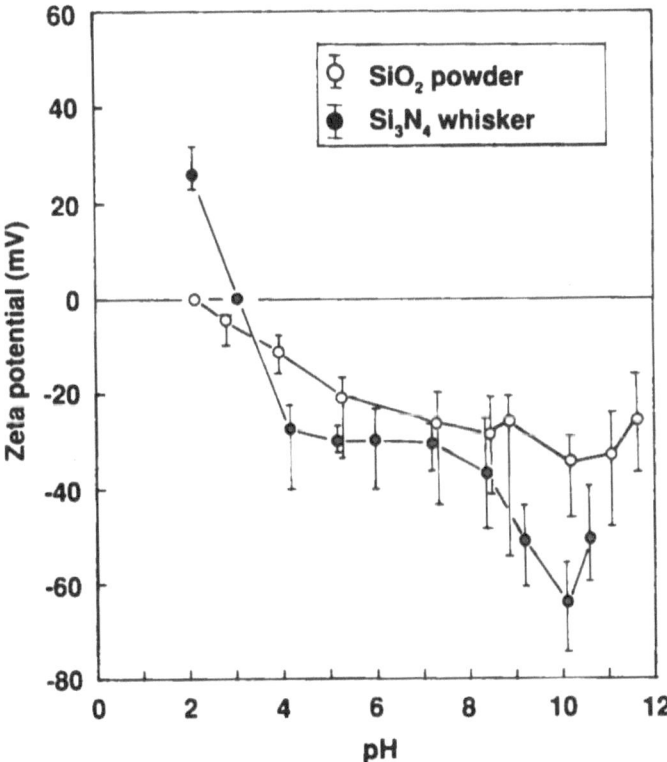

FIG. 6   Zeta potential as a function of pH for suspensions of silicon nitride whiskers and suspensions of silica powder. (From Ref. 6.)

ferent organizations, which would facilitate such an assessment. Even rough comparisons among composites reported in various publications do not seem worthwhile because of the variety of details used for preparing specimens for mechanical testing and the lack of standardization of strength and toughness tests. Hence, selected references including variations in processing that different researchers have judged to be important will be described.

B. Alumina Matrix Composites

Of all the types of composites discussed in this book, the silicon carbide whisker reinforced alumina system is certainly the most important commercially at this time, as a cutting tool material. Some processing details are given below.

Sacks et al [7] have reported considerable work on slurry processing for silicon carbide whisker/alumina particle slurries. The matrix materials were high purity α-alumina RCHP-DBM from Reynolds Metals Co. and AKP-50 from Sumitomo Chemical Co. Whiskers were SCW-1 from Tateho Chemical Industries and Silar SC-10 from Advanced Composite Materials Corporation. Treatments were made to remove large agglomerates from the alumina and to remove leachable impurities from the whiskers. The SCW-1 whiskers were separated into fractions having an average length of 7.8 μm (fraction A), 6.1 μm (fraction B), and 3.6 μm (fraction C). The alumina and whiskers were characterized by a variety of techniques including surface area determination by nitrogen gas adsorption, densities by helium-gas pycnometry, particle size by X-ray sedimentation, and scanning electron microscopy.

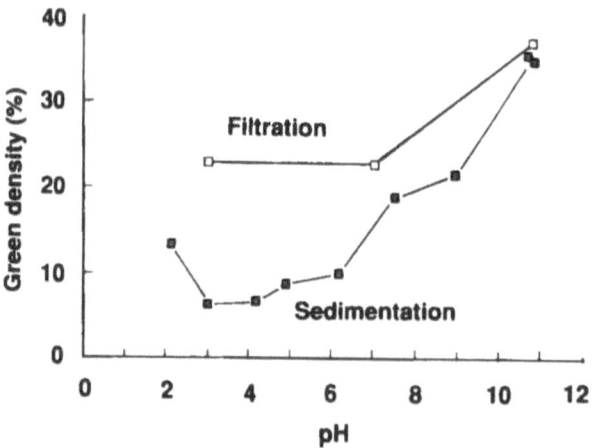

FIG. 7   Green densities of silicon carbide whisker, silica powder compacts formed by sedimentation and by filtration as functions of pH. (From Ref. 6.)

Composite samples were formed from well dispersed, mixed suspensions of whiskers and alumina particles. Suspensions of the individual components were prepared first, using ultrasonication to aid in breaking down agglomerates. The silicon carbide whisker suspension was then added to the alumina suspension. Overall solids concentrations ranged from 18–54 vol%. Green bodies were formed by slip casting. The slurry was poured into plastic or glass rings set on absorbant plaster blocks. Specimens were pressureless sintered in the range 1250° to 1600°C in a flowing nitrogen atmosphere.

Under the conditions of this work, the effect of pH on zeta potential was reasonably similar for both the whiskers and the alumina particles. Although zeta potential was highest at the lowest pH investigated, a somewhat higher pH, 4, was preferred to avoid complications created by large additions of ions to slurries. A polyelectrolyte additive was also found to decrease viscosity for concentrated whisker suspensions, indicating that improved dispersion was achieved.

By using the combination of pH adjustment and the polyelectrolyte addition, the alumina particles and whiskers could be dispersed at high solids concentrations while maintaining the relatively low viscosity needed for casting operations. The rheological behavior of a slurry of 95% alumina and 5% whiskers was similar to that with 100% alumina. At higher whisker volume fractions, viscosity increased, which was postulated to be due to the tendency of the whiskers to form network structures at high concentrations.

In addition to having well dispersed suspensions, segregation of particles during consolidation should be minimized to prevent inhomogeneities. In whisker/matrix particle systems, differences in density can result in differences in settling rate. Methods of minimizing segregation include consolidation of the suspensions at high rates and hindering settling by using suspensions with high solids loading. In the work of Sacks et al, fast consolidation was accomplished by slip casting, whereas slow consolidation was carried out by centrifugal casting. More segregation was observed with centrifugal casting at lower solids loadings, but not at higher solids loadings.

Although whisker content or length had relatively little effect on green density, sintered density decreased considerably with increasing whisker volume fraction and the effect was slightly less with the shortest whiskers. Sintering was inhibited by the whisker network structure. The effect was probably less with the shortest whiskers because of reduced network formation.

The dominant role of the whiskers in sintering is supported by the fact that the particle size of the major phase did not significantly effect composite densification. Alumina slurries without whiskers and with 10 vol% whiskers were cast using different aluminas: as-received RCPH, with a median particle diameter of 0.4 μm and AKP-50 (which had coarser particles removed by sedimentation), with a median particle diameter of 0.1 μm. Although effect of sintering temperature differed appreciably between the finer and coarser powders

in the absence of whiskers, there was no apparent effect of alumina particle size
when the whiskers were included (Figure 8). Sacks et al reported no mechanical
properties on their composites.

Other work on the effect of silicon carbide whisker length on alumina
matrix composites was carried out by Baek and Kim.[8] They used Tateho
SCW-1 whiskers varying in diameter from 0.1–0.5 μm and in length from 10–
40 μm. The alumina powder used was AES-11 from Sumitomo Chemical Co.
Mean particle diameter was 0.3 μm. Whiskers were dispersed in ethyl alcohol
and mechanically blended for times up to 60 min. During blending, average
whisker length (estimated using scanning electron microscopy) decreased from
an unblended average of about 18 μm to about 9 μm after 60 min of blending.

Whiskers were then blended with alumina powder in ethyl alcohol for 8
min and ultrasonically dispersed for another 60 min. Further whisker length
reduction by this process was not detected. The slurry was dried and granulated.
Hot pressing was carried out at 1800°C at 34.5 MPa for 1 h in an argon atmosphere. Whisker content of the composites was 20 vol%. Specimens for flexural strength and fracture toughness testing were prepared so that the long direction was perpendicular to the direction of hot pressing. Since whiskers tend
to be oriented in the plane perpendicular to the pressing direction, there was
some orientation in the long direction of the specimen. (Unless otherwise noted,
mechanical test specimen orientation in other studies on whisker reinforced ceramics to be described below was the same as in this work.)

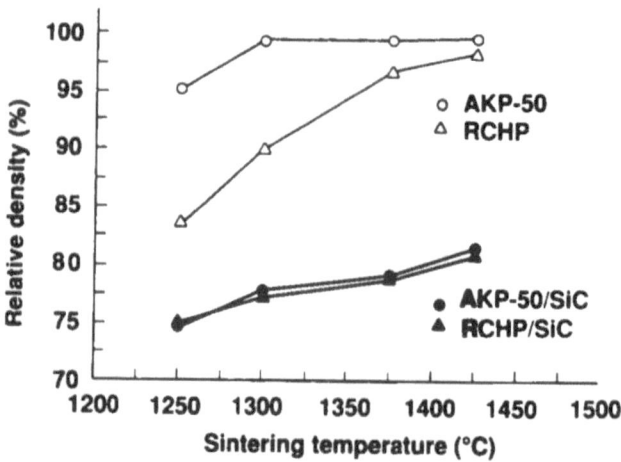

Fig. 8   Percent of theoretical density as a function of sintering temperature for a
0.4-μm diameter alumina (RCPH) and a 0.1-μm diameter alumina (AKP-50), with
and without 10 vol% silicon carbide whiskers. (From Ref. 7.)

Both flexural strength (4-point bending) and fracture toughness (single edge notched beam test and chevron notch test) increased with increasing whisker length (Figures 9 and 10). Strength was compared with that predicted using a mathematical model based on the rule-of-mixtures modified to account for variations in fiber length and orientation. The model tended to overestimate composite strength at the lower whisker lengths. The appreciable effect of measurement technique on measured fracture toughness values is clearly evident from Figure 10.

Transmission electron microscopy indicated that crack deflection was similar for all the composites, but that pullout lengths increased with increasing whisker length. Therefore, it was deduced that the whisker pullout mechanism rather than crack deflection contributes mainly to the toughening of hot pressed silicon carbide whisker reinforced alumina.

Becher et al [9] examined the effects of a number of factors on properties of silicon carbide whisker, oxide matrix composites. Whiskers were dispersed in water and agglomerates were removed by settling and decantation. Matrix

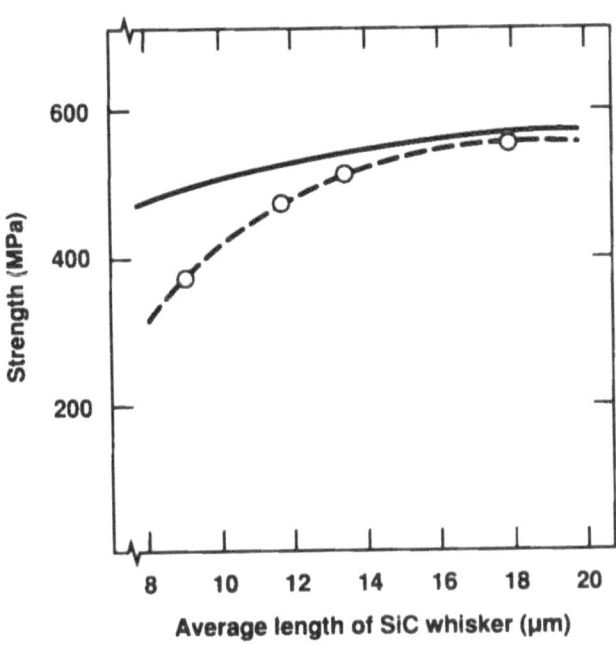

FIG. 9 Flexural strength as a function of whisker length for alumina reinforced with silicon carbide whiskers. The solid line is a prediction from a mathematical model described in the original reference. (From Ref. 8.)

powders were first milled and then similarly treated. Mixtures of whisker and matrix slurries were stabilized by pH adjustment and further blended in a high speed shear mill. The powder and whisker mixture was rapidly dried and gently ball milled to break up large agglomerates.

Specimens were prepared by vacuum hot pressing mixtures of silicon carbide whiskers and matrix powders. The maximum hot pressing temperature was varied according to the matrix material used, typically 1450° to 1850°C for alumina and 1450°C for mullite. (Two glass matrices were studied also, but work on these will not be described here.)

Fracture toughness measurements were made using the applied moment double cantilever beam method. Toughness results are reported in terms of a toughening increment, dK^{wt}, to the inherent matrix toughness. A variety of techniques were used to characterize microstructure and composition of the composites and whiskers. Theoretical predictions of toughening effects from various

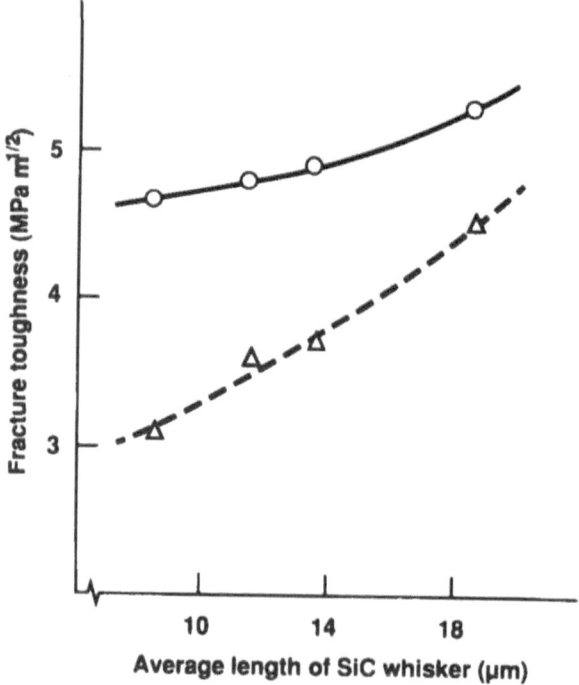

FIG. 10   Fracture toughness as a function of whisker length for alumina reinforced with silicon carbide whiskers. Data for the top curve were obtained by the single edge notched beam method. Data for the bottom curve were obtained by the chevron notch method. (From Ref. 8.)

factors were made using both the stress intensity in the presence of a bridging zone and on the strain energy change associated with a bridging zone.

Figure 11 shows toughness increase as a function of volume fraction for alumina and mullite matrix composites. The whiskers used for this experimental series had an average diameter of 0.8 μm, an average length of 30 μm, and a low surface oxygen content. Manufacturer was not specified. The experimental data agreed well with theoretical predictions.

The influence of the silicon carbide whisker characteristics on toughening was also investigated by Becher et al. Alumina composites were fabricated using silicon carbide whiskers from various sources. Although the goal was to examine the effect of whisker diameter, this was of course complicated by possible dif-

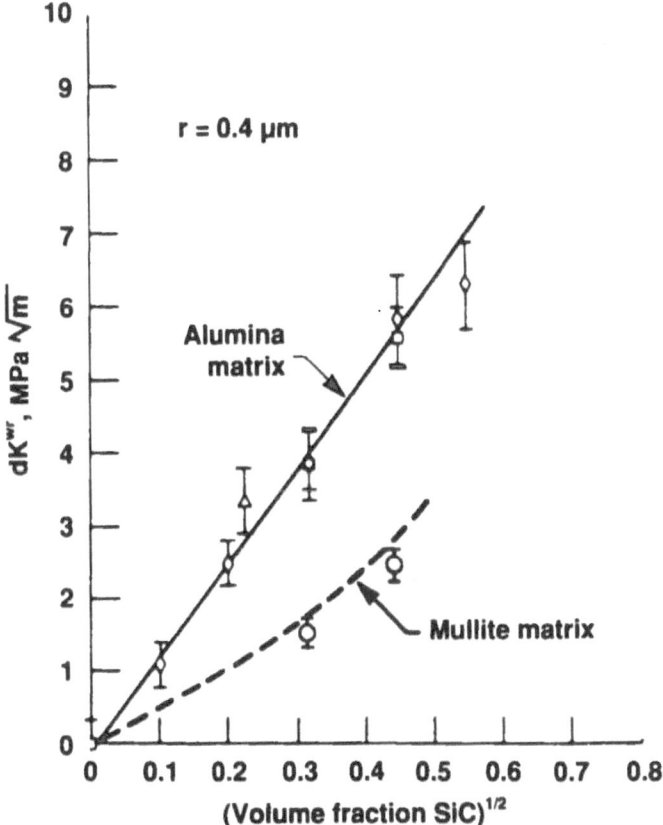

FIG. 11 Toughening contribution due to whisker reinforcement as a function of whisker volume fraction, for alumina and mullite. (From Ref. 9.)

ferences in other factors such as surface chemistry, which could modify the whisker/matrix interface. To minimize effects of characteristics such as surface roughness and the presence of defects, batches in which the whiskers contained tubular pores and/or were not straight, relatively smooth, and needlelike were excluded. The whiskers were also dispersed and decanted to attempt to remove agglomerates and fine particulates.

The cleaned, straight, solid whiskers were selected according to diameter. They were also grouped according to high surface oxygen content (>2 wt%) and low surface oxygen content (<1.5 wt%). High oxygen content was associated with a strong Si-O bond signal by X-ray photoelectron spectrometry. Low surface oxygen content was associated with high Si-C and C-C signals. TEM analysis also showed the presence of an amorphous oxygen-containing layer, especially for the high oxygen-content whiskers.

In order to study effects of surface oxygen content, which can affect interfacial bond strength and, hence, debonding and bridging, alterations in surface oxygen contents of some whisker batches were made. Larger diameter whiskers having a low oxygen content were oxidized at 1000°C for 1 h to achieve XPS signals that were similar to those of high oxygen-content whiskers. Conversely, one set of small-diameter whiskers having a high oxygen content was heat treated to reduce oxygen content and, hence, to produce XPS signals similar to the larger diameter low oxygen-content whiskers.

Composites were fabricated using whiskers of selected diameters and surface oxygen contents, and toughness values obtained. As shown in Figure 12, composite toughening increased with increasing whisker diameter, and composites containing whiskers having low surface oxygen contents were tougher. The increase in toughness with increasing whisker diameter was predicted by the theoretical model of Becher et al. The improvement in toughness with low surface oxygen content whiskers is consistent with formation of weaker interfaces during composite fabrication (indicating that chemical bonding at an interface is enhanced by the presence of a thin oxide layer on whisker). This was consistent with a dramatic decrease in whisker pullout length during fracture with decreased whisker surface oxygen content.

In agreement with other predictions of the theoretical model, overall toughness of the composites increased with matrix grain size and with increasing matrix Young's modulus.

Shih et al [10] studied the effect of whisker variations on alumina matrix composites. Composites containing 30 vol% whiskers were consolidated both by hot pressing and by hot pressing followed by hot isostatic pressing. Three types of whiskers produced in the United States were used. Table 1 gives chemical compositions and dimensions. Morphologies of the whiskers were shown previously in Chapter 5, Figure 23.

An alumina powder (CR-10 from Biakowsky Inc.) and a sintering additive, yttria, were wet milled in isopropyl alcohol. The whiskers were then introduced into the slurry, dispersed through pH control, and homogenized by additional mixing. The slurry was then dried, dry milled, and screened. After cold pressing at 10 MPa at room temperature, specimens were hot pressed at 1740°C in vacuum at a pressure of 27.8 MPa for 1h. Some of the hot pressed specimens were then further densified by hot isostatic pressing at 1675°C under 207 MPa nitrogen pressure. All densified specimens were then heat treated at 1200° to 1400°C for 1–2 h to minimize internal residual stresses.

Densities were measured using a helium autopycnometer. Various techniques were used to study microstructure and determine phases present. Flexural strength (4-point) and fracture toughness (single edge notched beam) were determined. Densities are given in Table 2. All specimens were quite dense, and hot isostatic pressing increased density only slightly. By X-ray diffraction, only

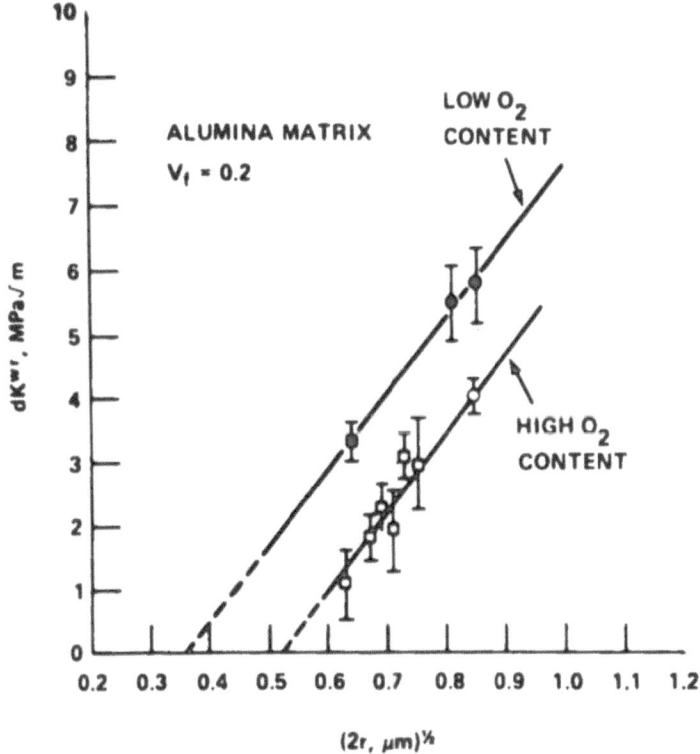

FIG. 12 Toughening contribution due to whisker reinforcement of alumina as a function of whisker diameter and whisker surface oxygen content. (From Ref. 9.)

TABLE 1. Chemical compositions and dimensions of silicon carbide whiskers used in processing studies on alumina matrix composites.

	Whisker I[a]	Whisker II[b]	Whisker III[c]
Al	120	200	>5000
B	<1	>1000	1
Ca	150	600	250
Co	<10	<1	<10
Cr	15	15	10
Fe	60	25	200
K	10	<1	5
Mg	90	200	5
Mn	70	1	5
Na	<10	20	<10
Ni	<10	20	<10
Zr	<10	10	<1
O (wt%)	1.25	2.85	0.73
Diameter (μm)	0.5-0.8	0.3-5.0	0.1-0.7

Units are ppm, except where noted.
[a]Advanced Composite Materials Corporation (SC-9)
[b]American Matrix Inc. (EX)
[c]Huber Inc. (XPW-2)
Source: Ref. 10.

phases originally detected in the whiskers and matrix material were detected after composite fabrication.

Mechanical properties are shown in Table 3. The "characteristic strength" listed is that at which the probability of failure was 0.63. After hot pressing, the mechanical properties were best with Type I whiskers, followed by Type II, and

TABLE 2. Density of silicon carbide whisker reinforced alumina as a function of sintering variables.

Whisker	Processing	Density g/cm³ (% of theor.)
I	HP	3.705 (99.1)
II	HP	3.699 (98.9)
III	HP	3.705 (99.1)
I	HP + HIP	3.726 (99.1)
II	HP + HIP	3.707 (99.2)

Source: Ref. 10.

Type III. Differences were not large with regard to strength, but Weibull modulus was appreciably higher with Type I whiskers than with the other two types. There were also appreciable differences in fracture toughness. Use of hot isostatic pressing after hot pressing raised strength, but had a detrimental effect on toughness. Weibull modulus was adversely affected by hot isostatic pressing with the Type I whisker.

Possible reasons for the differences in properties resulting from use of different whiskers were considered to be whisker diameter, aspect ratio, surface chemistry, surface morphology, and stability during processing. The increase in flexural strength resulting from hot isostatic pressing after hot pressing was attributed to elimination of porosity in the composite, while the reduction in fracture toughness was speculated to be due to increased bonding between the whiskers and matrix caused by hot isostatic pressing.

Iio et al [11] examined effects of whisker content and hot pressing temperature on silicon carbide whisker, alumina matrix composites. They used an alumina powder having an average particle diameter of 0.9 μm and silicon carbide whiskers having an average diameter of 0.6 μm and lengths of 10 to 80 μm. These were ball milled together in distilled water for 6 h. The slurry was then filtered with a 170-mesh sieve, dried, and granulated with a 60-mesh sieve. The mixed powder was hot pressed at 1850° or 1900°C for 0.5 h.

Flexural strength testing (3-point) and fracture toughness testing (chevron notch and indentation microfracture methods) were conducted. Figure 13 shows density, flexural strength, and fracture toughness values as functions of whisker content and hot pressing temperature. Density decreased appreciably at the 40% whisker level for specimens hot pressed at 1850°C, but not for specimens hot pressed at 1900°C. Flexural strength maximized with a 30% whisker content for specimens hot pressed at 1850°C, and similar values were obtained for specimens hot pressed at 1900°C. Fracture toughness increased with increasing

TABLE 3. Mechanical properties of silicon carbide whisker reinforced alumina as a function of sintering variables.

Whisker	Process	Flexural strength (MPa)	Weibull modulus	Fracture toughness (MPa · m$^{1/2}$)
I	HP	651	15.2	9.4 ± 0.4
II	HP	600	6.7	8.0 ± 0.4
III	HP	590	6.4	6.6 ± 0.8
I	HP + HIP	681	9.9	8.1 ± 0.6
II	HP + HIP	718	16.0	6.5 ± 0.5

Source: Ref. 10.

whisker content for specimens hot pressed at 1850°C, using both types of fracture toughness test methods. At a 40% whisker level, toughnesses of specimens
hot pressed at 1900°C were appreciably lower than those for specimens hot
pressed at 1850°C.

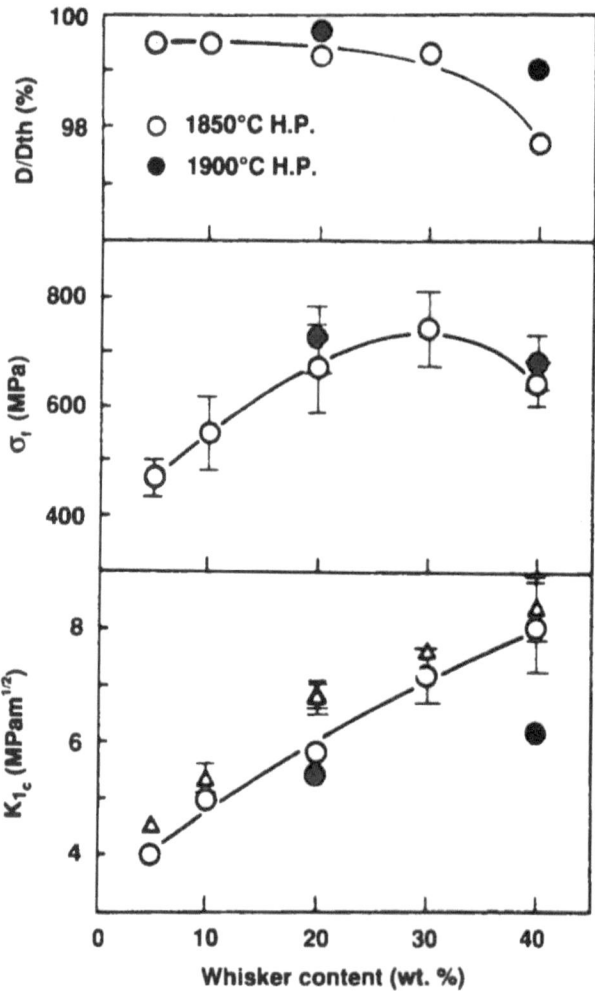

FIG. 13 Properties of silicon carbide whisker reinforced alumina as functions of
whisker content and hot pressing temperature. (From Ref. 11.)

Microstructural evaluations indicated that most of the fracture origins for specimens hot pressed at 1850°C were clusters of micropores due to incomplete densification, whereas those for specimens hot pressed at 1900°C were agglomerated whiskers. Whisker agglomeration with 40% whiskers and hot pressing was particularly severe, and this was speculated to be caused by the relatively short distances between whiskers at this volume loading and by alumina grain growth at the higher temperature.

Rhodes [12] reported results obtained on silicon carbide whisker reinforced alumina produced by hot pressing at Advanced Composites Materials Corporation, a major developer of this type of material. Processing conditions were not described in detail. As in much of the work described above, major gains in fracture toughness were made. Fracture toughness values about twice as great as values for unreinforced alumina have been obtained. Weibull modulus values in excess of 20 have been determined. This is in contrast to a Weibull modulus value of about 4.5 for unreinforced alumina.

A practical example of the increases in toughness and Weibull modulus is the increase in yield of finished, machined test bars from hot pressed billets of reinforced alumina, relative to unreinforced alumina. Unreinforced alumina was reported to have yields of only 10 to 50% of the available bars, due to specimen fracture caused by machining stresses during all stages of sawing and grinding operations. However, reinforced alumina composites have yielded over 98% of the specimens that could be produced from a billet.

As a consequence of the increase in toughness, thermal shock resistance of reinforced alumina was reported by Rhodes to be much improved over unreinforced alumina. Quenching of fifty specimens containing 25% whiskers from temperatures as high as 1100°C into boiling water resulted in no crack formation. Although more on wear properties of silicon carbide whisker reinforced alumina will be given in a later chapter on applications, it will be mentioned here that wear resistance is much greater than that of unreinforced alumina.

Although all the processes outlined above involved hot pressing, this process has limitations in shape making capabilities and is expensive. Hence, Rhodes reported on development work on pressureless sintering to produce this type of composite. Using pressureless sintering, complex shapes can be formed and green machined as required. Unfortunately, pressureless sintering alone has not yielded silicon carbide whisker reinforced alumina having very good properties. However, pressureless sintering to the extent that interconnected porosity exposed to the surface is eliminated permits subsequent hot isostatic pressing. As indicated earlier, hot isostatic pressing permits densification of specimens of complex shapes and is less expensive than hot pressing because a number of specimens can be simultaneously densified in a hot isostatic pressing chamber.

Flexural strengths of pressureless sintered and hot isostatically pressed specimens were lower than those of hot pressed specimens, but much higher than those of specimens that were only pressureless sintered. It was reported that fracture toughnesses of the sintered specimens often exceeded those of hot pressed specimens due to the combined toughening effect of microporosity and the whisker reinforcement.

Karunanithy [13] examined the cause of deterioration of silicon carbide whisker reinforced alumina after heat treatment at high temperatures. Whiskers were from Tateho (SCW-1-S1). They had average particle diameters of 0.5–1.5 μm and lengths of 5–20 μm. Alumina was Reynolds RC-HP-DBM having an average particle diameter of 0.5 μm. Stabilized, dispersed slurries having a total solids content of 60 vol% and whisker contents of 15–30 vol% were used. Slurries were adjusted to a pH of 4 to aid dispersion.

Whiskers were cleaned (with respect to metallic impurities) by washing in dilute hydrochloric acid. Excess acid was removed by water washing and the whiskers were dried. Powder and whiskers were shaken together for 2 h. A water slurry was prepared in a Waring blender for 5 min, and then blended in an ultrasonic mixer for 2 min, with adjustment of the pH. Slip casting, using a plaster mold, was used to prepare specimens. They were sintered at 1600°C for 2 h in a vacuum furnace. After sintering, thermal treatment was carried out in air at 1000°C for 1 h.

Microstructural studies of the heat treated specimens indicated that the whiskers had remained intact, but that a significant proportion of them appeared hollow in the core. These pores were 200–500 nm in size. A comparison of fracture surfaces before and after heat treatment showed that this hole formation is an important step in the whisker degradation. Only finer pores were observed in some of the whiskers after sintering but without composite heat treatment. These were attributed to partial decomposition during the sintering process.

C. Other Oxide Matrices

Less has been reported on whisker reinforcement of oxide ceramics other than alumina. Some results by Becher et al on mullite were included in the section on alumina. As another example, Liu et al [14] described work on silicon carbide reinforced mullite-zirconia composites. Starting materials included mullite (Dynamullit-351 mesh, from Dynamit Nobel AG), zirconia (SC-20, from Magnesium Electron, Ltd.) and Tokamax silicon carbide whiskers (from Tokai Carbon Company). Whisker diameters ranged from 0.3 to 1.0 μm and lengths from 30 to 60 μm.

The mullite and zirconia powders were milled in an attritor with alumina grinding media for 4 h in isopropyl alcohol and screened. Two powder processing routes were used. In the first method, the whiskers were tumbled in isopropyl alcohol for 20 h and then mixed with mullite or mullite-zirconia powder for 12 h under the same tumbling conditions.

In the second method, whiskers were ultrasonically dispersed in isopropyl alcohol and then the suspension was diluted with additional alcohol. After sedimentation for 20 min, the large particles and agglomerates that settled to the bottom were removed. The purified whiskers were tumbled for 12 h in isopropyl alcohol with 3-mm plastic balls and then mixed with mullite or mullite-zirconia powder for 4 h using the same tumbling procedure.

To avoid separation of the whiskers from the matrix powders, a rotary-vacuum evaporator was used to dry the slurry at 80°C. The dried whisker/powder mixtures were screened through a 315-μm sieve, cold pressed, and then hot pressed in flowing argon. Hot pressing temperatures of 1500°, 1550°, and 1600°C were used. Other conditions were not given. Compositions produced were mullite, mullite + 10 vol% zirconia, mullite + 20 vol% whiskers, and mullite + 10 vol% zirconia + 20 vol% whiskers. Flexural strengths (4-point) and phase and microstructural analyses were performed.

Using the first processing method, relatively little improvement in flexural strength was attained relative to the nonreinforced materials. Significant improvements were made using the second processing method. For example, with a hot pressing temperature of 1550°C, mullite had a strength of about 250 MPa, while mullite with whiskers had a strength of about 325 MPa when produced by the first method and about 424 MPa when produced by the second method. At the same hot pressing temperature, both mullite-zirconia and whisker reinforced mullite-zirconia produced by the first method had strengths of about 400 MPa, and whisker reinforced mullite-zirconia produced by the second method had a strength of about 575 MPa.

Microstructural examination of crack surfaces of composites produced by the first method identified two types of fracture origins. One was from extensive porous regions resulting from whisker agglomerates. The other was a result of large inclusions. Typical size of each type of flaw was about 100 μm. In contrast, these types of failure origins were not found for composites produced by the second method.

It was concluded that the whiskers must have introduced hard agglomerates and large inclusions that were not dispersed by simple tumbling without balls or by ultrasonic treatment. Sedimentation successfully removed large particles and hard whisker agglomerates. However, separation was complicated if soft agglomerates of whiskers were also formed during processing. The soft agglomerates could be broken up by tumbling or ultrasonic vibration. Rotary vacuum evaporation was judged to be a good method of preventing separation of whiskers and powder during drying, since the stresses resulting from the rotation can overcome the whisker-to-whisker and particle-to-particle attractive forces. It was concluded also that whisker aspect ratios greater than 50 tended to cause bundling and clumping, while aspect ratios of about 10-20 avoided this problem while providing sufficient whisker pullout for toughening.

Investigations on whisker reinforced glass and glass-ceramics were reported by Gadkaree and Chyung.[15] Several of the whisker types previously described were used. Glass-ceramic compositions were based on barium "stuffed" cordierite ($MgO \cdot 4Al_2O_3 \cdot 10SiO_2$) and on barium osumilite ($2BaO \cdot 4MgO \cdot 6Al_2O_3 \cdot 18SiO_2$). Whisker/glass powder mixtures were either prepared by a proprietary process by a whisker manufacturer, or by the following procedure: deagglomeration of whiskers by blending for 5 min in a water/isopropyl alcohol mixture, addition of glass, further blending for 5 min, filtering, and drying. Hot pressing was carried out in nitrogen.

Flexural strengths and fracture toughnesses were determined. Results were in general agreement with many of those obtained above with whisker reinforced crystalline matrix materials. Severalfold improvements in strength and fracture toughness were obtained by incorporating silicon carbide whiskers into the glasses and glass-ceramics. Incorporation of the whiskers in the glass particle slurries produced a large viscosity increase, necessitating much higher composite processing temperatures than with the matrix alone. The commercially available whiskers differed greatly in their ability to reinforce glasses and glass-ceramics. Increase in thermal expansion mismatch and the presence of alkali metal oxides in the matrix appeared to have an adverse effect on composite quality. For glass-ceramics, elimination of the glassy phase and increasing the softening point of the matrix resulted in better high temperature properties.

D. Silicon Nitride Matrix Composites

In addition to interest in alumina and other oxide base ceramics for whisker reinforcement, there is considerable activity in non-oxide base matrices, particularly silicon nitride. As with the oxide matrix composites, it does not appear useful at this time to attempt an assessment of best overall processing. Hence, selected work dealing with specific aspects of processing will be discussed independently.

Zheng et al [16] studied reinforcement of silicon nitride with silicon carbide whiskers and platelets. Suppliers were not identified. Whiskers had diameters of 0.1–1.0 μm and aspect ratios of 50–200. They consisted of 99 wt% silicon carbide, with less than 0.5 wt% silica, and with traces of chromium, cobalt, and iron. Platelets had thicknesses of 1–10 μm and diameters of 20–70 μm. They consisted of 99.9% α-silicon carbide. Silicon nitride powder was Starck LC 12, having a ratio of the alpha crystalline form to beta crystalline form of 95/5 and an average particle size of 0.8 μm. Since silicon nitride does not sinter well without a liquid phase, magnesium oxide and neodymium oxide in a 1.67/1 ratio were added for this purpose. In some cases, alumina additions were made also.

Specimens containing 0 to 30 vol% whiskers were produced. The slurry processing procedure was given in only limited detail. The whiskers or platelets were blended with the powder in isopropyl alcohol. Forming was by slip casting followed by cold isostatic pressing. After thermal removal of binder, pressureless sintering was carried out at 1600°, 1650°, and 1700°C for 1 h in nitrogen under a powder bed. Densities were measured and microstructural analyses were conducted. Fracture toughness (single edge notched beam) was measured.

Green densities were increased by platelet additions, but reduced by whisker additions. Sintered densities decreased appreciably with increasing whisker contents. Density with 30 vol% whiskers was only about 50% that of the matrix material alone. Density differences caused by different sintering temperatures were small relative to those from differences in whisker content. Density also decreased with increasing platelet content, but the decrease was smaller. With the alumina addition in combination with the magnesium oxide/niodymium oxide addition, density at a 20 vol% platelet level was still 95% of that of the matrix without any whisker or platelet content. The differences were attributed to more efficient particle packing and less inhibition of shrinkage during sintering with the platelets. In addition, whiskers reacted more with the liquid sintering phase, causing more mass loss. This was speculated to be due to the lower purity of the whiskers.

As might be expected from the very low densities of the whisker containing composites, fracture toughness was not significantly improved over the matrix material, which has a toughness of about 4.5 MPa·m$^{1/2}$. However, platelet containing composites had increased fracture toughness. At 20 vol% platelets, toughness was about 6.5 MPa·m$^{1/2}$.

Hot pressed specimens were also prepared (1550°C, 15 MPa). Densities of over 95% of theoretical could be obtained with 10 vol% platelets and over 90% of theoretical with 10 vol% whiskers. Densities with higher whisker contents were much lower. In contrast to the sintered specimens, fracture toughness increased with increasing whisker content for the hot pressed specimens.

Crimp and Piller [17] described dispersion studies on silicon carbide whiskers (American Matrix, Inc.), silicon nitride powder (UBE Industries), and sintering aid powders (alumina and yttria). Electrophoresis studies were conducted. Aqueous dispersions were prepared by ultrasonically dispersing each material to produce 10 wt% whiskers in silicon nitride with 4 wt% yttria and 2 wt% alumina. The pH was adjusted using nitric acid or potassium hydroxide.

Isoelectric point pH values for the various materials were 6.3 for the silicon nitride, 2.0 for the whiskers, 8.2 for the yttria, and 3.5 for the alumina. Examination of the zeta potential curves resulted in a selection of a pH of 3 for the slurry. At this pH, potentials of the matrix powders were all more positive than 25 mV and the potential of the whiskers was more negative than −25 mV. Hence, there was mutual repulsion among the matrix powders but attraction between the matrix powders and the whiskers. SEM analysis of freeze dried powder indicated good dispersion of whiskers and powder. Dispersion at a pH of 5 resulted in poor suspension of whiskers and particles. Dispersion at a pH of 8, which gave all materials a negative surface potential resulted in good suspension but did not produce mixing between matrix powders and whiskers as good as that at a pH of 3.

Stedman et al [18] investigated whisker length degradation during processing of silicon carbide whisker reinforced silicon nitride by injection molding. The whiskers were SCW-1 from Tateho Chemical Industries. The silicon nitride powder was grade Nu-10 from Showa Denka Co. Ltd. Yttria and alumina were used as sintering aids. Polypropylene (GY545M, from ICI), microcrystalline wax (Okerin 1865Q, from Astor Chemicals Ltd.), and stearic acid (from BDH Chemicals) were used as additives.

Six blending procedures, summarized in Figure 14, were carried out. Material produced by Route 1 caused seizure of the capillary rheometer, indicating that it was not suitable for use in injection molding. Material prepared by Route 2, which differed from that of Route 1 only to the extent that the mixture was slurried with carbon tetrachloride instead of acetone before drying, passed the capillary rheometer but had uneven flow, which suggested that the mixture was not homogeneous. The improvement with Route 2 compared with Route 1 was considered to be due to redistribution of the stearic acid by the carbon tetrachloride, which is a solvent for stearic acid.

In Route 3, the whiskers were deliberately precoated with part of the stearic acid used in the blend and added to the mixture just before injection molding. The object was to avoid processing the whiskers in the twin screw extruder, which was believed to cause considerable whisker degradation. Some sticking of the screw of the rheometer occurred with this blend, suggesting that the whiskers were not well dispersed.

In Route 4, the whiskers and powder were precoated with stearic acid and reliance was placed on the plasticizing action of the single screw injection molding machine for mixing the suspension. However, this was unsuccessful; the screw seized in the barrel. In Route 5, the whiskers and powder were precoated with stearic acid before blending with the remaining organic materials, and the twin screw machine was used for final blending. Acceptable product was made using this route. In Route 6, preblending was carried out in a high speed mixer before compounding with the twin screw extruder. It passed the extruder without difficulty and an acceptable extrudate was produced.

Whisker lengths were measured on as-received samples and on whiskers extracted from the injection molding blends after ashing to remove the organic materials. Samples were ultrasonically dispersed in a glycerol-water mixture containing an ionic dispersant and drops of the suspensions were placed between glass plates and average whisker lengths determined using an optical image analysis system.

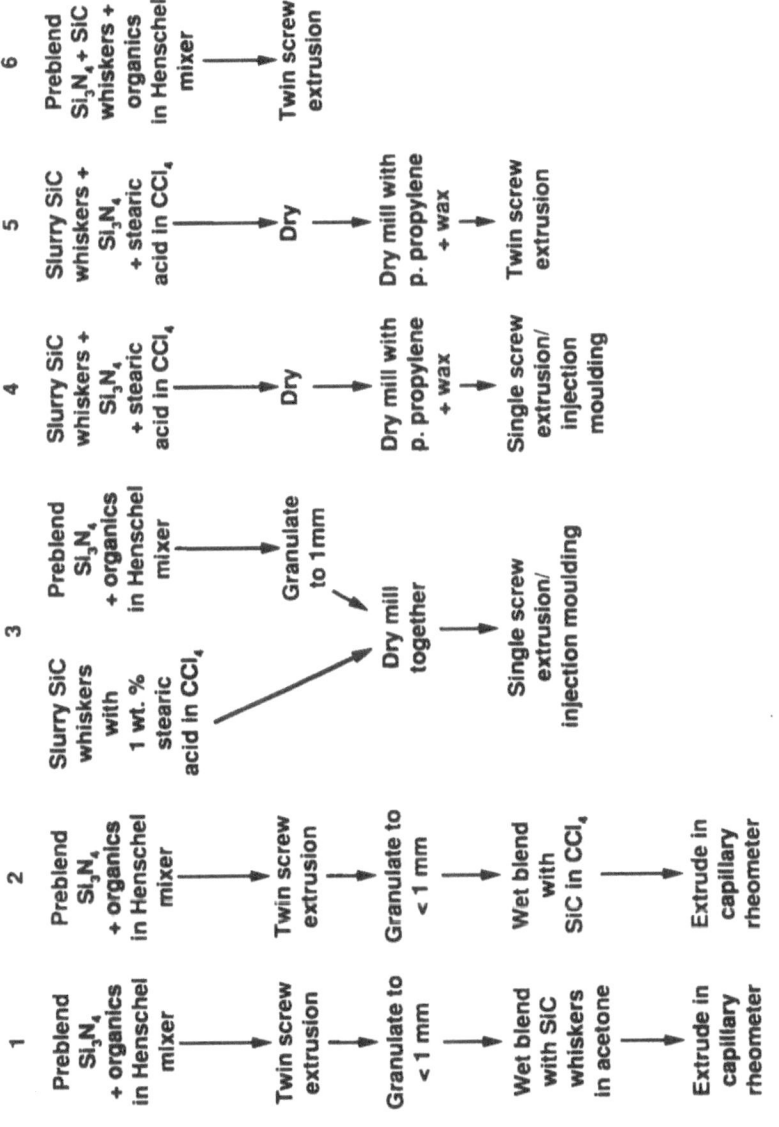

FIG. 14 Summaries of six processing routes investigated for preparing silicon carbide whisker, silicon nitride powder slurries for injection molding. Results for each route are described in the text. (From Ref. 18.)

In most cases, whisker degradation was not severe. Since Routes 5 and 6 resulted in the most suitable product for injection molding, degradation by these routes was of most interest. In Route 5, average length was decreased by only 1–2 μm, depending on screw speed (from an average of 9 μm for the as-received whiskers). Degradation was greater using Route 6. Using Route 5, whisker degradation increased appreciably with increasing volume fraction of whiskers, probably due to increased whisker-whisker interactions. Other detailed work on the rheology of silicon carbide whisker, silicon nitride slurries intended for injection molding was reported by the same authors.[19]

Ohji et al [20] investigated the high temperature toughness and tensile strength of silicon carbide whisker reinforced silicon nitride. Hot pressed silicon nitride specimens containing 20 wt% silicon carbide whiskers were produced. Whisker lengths ranged from 10 to 100 μm, and diameters ranged from 0.1 to 1.0 μm. Slurry processing details were not given. Specimens were hot pressed at 1750°C with 5 wt% yttria and 5 wt% spinel as sintering aids.

Mechanical testing was carried out from room temperature to 1300°C. Fracture toughness was determined using the chevron notched beam technique. Tensile testing was carried out also. Mechanical test results as a function of displacement rate were studied.

The tensile test results as a function of temperature were not reported in detail. It was mentioned that tensile strength showed substantial degradation above 1000°C. Tensile strength at 1300°C was 290 MPa (using a displacement rate of 0.1 mm/min), which was less than one-half that at room temperature.

At room temperature through 1000°C, fracture energy was nearly independent of displacement rate. At 1200° and 1300°C, however, fracture energy was largely dependent on rate. Large increments of fracture resistance with increasing crack length (known as R-curve behavior) were observed as displacement rate was decreased from 0.1 to 0.0001 mm/min at 1300°C. The sharply rising R-curve behavior was speculated to be caused by bridging effects of the whiskers. Apparently the whiskers can pull out more readily at the elevated temperatures due to softening of the glassy phases at the boundaries.

Bradley et al [21] determined effects of two whisker types on stability during processing of silicon carbide whisker reinforced silicon nitride. Whiskers were SC-9 from Advanced Composite Materials Corp. and SCW-1 from Tateho Chemical Industries Co. Both types of whiskers were produced by the rice hull process and had similar bulk compositions. However, the SCW-1 whiskers were found to have greater surface iron contents and lower surface oxygen contents than the SC-9 whiskers. Surface oxides for both types of whiskers resembled a Si-O-C glass in which the carbon is dispersed on an atomic scale. The SC-9 whiskers were straighter and tended to have less branched whiskers, distorted structures, and particulate debris.

Whiskers were blended with silicon nitride powder (SN-502 from GTE) along with 6 wt% yttria and 2 wt% alumina. Slurries were aged for 7 to 10 h before casting into billets. Carrier liquid was presumably water, although this was not reported. Billets were dried for one day in air at ambient temperature followed by one to two days at 40° to 50°C. The billets were then presintered at a temperature of 1400°C for 1 h to drive off remaining volatile material. The composites were sealed in niobium cans and hot isostatically pressed for 2 h at 1800°C and 190 MPa. Composite specimens containing the SC-9 whiskers had whisker contents of about 20 wt% and those containing SCW-1 whiskers had whisker contents of about 30 wt%. Microstructural characterization tests were carried out. Flexural strength was determined and fracture toughness was measured by the chevron notched beam method.

There were significant differences between the two materials. In addition to overall microstructural variations, there were differences in degradation of the whiskers. After the presintering stage, there appeared to be some degradation of the SC-9 whiskers, since some of them had different morphologies than the starting whiskers. At the same stage, the SCW-1 whiskers appeared to have suffered significantly more degradation. Some seemed to be losing their morphological character and appeared to have reacted with the silicon nitride. Many whiskers had a hollow morphology and several exhibited axial defects. Microanalytical results suggested that the SCW-1 whiskers were being nitrided and oxidized during the presintering operation.

After hot isostatic pressing, specimens containing both types of whiskers had near-theoretical densities. Properties of mechanical test specimens containing SC-9 whiskers were somewhat better than those of the matrix alone. Matrix specimen flexural strength was 815 ± 50 MPa and K_{Ic} was 5.90 ± 0.25 MPa·m$^{1/2}$. Composite flexural strength was 906 ± 68 MPa and K_{Ic} was 6.45 ± 0.58 MPa·m$^{1/2}$. However, properties of specimens containing SCW-1 whiskers were significantly poorer than those of the matrix material alone. Flexural strength was only 440 ± 49 MPa and K_{Ic} was 5.08 ± 0.23 MPa·m$^{1/2}$. As with the presintered specimens, there was much more evidence of whisker degradation with the SCW-1 whiskers. Many whiskers had apparently totally reacted, since counting methods indicated a volume fraction of only about 17%, compared with the nominal 30%. It was hypothesized that the higher surface iron content of the SCW-1 whiskers compared with the SC-9 whiskers (0.9 at% vs. 0.2 at%) catalyzed nitridation and oxidation reactions, causing the greater degradation.

Hoffmann et al [22] determined the influence of whisker loading on the rheology of suspensions and on the sintering behavior for composites containing up to 20 vol% silicon carbide whiskers in silicon nitride. Whiskers were SCW-1-S from Tateho Chemical Industries and silicon nitride powder was LC 12 from H. C. Starck. The silicon nitride powder was oxidized in air at 600°C for 8 h to generate a silica layer on the surface. Yttria and alumina were used as sintering aids. The silicon carbide whiskers were purified by sedimentation in aqueous suspensions at pH 9 and stabilized by polyacrylate.

Aqueous suspensions with a solids content of 45 vol% were prepared at pH 9 by deagglomeration of the powder in a ball mill for 1 h with the addition of a deflocculant. The whiskers were then added and the slurries were homogenized by tumbling for 20 h. Rheologies of slurries containing 0 to 20 vol% whiskers were determined using rotation viscometry. Specimens were cast in plaster molds or pressure cast in molds of porous stainless steel at 1 MPa. The cast specimens were dried in a controlled atmosphere for three days at 40°C and 90% relative humidity.

Sintering behavior was studied on samples having defined whisker orientations. Sintering was carried out in an atmosphere of 0.13 MPa of nitrogen using a heat-up rate of 45°C/min to a temperature of 1820°C, followed by a 5-h soak at that temperature. Microstructural characterizations were made.

There was no influence of whisker loading on the viscosity of the slurries at up to 15 vol% whisker content and viscosity was only slightly higher with 20 vol% whiskers. During slip casting, whiskers were preferentially aligned parallel to the mold surface. Green densities were 64 to 69% of theoretical. Highly anisotropic shrinkage occurred during pressureless sintering, with a shrinkage of up to 21% perpendicular to the plane of fiber alignment but only 7% parallel to this plane. Sintered densities decreased from 98% of theoretical with no whiskers to 88% of theoretical with 20 vol% whiskers.

Chu and Singh [23] prepared silicon nitride whisker reinforced silicon nitride with up to 35 vol% whiskers, and determined mechanical properties and microstructural details. UBE SN-WB whiskers and UBE-SN-E10 powder were used. Magnesium nitrate hexahydrate was added in a quantity sufficient to produce 2.3 wt% magnesium oxide as a sintering aid. The silicon nitride powder and sintering aid were ball milled in ethyl alcohol for 16 h. Whiskers having an average diameter of 0.6 μm and average length of 45 μm were then added and ball milling was continued for another 3 h. Ball milling reduced average whisker length to 15 μm. The mixture was blended in a high-speed laboratory blender for 5 min and dried in a thin layer to avoid whisker settling. The dried blends were hot pressed at 1650° to 1750°C under 30 MPa pressure for 2 h in nitrogen. Density was at least 99% of theoretical for all specimens.

Young's modulus was determined by the sonic method and hardness by the Vickers indentation method. Flexural strength (4-point) and fracture toughness (indentation method) were determined. Young's modulus and hardness were independent of whisker content. Fracture toughness increased from 6.5 MPa·m$^{1/2}$ to 8.8 MPa·m$^{1/2}$ with 5 vol% whiskers, decreased with additional whisker contents up to 20%, and then increased again. This was hypothesized to be due to competing factors of increased toughening with increasing whisker content (due to whisker pullout and crack deflection mechanisms), but decreased toughening due to a decrease in matrix grain size with increasing whisker content. The decrease in grain size with increasing whisker content could be seen in the microstructural analysis.

Flexural strength variation with whisker content was similar to that for fracture toughness. Strength increased from 673 MPa to 802 MPa with 5 vol% whiskers, but decreased and then increased with increasing whisker contents.

Homeny and Neergaard [24] prepared composites containing 30 vol% silicon nitride whiskers in a silicon carbide matrix. Whiskers were UBE-SNWB. Average diameter was 1.0 μm and average length was 30 μm. Powder was UBE-SN, having an average diameter of 0.2 μm.

The whiskers and powder were combined in ethyl alcohol. Magnesium nitrate hexahydrate was added in a quantity sufficient to provide 1 or 5% magnesium oxide as a sintering aid. The mixture was blended for 5 min at high speed to provide good distribution of the components. Remaining powder agglomerates and whisker bundles were broken up by ultrasonic dispersion for 5 min. The resulting slurries were dried and sized for hot pressing, which was carried out for 2 h at 30 MPa at temperatures ranging from 1750° to 1825°C. Flexural strengths (4-point) and fracture toughnesses (controlled flaw indentation technique) were determined, and microstructural characterizations were made.

Table 4 shows composite properties. Flexural strengths were not significantly increased by whisker additions. Fracture toughness averaged 8 MPa·m$^{1/2}$ with 30 vol% whiskers, about twice as high as the average for the matrix alone. The flaw size values responsible for failure shown in the table were calculated from the flexural strength and fracture toughness values. Microstructural analyses indicated strong bonding between the whiskers and matrix, so whisker pullout as a toughening mechanism was probably limited. The primary toughening mechanism appeared to be bridging in the vicinity of the crack tip.

Sialon, a ceramic material containing both oxide material and silicon nitride (the most common form having the formula, $Si_{6-z}Al_zO_zN_{8-z}$, where z ranges from 0–4), will arbitrarily be included in this section. Bower et al [25] investigated methods for fabrication of one sialon composition (termed β'-sialon) reinforced with silicon carbide whiskers and with chopped carbon fibers. Starting matrix material resulting in β'-sialon upon sintering was supplied by the Cookson Group. Tateho silicon carbide whiskers and chopped carbon fibers from Osaka Gas Company were used. The matrix and whiskers or fibers were milled in isopropyl alcohol for up to 100 h. The slurry was then dried and uniaxially pressed.

The whisker containing specimens were pressureless sintered or overpressure sintered and the fiber containing specimens were pressureless sintered or hot pressed. Since carbon monoxide can inhibit reactions between sialon and either carbon or silicon carbide, it was used as a component of many of the sintering atmospheres.

In pressureless sintering, the pressed pellets were packed in a bed of β'-sialon. Sintering temperatures were 1550°C for carbon fibers in β'-sialon and 1700°C for silicon carbide whiskers in β'-sialon. Sintering time was 1 h. Hot pressing was carried out at 30 MPa for 1 h at temperatures of 1550° or 1700°C. In overpressure sintering, pressed pellets were supported on a bed of β'-sialon. The atmosphere was nitrogen or 50/50 nitrogen/carbon monoxide. Sintering was conducted at 0.1 MPa for 1 h. Temperatures were 1550°C with carbon fiber reinforcement and 1700°C with silicon carbide whisker reinforcement. Microstructural analyses were conducted on all specimens.

In carbon fiber containing specimens produced by pressureless sintering in nitrogen, all the carbon fibers reacted, forming regions of porosity. Similar specimens sintered in nitrogen containing 8 vol% carbon monoxide indicated only partial reaction of the fibers. Specimens sintered in nitrogen containing 35% carbon monoxide showed no reaction between the fibers and matrix. The latter specimens had densities greater than 90% of theoretical.

Specimens containing silicon carbide whiskers had low sintered densities with both pure nitrogen and nitrogen/carbon monoxide atmospheres. Density decreased with increasing whisker content. With a 30 vol% whisker content, density was only about two-thirds that of the matrix alone. It was concluded

TABLE 4. Physical and mechanical properties of unreinforced silicon nitride and silicon nitride reinforced with 30 vol% β'-silicon nitride fibers.

Hot-pressing temperature (°C)	MgO content (wt%)	Theoretical density (%)	Flexural strength (MPa)	Fracture toughness (MPa · m$^{1/2}$)	Flaw size (μm)
0 vol% whiskers					
1750	1	99.1	650 ± 50	4.0 ± 0.7	22.8
30 vol% whiskers					
1800	1	95.9	680 ± 60	8.6 ± 1.8	96.5
1825	1	97.3	670 ± 50	7.6 ± 1.2	77.6
1800	5	99.9	640 ± 60	8.1 ± 1.5	96.6
1825	5	99.3	620 ± 40	8.0 ± 1.6	100.5

Source: Ref. 24.

that methods for better dispersion of whiskers would need to be developed for sintered silicon carbide whisker/sialon matrix composites.

Hot pressing of β'-sialon containing 10 vol% carbon fibers resulted in a material having a density of over 97% of theoretical, and no apparent reaction between fibers and matrix. With overpressure sintering of silicon carbide whisker/β'-sialon materials, complete reaction of the whiskers occurred, whereas reaction was inhibited in the nitrogen/carbon monoxide atmosphere.

IV. SUMMARY

Silicon carbide whisker reinforced alumina is the most advanced fiber or whisker reinforced ceramic from a commercial standpoint. Detailed processing information for the commercial product is not readily available, but there is considerable reported work from academic and other nonprofit institutions. Details of raw materials, additives, slurry suspension medium, blending/milling conditions, and sintering conditions differ appreciably in reported works.

Since the extent of composite evaluation and details of the methods of evaluating these composites also differ appreciably, no attempt has been made here to suggest preferred raw materials or processing methods. In general, it appears that water or alcohols can function equally well as the suspension medium. The importance of whisker dispersion is clear from many of the investigations. It is also apparent that pressureless sintering alone results in composites having significantly inferior properties to those produced by hot pressing, because of the presence of the nonshrinking whisker network.

The second most studied system has been silicon carbide whisker reinforced silicon nitride. As with an alumina matrix, processing details vary widely in reported studies, and it is difficult to assess optimum conditions at this time. Other matrices such as mullite and sialon have also been investigated with silicon carbide whisker reinforcement, and silicon nitride whiskers have been investigated with several matrices.

REFERENCES

1. Onoda, G. Y., Jr., The rheology of organic binder systems. In *Ceramic Processing before Firing*, G. Y. Onoda and L. L. Hench, Jr. (Ed.), John Wiley and Sons, New York, NY, 1978, 235–251.
2. Sheppard, L. M., The changing demand for ceramic additives, *Ceramic Bulletin*, vol. 69, no. 5, 802–806 (1990).
3. Chan, K. K., and Shanefield, D. L., Growth forecast for high-value ceramic additives, *Ceramic Bulletin*, vol. 68, no. 4, 854–856 (1989).
4. Waack, R., Venkataswamy, K., Novich, B. E., Halloran, J. W., Egozy, A. R., Hodge, J. D., and Tormey, E. S., Polymerizable binder solution for low viscosity, highly loaded particulate slurries and methods for making green articles therefrom,

International Patent Application Number: PCT/US88/ 01232; International Publication Number: WO 88/07505; International Publication Date: October 6, 1988.

5. English, L. K., Fabricating with composite materials: a primer, *Materials Engineering*, vol. 107, no. 10, 41–45 (1990).

6. Hirata, Y., Nakagama, S., and Ishihara, Y., Dispersion and consolidation of silicon nitride whisker in aqueous suspension, *J. Mater. Res.*, vol. 5, no. 3, 640–646 (March 1990).

7. Sacks, M. D., Lee, H.-W., and Rojas, O. E., Suspension processing of Al₂O₃/SiC whisker composites, *J. Am. Ceram. Soc.*, vol. 71, no. 5, 370–379 (1988).

8. Baek, Y. K., and Kim, C. H., The effect of whisker length on the mechanical properties of alumina-SiC whisker composites, *Journal of Materials Science*, vol. 24, 1589–1593 (1989).

9. Becher, P. F., Hsueh, C.-H., Angelini, P., and Tiegs, T. N., Toughening behavior in whisker-reinforced ceramic matrix composites, *J. Am. Ceram. Soc.*, vol. 71, no. 12, 1050–1061 (1988).

10. Shih, C. J., Yang, J.-M., and Ezis, A., Processing and performance of several SiC whisker-reinforced Al₂O₃ matrix composites, *Materials and Manufacturing Processes*, vol. 5, no. 1, 35–49 (1990).

11. Iio, S., Watanabe, M., Matsubara, M., and Matsuo, Y., Mechanical properties of alumina/silicon carbide whisker composites, *J. Am. Ceram. Soc.*, vol. 72, no. 10, 1880–1884 (1989).

12. Rhodes, J. F., Whisker reinforced ceramic composites. In *Proceedings of the Fifth Annual Conference on Materials Technology*, Materials Technology Center, Southern Illinois University, Carbondale, IL, 1988, 205–219.

13. Karunanithy, S., Chemical processes that degrade composites of alumina with SiC whiskers, *Materials Science and Engineering*, vol. A112, 225–231 (1989).

14. Liu, H. Y., Claussen, N., Hoffmann, M. J., and Petzow, G., Fracture sources and processing improvements in SiC-whisker-reinforced mullite-zirconia composites, *Journal of the European Ceramic Society*, vol. 7, 41–47 (1991).

15. Gadkaree, K. P., and Chyung, K., Silicon-carbide-whisker-reinforced glass and glass-ceramic composites, *Ceramic Bulletin*, vol. 65, no. 2, 370–376 (1986).

16. Zheng, X. Y., Zeng, F. P., Pomeroy, M. J., and Hampshire, S., Reinforcement of silicon nitride ceramics with whiskers and platelets. In *Fabrication Technology*, R. W. Davidge and D. P. Thompson (Ed.), British Ceramic Proceedings, No. 45, The Institute of Ceramics, Stokes-on-Trent, UK, 1990, 187–198.

17. Crimp, M. J., and Piller, R. C., Dispersion of SiC whiskers in a Si₃N₄ matrix using pH control. In *Fabrication Technology*, R. W. Davidge and D. P. Thompson (Ed.), British Ceramic Proceedings, No. 45, The Institute of Ceramics, Stokes-on-Trent, UK, 1990, 199–204.

18. Stedman, S. J., Evans, J. R. G., and Woodthorpe, J., Whisker length degradation during the preparation of composite ceramics for injection moulding, *Journal of Materials Science*, vol. 25, 1025–1032 (1990).

19. Stedman, S. J., Evans, J. R. G., and Woodthorpe, J., Rheology of composite ceramic injection moulding suspensions, *Journal of Materials Science*, vol. 25, 1833–1841 (1990).

20. Ohji, T., Goto, Y., and Tsuge, A., High-temperature toughness and tensile strength of whisker-reinforced silicon nitride, *J. Am. Ceram. Soc.*, vol. 74, no. 4, 739–745 (1991).
21. Bradley, S. A., Karasek, K. R., Martin, M. R., Yeh, H. C., and Schenle, J. L., Silicon carbide whisker stability during processing of silicon nitride matrix composites, *J. Am. Ceram. Soc.*, vol. 72, no. 4, 628–636 (1989).
22. Hoffman, M. J., Nagel, A., Greil, P., and Petzow, G., Slip casting of SiC-whisker-reinforced Si_3N_4, *J. Am. Ceram. Soc.*, vol. 72, no. 5, 765–769 (1989).
23. Chu, C.-Y., and Singh, J. P., Mechanical properties and microstructures of Si_3N_4-whisker-reinforced Si_3N_4 matrix composites, *Ceram. Eng. Sci. Proc.*, vol. 11, no. 7–8, 709–720 (1990).
24. Homeny, J., and Neergaard, L. J., Mechanical properties of β'-Si_3N_4-whisker/Si_3N_4-matrix composites, *J. Am. Ceram. Soc.*, vol. 72, no. 11, 3493–3496 (1990).
25. Bower, R. M., Edrees, H. J., and Hendry, A., Techniques for fabrication of β'-sialon ceramic matrix composites. In *Fabrication Technology*, R. W. Davidge and D. P. Thompson (Ed.), British Ceramic Proceedings, No. 45, The Institute of Ceramics, Stokes-on-Trent, UK, 1990, 169–177.

11

Slurry Processed Continuous Fiber Reinforced Ceramics

I. BACKGROUND

Several relatively well developed processes for fabricating continuous fiber reinforced ceramics also utilize slurry processing. After forming a fibrous preform by this method, one method of consolidating and densifying the matrix is hot pressing. This process has been developed to the greatest extent for production of glass and glass-ceramic matrix composites. (Glass-ceramics are glass systems that can be partially crystallized by heat treatment, or "ceraming," which changes material properties appreciably.) Work on carbon fiber reinforced glasses at Harwell in England in the early 1970s resulted in strengths of about 700 MPa, but problems with oxidation of the fibers and the poor perception of carbon reinforced composites in general resulted in suspension of work.[1]

The major development of this method was subsequently done by Prewo and co-workers at United Technologies Research Corporation (UTRC). They too have worked to some extent with carbon fibers as the reinforcement, but emphasis has been on the more oxidation resistant silicon carbide base fibers. Considerable development of glass and glass-ceramic matrix composites has also been done at Corning Glass Works, and work has been reported from other organizations as well.

More limited work has been carried out on production of continuous fiber, crystalline matrix composites by this fabrication route.

II. GLASS AND GLASS-CERAMIC MATRIX COMPOSITES

A. General Fabrication Description

Figure 1 is a general schematic diagram of the process as practiced at UTRC.[2] Fiber tows are passed through a slurry-containing tank, wound onto a drum (being translated axially to achieve the desired fiber spacing), and dried. A binder included in the slurry permits removal of sheets of matrix material with uniaxially aligned fibers. The sheets can be stacked to give a uniaxial or a two-dimensional orientation. After burn-off of the binder, the matrix is consolidated by hot pressing. Specific details have undoubtedly varied over the years. Since hot pressing of a glass matrix composite involves considerable viscous flow, densities of near 100% of theoretical are obtained. Figure 2 shows a cross section of a glass matrix composite reinforced with Nicalon fibers.

In some of the earliest work reported from UTRC, Prewo and Brennan [3] described processing and properties of borosilicate glass (Corning Code No. 7740) reinforced with Nicalon fibers and with SCS-6 fibers. In the work using SCS-6 fibers, procedure differed somewhat from that shown in the schematic diagram. The fibers were wound onto a drum, bonded together periodically with polystyrene, and cut from the drum to form tapes. The tape layers were alternated with layers of glass powder. The amount of powder was varied to make composites with 35 vol% and 65 vol% fibers. The lay-ups were hot pressed for 20 min at a temperature of 1150°C and a pressure of 6.9 MPa in an argon atmosphere. This hot pressing schedule resulted in complete densification of the glass with very little bubble or void formation.

The Nicalon fiber reinforced borosilicate glass specimens were fabricated using a procedure that had been developed earlier for graphite fiber/borosilicate glass matrix composites, which is like that shown in the schematic diagram. Nicalon fiber tow was passed through a slurry of glass powder and isopropyl alcohol, dried, and cut into tapes to fit the hot-pressing die. Sufficient tape was stacked in the die to provide a desired thickness, and hot pressed in vacuum for 1 h at a temperature of 1400°C and a pressure of 14 MPa.

B. Properties and Further Process Development

Flexural tests (3-point) were conducted on the specimens described by Prewo and Brennan. With SCS-6 fiber reinforcement, strength at ambient temperature averaged about 830 MPa with 65 vol% fibers and 650 MPa with 35 vol% fibers. Tests were also conducted at elevated temperatures to 700°C (Figure 3). With 65 vol% fibers, flexural strength increased with increasing temperature to 600°C, then decreased. Behavior was probably similar with 35 vol% fibers, although only one elevated temperature was investigated. At 700°C, the matrix completely deformed plastically. With 40 vol% Nicalon fibers, room temperature strength

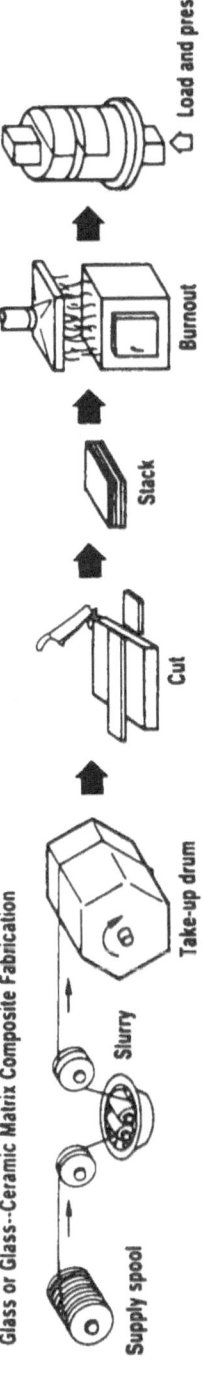

Glass or Glass--Ceramic Matrix Composite Fabrication

Supply spool

Slurry

Take-up drum

Cut

Stack

Burnout

Load and press

FIG. 1 Schematic representation of the URTC process for glass and glass-ceramic matrix composites. (From Ref. 2.)

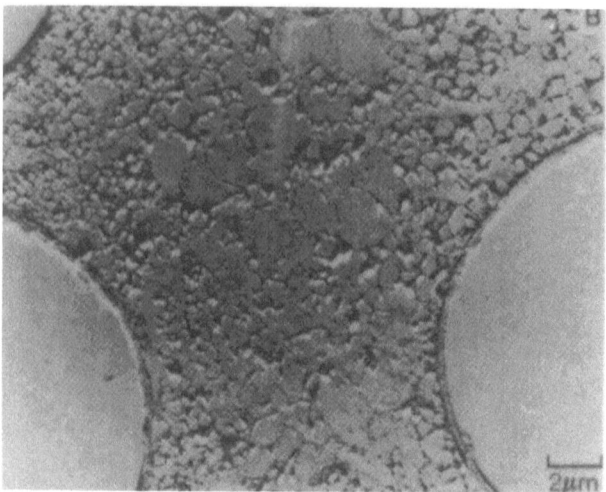

FIG. 2   Cross section of a glass matrix composite reinforced with Nicalon™ fibers. (From Ref. 2.)

was lower, 290 MPa. Effect of test temperature was similar to that with SCS-6 fibers.

It was proposed that the increase in observed flexural strength to 600°C was due to the fact that increasing the test temperature lowered the viscosity and effective yield stress of the matrix, which permitted more specimen yielding.

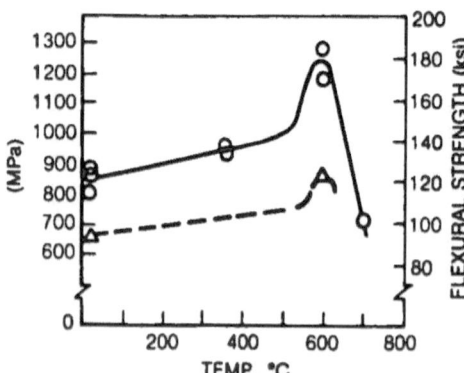

FIG. 3   Three-point flexural strength in air as a function of temperature for SCS-6 fiber reinforced Code 7740 borosilicate glass. Upper curve is for 65 vol% fibers; lower curve is for 35 vol% fibers. (From Ref. 3.)

As a result, the actual stress on the specimen tensile surface was relieved and the calculated maximum flexural strength could exceed that calculated from a simple beam equation. Therefore, the measured increase in load-carrying capacity does not necessarily mean an increase in material strength.

Fracture toughness was measured at room temperature and 600°C in air using a notched flexural test specimen in 3-point bending. For the 35 vol% SCS-6 reinforced material, K_{Ic} was 18.8 MPa·m$^{1/2}$ at room temperature and 14.3 MPa·m$^{1/2}$ at 600°C. Specimens were still in one piece after testing. For Nicalon fiber reinforced glass, fracture toughness values were lower, 11.5 MPa·m$^{1/2}$ at room temperature and 7.0 MPa·m$^{1/2}$ at 600°C. These are still much higher than those of unreinforced glassy materials. Young's moduli were 185 GPa and 290 GPa with 35% and 65% SCS-6 fibers, respectively, and 120 GPa with Nicalon fibers.

During room temperature testing, it was noted that composite failure invariably occurred by splitting along the fiber/matrix interface, indicating a relatively weak interface. Fiber pullout and crack deflection along the fiber/matrix interfaces were less prevalent for the Nicalon fiber reinforced material than for the SCS-6 fiber reinforced system. While strength for the SCS-6 fiber reinforced material was considerably greater, the flexibility of the smaller diameter Nicalon fibers increases their potential for fabricating intricate shapes.

The encouraging results on fabrication of continuous fiber reinforced glass matrix composites by slurry processing and hot pressing provided impetus for considerable further research at UTRC and elsewhere. Higher temperature glasses as well as glass-ceramics have been used as matrix materials, and processing conditions have been studied in considerable detail. Some examples are given below.

Prewo [4] reported on work using Nicalon fibers with two glass-ceramic compositions designated LAS-I and LAS-II. Both these materials consist predominantly of lithium aluminosilicate (Corning Code 9608), but were tailored to be more compatible with the fibers. Fabrication can be adequately represented by Figure 1 although additional heat treatment (ceraming) was required to convert the glassy matrix to a partly crystalline one. Fiber tapes were oriented to produce uniaxial fiber reinforcement. Fiber volume fraction was about 46%.

In order to better relate composite properties to fiber properties, individual fibers were tested as-received and after extraction from fabricated composites. The as-received fibers were removed from the tows by first burning off the vinyl acetate size in a flame. Hydrofluoric acid was used to extract fibers from composites. Individual 7.5-cm lengths of fibers were mounted on glass slides and average diameters determined. Fibers were mounted such that gage length was 2.5 cm. For weak fibers, the location of fracture was easily identified, but for the strongest fibers multiple fractures frequently occurred. Average ultimate tensile strength for as-received fibers ranged from 940 to 3665 MPa, with an average of 2300 MPa. Extracted fibers were significantly weaker. A typical Weibull plot for as-received and extracted fibers is shown in Figure 4.

Table 1 gives more complete information on fiber properties. Fibers extracted from composites having the LAS-I and LAS-II compositions were reduced in strength by 40% and 30%, respectively. There was no additional loss in strength due to the ceraming treatment. The cause for the strength reduction was not determined, although the fibers were wavy, rather than straight, after hot pressing, which could have contributed to the strength loss.

Composite tensile testing was carried out using flat specimens on which epoxy tabs were bonded. Strain was measured using strain gages on both of the specimen faces. Three-point and 4-point flexure tests were performed in both air and argon atmospheres up to a temperature of 1100°C.

The tensile stress-strain curve for a ceramed LAS-I matrix specimen was linear, with an ultimate tensile strength of about 450 MPa and strain to failure of about 0.35%. In contrast, the stress-strain curves for the LAS-II matrix specimens were initially linear, but deviated from linearity as stress was increased. There was a relatively small difference between as-pressed and ceramed specimens. The break points in the stress-strain curves are believed to be associated with the beginning of matrix microcracking.

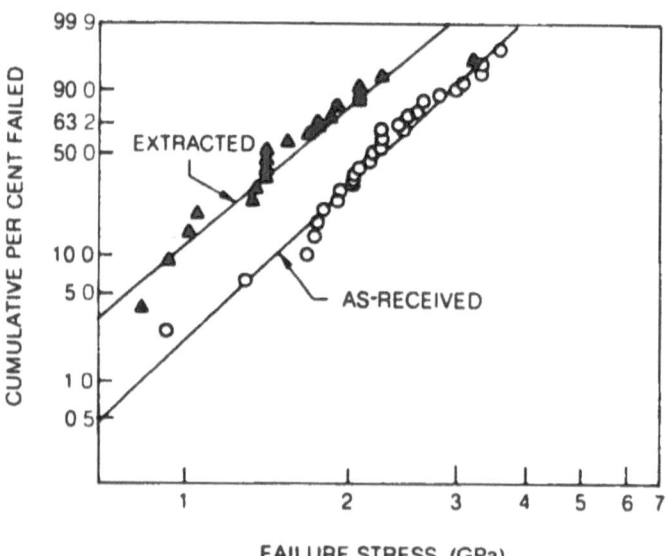

Fig. 4 Typical Weibull plots for as-received Nicalon™ fibers and fibers extracted from an LAS matrix composite using hydrofluoric acid. (From Ref. 4.)

TABLE 1. Tensile strength data for Nicalon™ fibers as-received and cleaned and for fibers extracted from glass-ceramic matrix composites.

Fiber spool no.	Composite matrix	Fiber condition	No. fibers tested	Average diameter (μm)	Average UTS (MPa)	Standard deviation (MPa)	Weibull Parameter, m
967	—	As received & cleaned	26	14	1820	560	3.6
	—	HF acid soaked	19	15	1810	630	3.2
	LAS-I	Extracted from composite 2160 (as-pressed)	20	16	1070	250	4.8
840	—	As-received & cleaned	19	15	2020	430	5.4
	LAS-I	Extracted from composite 2157-6 (ceramed)	20	14	1200	450	2.9
4585	—	As-received & cleaned	30	15	2300	610	4.2
	LAS-II	Extracted from composite 2369-7 (as-pressed)	20	17	1580	460	3.8
	LAS-II	Extracted from composite 2369-7c (as-pressed)	20	17	1520	600	2.7
	LAS-II	Extracted from composite 2376-2 (as-pressed)	20	15	1470	430	3.9
	LAS-II	Extracted from composite 2376-5c (ceramed)	20	15	1450	520	3.1

Note: Details on fiber spool number and composite number are given in the original reference.
Source: Ref. 3.

Flexural strengths (4-point) of LAS-I matrix specimens averaged about 750 MPa, which is about 1.7 times greater than the tensile strength. There was no significant difference between as-pressed and ceramed specimens. All specimens failed first by tensile cracks on their tensile sides and resisted complete cracking by diversion of cracks. In contrast, the LAS-II matrix composites had significant differences before and after ceraming. The as-pressed specimens averaged about 1375 MPa in strength, about twice their tensile strength. On the other hand, the ceramed specimens averaged less than 1000 MPa. It was noted that in contrast to the other specimens, these ceramed LAS-II matrix specimens failed in compression rather than in tension.

Elevated temperature 3-point and 4-point flexural testing was conducted also. Flexural strengths of LAS-I matrix specimens in air increased with temperature up to 1100°C, and stiffness decreased. This was accompanied by a more noticeable rounding of the stress-strain curve with increasing temperature and, at 1100°C, a permanent bowing of the specimen. As described previously, this can be attributed to matrix softening. Flexural strength in argon also increased with increasing temperature, although the appearances of the fracture surfaces differed from those tested in air.

In contrast, flexural strengths of the LAS-II matrix specimens decreased significantly at elevated temperatures in air. This was accompanied by a change in fracture surface appearance. Strengths were much higher in argon. For example, in 4-point flexure at 900°C, strength was 1180 MPa in argon, but only 275 MPa in air. In the latter case, there was no fiber pullout at the tensile side of the fracture. It is likely that in air, after matrix microcracking had initiated, accelerated fiber degradation or an increase in the fiber/matrix bond occurred, either of which could decrease strength and toughness by the bridging and pull-out mechanisms.

Prewo also contrasted the tensile stress-strain behavior of an LAS-II matrix composite with that of an epoxy matrix composite containing Nicalon fibers (Figure 5). As indicated earlier, the break in the curve for the LAS-II matrix composite can be attributed to matrix microcracking and subsequent fiber/matrix debonding. In contrast, the epoxy matrix curve is linear since the epoxy matrix failure strain is greater than that of the fibers. In addition, the ultimate strength of the epoxy matrix composite is greater, probably due to the less severe fabrication conditions, resulting in a greater retention of fiber strength.

Although the flexural strength of the LAS-II matrix composite was apparently greater than the flexural strength of the epoxy matrix composite, Prewo pointed out that the calculated maximum flexural stress for the LAS-II composite is not valid due to the nonlinear tensile behavior, which causes the neutral axis to be shifted toward the compression side of the specimen, and negates the simple elastic beam calculation used to calculate flexural strength. Nonetheless, the fact remains that the glass-ceramic matrix composite was able to carry a significantly higher load in flexure than the epoxy matrix composite.

Mah et al [5] studied high temperature properties of UTRC unidirectional Nicalon/LAS-II composites in a variety of atmospheres. Included were 7-mPa vacuum, 105-Pa argon, air, and various partial pressures of oxygen in argon. Four-point flexural tests were carried out at 900° and 1000°C. (Tensile tests were also conducted, but these were more limited in scope than the flexural testing and will not be discussed here.) The failure behavior and ultimate strengths of specimens tested at 900°C under vacuum and in argon were the same as those tested at room temperature.

Specimens were also tested at 900°C in oxygen partial pressures of 10^2, 10^3, and 10^4 Pa in argon. The furnace chamber was evacuated to 0.1 mPa before the gas mixture was introduced and heating started. Total pressure was kept at 10^5 Pa. Table 2 shows flexural strengths after testing in the different atmospheres. A progressive decrease in ultimate strength occurred with increasing partial pressure of oxygen. All the specimens failed in a brittle manner from the tensile surface, except the one tested in the lowest oxygen partial pressure. The latter specimen failed in the same manner as as-received specimens. The degradation happened rapidly, since the specimens were rapidly heated to the test temperature and stressed. Load on the sample was required, since a specimen treated at 900°C in oxygen for 20 min had the same failure behavior and ultimate tensile strength as an as-received specimen.

Upon microstructural analysis, only one crack was observed in each specimen tested at the elevated temperatures. Fibers adjacent to the surface were broken, and no fibers bridging the matrix were observed. The fracture surface consisted of an outer region about 150 μm thick exhibiting completely brittle failure (no fiber pullout) and an inner region showing fiber pullout. It was noted

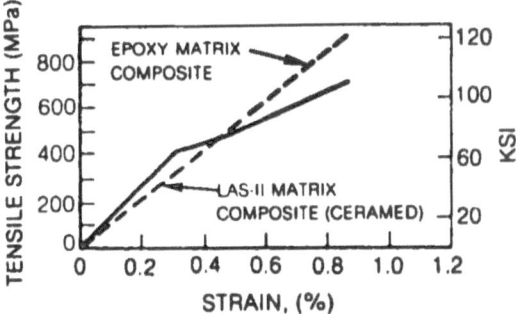

FIG. 5 Tensile stress-strain curves for Nicalon™ fiber reinforced epoxy and Nicalon™ fiber reinforced LAS-II. (From Ref. 4.)

TABLE 2.   High temperature flexural strengths of Nicalon™ fiber reinforced LAS-II in various atmospheres.

Oxygen partial pressure, Pa (atm)	Flexural strength based on maximum load, MN/m² (ksi)
2×10^4 (2×10^{-1})[a]	521 (75.5)
1×10^4 (1×10^{-1})	586 (85.0)
1×10^3 (1×10^{-2})	664 (96.3)
1×10^2 (1×10^{-3})	1010 (146.5)

[a]In air.
Source: Ref. 5.

that similar specimens previously tested at room temperature exhibited equally spaced multiple cracks with unbroken fibers bridging the matrix.

Allaire et al [6] discussed work involving several fibers and matrices. Nicalon fibers as well as A-4 and HM graphite fibers from Hercules and P55 and P75 fibers from Union Carbide (now, Amoco Performance Products) were used. Matrix compositions included Corning Code 7740, which is an alkali borosilicate (ABS) glass, Corning Code 1723, which is an alkaline earth aluminosilicate (AEAS) glass, and a calcium aluminosilicate (CAS) glass-ceramic. Fabrication procedure was similar to that described earlier.

Four-point flexural tests were conducted. Mechanical properties are shown in Table 3. Composite properties were largely governed by fiber properties. High modulus fibers (HM, P55, and P75) yielded high modulus composites, while lower modulus fibers yielded lower modulus composites. Composite strengths

TABLE 3.   Mechanical properties of glasses and glass-ceramics reinforced with Nicalon™ or carbon fibers.

Fiber	Matrix	Microcrack yield stress, MPa (Kpsi)	Ultimate stress, MPa (Kpsi)	Modulus, GPa (Mpsi)
Hercules A-4	ABS	670 (97)	1104 (160)	131.1 (19.0)
Hercules HM	ABS	426 (67)	649 (94)	165.6 (24.0)
Union Carbide P55	ABS	518 (75)	731 (106)	140.1 (20.3)
Union Carbide P75	ABS	366 (53)	649 (94)	198.7 (28.8)
Nicalon	AEAS	511 (74)	1428 (207)	135.9 (19.7)
Nicalon	CAS	266 (39)	752 (109)	124.2 (18.00)

Source: Ref. 6.

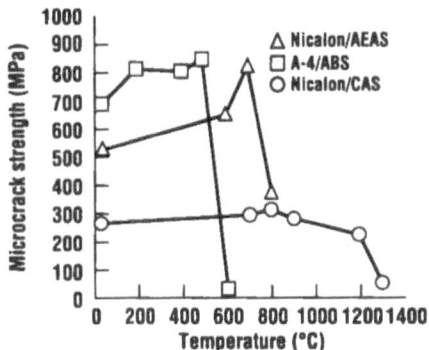

FIG. 6 Tensile microcrack yield stress as a function of temperature for glass and glass-ceramic matrix composites. (From Ref. 6.)

were also dependent upon fiber strengths. High strength fibers (A-4) yielded high strength composites, while lower strength fibers (HM, P55, and P75) yielded lower strength composites. Nicalon fiber reinforced glass (AEAS) had a higher microcrack yield stress and a higher ultimate stress than the Nicalon fiber/glass-ceramic matrix composite.

Strengths at elevated temperatures in air were measured also (Figures 6 and 7). The A-4/ABS composite maintained a microcrack yield stress of over 650 MPa and an ultimate stress of over 950 MPa to at least 500°C. This temperature is far above the upper use temperature for the best polymer matrix

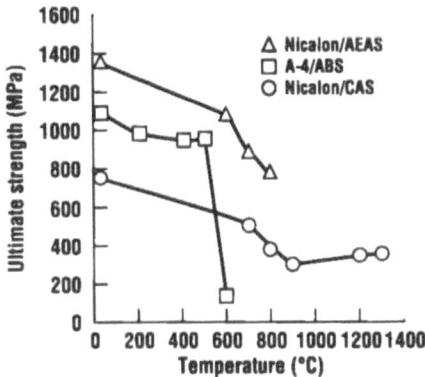

FIG. 7 Ultimate tensile stress as a function of temperature for glass and glass-ceramic matrix composites. (From Ref. 6.)

composites. Nicalon fiber reinforced materials maintained good properties to even higher temperatures. Nicalon/AEAS had a microcrack yield stress of about 800 MPa and an ultimate stress of about 900 MPa at 750°C. Nicalon/CAS strength was lower but respectable strength was maintained to at least 1200°C.

The thermal exposure time for the tests shown in Figures 6 and 7 was only 15 min. Long term stabilities of graphite/glass composites were assessed by exposing specimens to a 400°C temperature in air for various lengths of time, and then measuring mechanical properties at room temperature. Results are shown in Table 4. Even after 72 h, there was no significant loss in properties. The dense glassy matrix protected the carbon fibers from oxidation. However, this would obviously not be the case if the composite were first stressed beyond the point of matrix microcracking, which would permit exposure of fibers to the air environment. Even with the more oxidation stable Nicalon fibers, matrix microcracking can result in greatly reduced elevated temperature properties in air because the oxidation of the carbon layer generated during processing will be destroyed and interfacial reactions that increase strength of the fiber/matrix bond can occur.

Mechanical properties of the Nicalon/BAS (barium aluminosilicate) system were investigated by Kim and Katz.[7] Material with unidirectionally aligned fibers was supplied by Corning Glass Works. Tensile and 4-point flexural tests were conducted. End tabs of fiber glass/epoxy were used for the tensile specimens.

As in other reports on this type of material, tensile stress-strain curves had a proportional limit, followed by a nonlinear region. Average tensile strength was 416 MPa and flexural strength was 703 MPa. Stress at the proportional limit, i.e., the matrix microcrack stress, was 254 MPa in tension and 362 MPa in flexure. An interesting finding was that in tension just beyond the proportional limit the axial strain increased while the transverse strain decreased. That is, the transverse strain changed its direction. This was also true for the tension surface

TABLE 4. Mechanical properties based on flexural testing of glasses and glass-ceramics reinforced with Nicalon™ or carbon fibers after thermal treatments in air at 400°C.

Exposure, hours	Microcrack yield stress, MPa (Kpsi)	Microcrack elongation, %	Ultimate stress, MPa (Kpsi)	Elongation, %	Modulus, GPa (Mpsi)
0	697 (101)	0.56	1076 (156)	0.96	125.6 (18.2)
24	649 (94)	0.50	1180 (171)	1.00	127.7 (18.7)
72	683 (99)	0.53	987 (143)	0.99	129.0 (18.7)

Source: Ref. 6.

in flexural testing. The mechanism for the transverse strain reversal was not determined, but was postulated to be due to axial cracks and debonding between fibers and matrix. Similar Young's modulus values were determined from both tests, 149 GPa in tension and 141 GPa in flexure. Poisson's ratios calculated from the testing were 0.26 and 0.25, from tensile and flexural tests, respectively.

Fracture toughnesses of UTRC-produced Nicalon/LAS were determined by Wu et al [8] using what is termed the applied moment double cantilever beam method. Flexural strengths were determined also. They found that in fracture toughness testing of 0/90 specimens, failure always started with complete pullout of a section of fibers about 1 cm long along the crack path. This was followed by more conventional fracture, with individual fibers having different pullout lengths. When cracks were made to propagate perpendicular to fibers, average toughnesses of over 50 MPa·m$^{1/2}$ were determined. When cracks were made to propagate parallel to fibers, toughness was about 3 MPa·m$^{1/2}$, which is below that of matrix material alone. Flexural strengths averaged 240 MPa for 0/90 composites and 500 MPa for uniaxial composites in the longitudinal direction.

Effects of fiber orientation on properties of Nicalon/CAS composites fabricated at Corning Glass Works were investigated by Wang and Parvisi-Majidi.[9] Specimens included one with eight plies of uniaxially aligned fibers and several with three 0° plies toward each outer surface but different numbers of 90° plies in the interior. Composite stress-strain behavior was determined under tensile loading. The sequence of failure events in cross-ply composites included transverse and longitudinal cracking in the 90° plies and matrix microcracking in the 0° ply. The first transverse cracking strain was greatest for a composite having only one 90° ply, intermediate with two 90° plies, and least with three 90° plies. Delamination-type cracking of the 90° plies was observed, and postulated to be due to a Poisson's contraction effect.

Fatigue properties of Nicalon/AEAS (Corning Code 1723) were studied by Zawada et al.[10] Composite fabrication process was similar to that illustrated by UTRC. Vacuum hot pressing was carried out at 1090°C and 10.35 MPa. Fiber volume fraction was about 45%. To identify stress-strain behavior prior to fatigue testing, tensile strength tests using straight sided, tabbed specimens were carried out. Acoustic emission was used to determine the onset of matrix microcracking during tensile testing. The matrix microcracking stress level corresponds to the stress reached at the first major acoustic event measured during tensile testing. Extent of microcracking after various loads was also investigated microscopically.

Tension-tension fatigue was performed in air at room temperature. A unique hydraulic gripping system was used, and great care was taken to assure alignment and stability of the data acquisition system. The fatigue tests were conducted under load control using a frequency of 10 Hz. Untabbed, straight sided specimens were used. Tests were run for up to 10^6 cycles. Some of the specimens that survived the 10^6 cycles without failure were then tested in tension to determine residual tensile strength and Young's modulus.

First matrix microcracking for unidirectional material was found to occur at about 300 MPa, while the proportional limit was at about 400 MPa. Ultimate tensile strength was 680 MPa. For cross-plied material, matrix microcracking initiation and the proportional limit of the stress-strain curve both occurred at about 70 MPa. The curve then regained linearity until a second proportional limit was reached at about 200 MPa. The reason for this behavior was speculated to be cracks initiating and propagating in the 90° plies before they initiated and propagated in the 0° plies. Ultimate tensile strength was 318 MPa. Results of these tension tests were used to select fatigue test levels to evaluate both the importance of the matrix microcracking level and the proportional limit values.

Unidirectional material was fatigue tested at maximum stress levels ranging from 73% to 250% of the matrix microcracking stress value. At low stress levels, the modulus decreased slightly until the tests were terminated. At a stress level of 500 MPa, modulus dropped off to about 85% of its initial value (indicating damage accumulation) during the first 10^3 cycles and then recovered slightly after about 10^4 cycles. The specimen failed prior to 10^6 cycles.

The cross-plied material, which was fatigue tested at 79% to 400% of its matrix microcracking stress value, exhibited similar modulus behavior. At a maximum stress level of 55 MPa, the modulus first remained constant, then decreased gradually. At a stress level of 150 MPa, the modulus dropped to about 70% of its initial value within ten cycles, but recovered somewhat. Behavior was similar with a stress level of 201 MPa, but the specimen failed prior to 10^6 cycles. The reason for the rise in modulus at an intermediate number of cycles for both unidirectional and cross-ply composites might have involved creation of debris that precluded full closure of cracks during unloading, and/or realignment of fibers debonded from the matrix into a stiffer orientation as the stress on the specimen continued to be cycled.

By comparing tensile stress-strain behavior with the fatigue test data, it was seen that fatigue failure of unidirectional specimens occurred only when the maximum fatigue stress was above the proportional limit, even though the matrix began to microcrack at a much lower level. For cross-ply specimens, the fatigue stress limit closely matched the onset of the second proportional limit described above. In both cases, then, it can be implied that fatigue at stress levels high enough to produce nonlinear stress-strain behavior in the 0° plies was required to develop the type of damage that led to failure.

After fatigue testing, there was little relationship between residual strength and the maximum stress at which the material was cycled, with unidirectional composites. However, modulus decreased with increasing maximum fatigue stress, to 80% of its initial value, when cycled to 400 MPa. For the cross-plied material, residual tensile strength and modulus both decreased with increasing maximum fatigue stress. At a maximum fatigue stress of 175 MPa, tensile strength and modulus were 75% and 54%, respectively, of their initial values.

Although relatively weak interfacial bonds have been employed in most successful ceramic matrix composite systems to obtain appreciable toughening, Michalske and Hellman [11] deliberately produced glass matrix composites expected to have strong interfacial bonding in an attempt to form useful composites without matrix microcracking and fiber debonding and pullout prior to failure. Alumina Fiber FP was used with four silica glass compositions having thermal expansion coefficients above and below that of alumina. Processing conditions similar to those utilized in other work described above were used to produce unidirectional composites having volume fractions of about 45%. Composite flexural strengths (4-point) and fracture toughnesses (chevron notched beam) were determined.

Despite the apparent lack of toughening due to the factors described previously, K_{Ic} values for the composites ranged from 3.0 to 4.0 MPa·m$^{1/2}$, compared with typical alumina and glass values of 2.6 and 0.7–0.8, respectively. Flexural strengths were moderate, ranging from 158 to 311 MPa. Toughness increased with increasing residual compressive stress in the matrix (greatest positive difference between fiber CTE and matrix CTE), but strength was greatest with the minimal CTE difference between the fibers and matrix. Microscopic evaluation showed nearly planar fracture surfaces, indicating that interfacial bonds were indeed strong. The observed toughness increase was attributed to shielding the reinforcing fibers from direct interaction with matrix cracks due to a modulus mismatch between the glass matrices and the alumina fibers.

Gadkaree et al [12] studied Nicalon/Code 1723 AEAS glass matrix composites with silicon carbide whiskers (SC-9) incorporated into the matrix. Although continuous fiber reinforced glass and glass-ceramic matrix composites can have very high ultimate strengths, matrix microcracking occurs at a much lower stress level, leaving the composite susceptible to environmental attack at elevated temperature. This work was aimed at increasing the matrix microcrack stress by toughening the matrix material.

Incorporation of the whiskers into the matrix utilized techniques similar to those described for whisker reinforcement, and incorporation of Nicalon fibers into this material was similar to procedures described previously in this section. Composites were fabricated with whisker contents varying from 0 to 24 wt%. Microcrack stress and ultimate stress were measured in 4-point flexural testing. Tensile strength and other measurements were made also.

Variations in microcrack stress and ultimate strength as a function of whisker content are shown in Figure 8. Ultimate strength decreased with increasing whisker content, while matrix microcrack stress optimized at a 10% whisker content. (These results were in contrast to theoretical calculations that ultimate strength should be constant and matrix microcrack stress should increase with increasing whisker content.) Fracture morphology changed to one of long pullout lengths with no whiskers to almost no pullout with 24% whiskers. Qualitatively similar results were obtained in tensile testing.

Reasons for the decreasing ultimate strength and the peak in the matrix microcrack stress with whisker content were sought. Microanalytical studies revealed no significant change in fiber/matrix interfaces that would have changed bond strength. Another possible cause of the decrease in ultimate strength with an increase in whisker content that was investigated was decrease in fiber volume fraction with increasing whisker content due to an increased slurry viscosity with increasing whisker content. Fiber volume fraction decreased from 45% to 36% upon increasing whisker content from 0 to 24%, which could account for only part of the reduction in strength. The largest factor was concluded to be fiber damage due to the whiskers. SEM examination indicated such damage with 24% whiskers. An increased presence of flaws in a shorter spacing could account for the lack of fiber pullout during fracture.

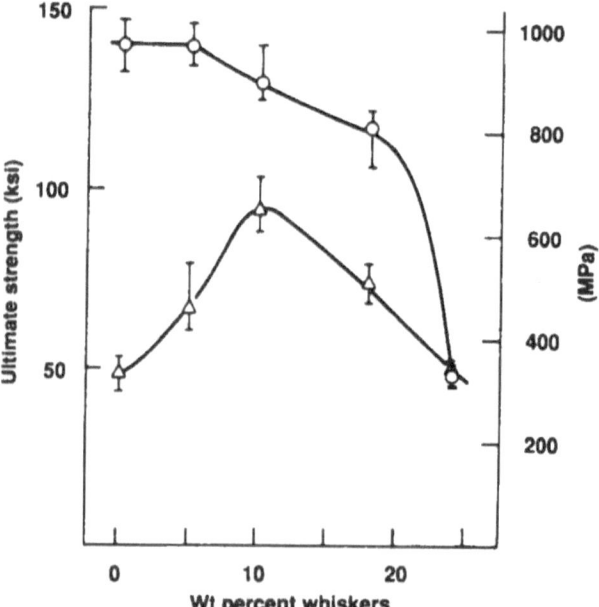

Fɪɢ. 8 Variations in microcrack yield stress (lower curve) and ultimate strength (upper curve) as a function of whisker content for Code 1723 AEAS reinforced with both Nicalon™ fibers and silicon carbide whiskers. (From Ref. 12.)

Although results were poorer than predicted, it was demonstrated that the stress level causing microcracking can be increased by whisker additions, increasing the possibility of composite use at higher strain levels without exposing fibers to the atmosphere, which would be of particular benefit at elevated temperatures. In addition, interlaminar shear strength increased from 10.3 MPa to 53.7 MPa and transverse strength increased from 44.8 MPa to 134.8 MPa with increasing whisker content from 0 to 24%.

III. HOT PRESSED CRYSTALLINE MATRIX COMPOSITES

A. Silicon Nitride Matrix

Fabrication of continuous fiber reinforced crystalline matrix systems by slurry processing and hot pressing has been carried out also. Major emphasis has been on silicon nitride. There is some similarity between silicon nitride matrix composites produced by hot pressing and the glass matrix composites discussed above, since considerable quantities of sintering aids to provide a glassy intergranular phase are used. In addition, the glass-ceramic matrices have considerably crystallinity, so that there is some similarity in this regard. Nonetheless, the hot pressed crystalline matrix composites are sufficiently different to warrent a separate treatment.

Work at Textron Speciality Materials of Lowell, MA, on silicon nitride reinforced with SCS-6 fibers has been reported by Thomson and LeCostaouec.[13] The matrix powder consisted of silicon nitride with 5 wt% yttria and 1.25 wt% magnesia. The powder was ball milled and dispersed in water including a binder to form a sprayable slurry. Total milling time was typically 72 h. SCS-6 fiber was wound onto a drum with a uniform spacing of about 40 fibers per cm.

The matrix powder was then sprayed onto the collimated fibers to form a tape. A slow buildup of matrix was used to prevent cracking during drying. A polymeric nonwoven backing on the tape improved handling. Tape was cut to size by die punching and plies laid up to the desired orientation in the hot press mold. Binder was thermally removed and the composite hot pressed to full theoretical density. An earlier news release [14] reported that processing methods were developed in cooperation with the University of Lowell (MA).

Flexural strength tests have been conducted at room temperature and at elevated temperatures and tensile strength tests conducted at room temperature. Interlaminar shear strength was determined also. Thermal shock resistance was determined by quenching in oil from 1000° and 1200°C. Ballistic impact was determined. Details of the tests were not reported.

Flexural strengths at room temperature for unidirectional material were above 650 MPa. Toughness was qualitatively demonstrated by a load carrying capability beyond the proportional limit. At 1350°C, strength was 380 MPa.

After exposure for 100 h at 1260°C in air, strength at room temperature was 480 MPa. Ultimate tensile strength at room temperature for a 0/90 cross-plied specimen was about 350 MPa (Figure 9). Young's modulus was 250 GPa. Room temperature interlaminar shear strengths on unidirectional composites were 117 MPa or higher.

On average, flexural strength and Young's modulus values were retained after thermal shocking the specimens from either 1000° or 1200°C. Additional thermal shock testing was done at NASA-Lewis Research Center. This test rig uses the exhaust from a hydrogen/oxygen rocket engine. Heat-up rates of several thousand degrees Celsius per second can be achieved. The maximum temperature is adjusted by varying the hydrogen/oxygen ratio. For these specimens, cycle times of 1–3 s for up to three cycles were used. All specimens having near theoretical density performed well. At 1300°C both unidirectional and cross-plied specimens were either visually undamaged or showed slight oxidation and cracking of the leading edge. At 1500°C, specimens had some scale formation and edge cracking, and at 1700°C more scale formation occurred. At 1900° and 2300°C, there was a glassy formation on the specimen surfaces.

During high velocity ballistic testing, cross-plied composites demonstrated high toughness. Only localized damage was observed in the area of the projectile penetration. The toughness and potential multiple hit capability were said to make it a good candidate for containment applications in a gas turbine engine.

B. Silicon Carbide Matrix

Fiber reinforced composites having silicon carbide matrices formed by hot pressing have been fabricated by Miyoshi et al.[15] Although continuous SCS-6 silicon carbide fibers and Toray PAN-derived carbon fibers were used, they were chopped into shorter lengths in this work.

Characteristics of the silicon carbide powder were not reported. Aluminum nitride (2 wt%) was added as a sintering aid. In addition, a small amount of phenolic resin was added as a carbon source. The silicon carbide powder and additives were thoroughly mixed and then combined with the filaments. The SCS-6 fibers were unidirectionally arranged into planar sheets. Twelve layers of fibers were alternately stacked with 13 layers of matrix powder and cold pressed at 50 MPa. The carbon fibers were cut into 5-mm lengths and mixed with the powders using a mortar and pestle. Average length of the carbon fibers was about 0.5 mm after mixing. These mixtures were also cold pressed first. Hot pressing was carried out at 2100°C for 1 h in vacuum, which resulted in near theoretical density. Volume fraction of fibers was up to 30%.

Test specimens were cut from the SCS-6 reinforced material normal to the pressing direction so that fiber orientation was in the long direction. Specimens of the carbon reinforced material were also cut so that the long direction was normal to the pressing direction; orientation in this plane was multidirectional because the fibers had been added directly to the matrix material (as in whisker reinforced material). Four-point flexural tests and single edge notched beam fracture toughness tests were conducted, and Young's modulus values were calculated from the stress-strain curves. Other characterization methods were used also.

The carbon addition had a large effect on composite properties. Without the addition, fracture toughness for the SCS-6 reinforced material was about 4.5 MPa·m$^{1/2}$, not appreciably higher than that for monolithic silicon carbide (3–4 MPa·m$^{1/2}$). Scanning electron microscopy of fracture surfaces showed a flat fracture surface, indicative of strong interfacial bonding. It was postulated that the carbon layer on the SCS-6 fibers had been destroyed during hot pressing by reaction with a silica surface layer on the silicon carbide powders.

With the carbon addition, fiber pullout could be observed on the fracture surfaces and fracture toughness was increased. With a 2.5 wt% carbon addition, fracture toughness increased from about 3.5 MPa·m$^{1/2}$ with no whisker addition to 7.5 MPa·m$^{1/2}$ with 30 vol% whiskers. However, flexural strength decreased from about 630 MPa with no whisker addition to 500 MPa with 30 vol% whiskers. Stress-strain curves had about three times the deflection to failure compared with unreinforced silicon carbide. It was postulated that the carbon addition helps reduce reaction of the carbon layer on the fibers with the silica surface layer on the silicon carbide powder.

With carbon fiber reinforcement, fracture toughness at room temperature was only about 3.5 MPa·m$^{1/2}$. Fractured surfaces had gaps between the fibers

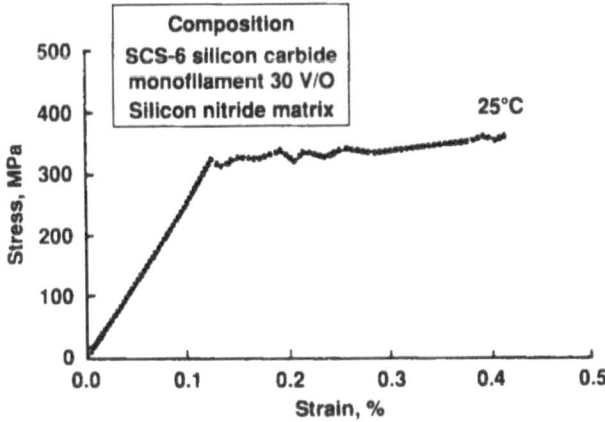

Fig. 9 Room temperature tensile stress-strain curve for SCS-6 fiber reinforced silicon nitride produced by hot pressing. (From Ref. 13.)

and matrix. It was suggested that the greater CTE of silicon carbide than carbon ($5.5 \times 10^{-6}/°C$ vs. $4.0 \times 10^{-6}/°C$) resulted in these gaps upon cooling from the hot pressing temperature. Supporting this hypothesis, fracture toughness increased with increasing temperature, reaching about 8 MPa·m$^{1/2}$ at 1450°C.

IV. SUMMARY

Continuous fiber reinforced glass and glass-ceramic materials have undergone rather extensive development. Nicalon fibers have been used most often, but carbon and other fibers have been used as well. Because these materials are hot pressed at temperatures at which the matrix powders can undergo viscous deformation, porosities can be very low. Flexural strengths of well over 1000 MPa and fracture toughnesses of over 50 MPa·m$^{1/2}$ have been reported. These compare very favorably with other engineering materials.

Because of the low porosities of these materials, the fibers and fiber/matrix interfaces can be shielded from atmospheric oxygen. Hence, under some conditions, mechanical properties at elevated temperatures are maintained or even enhanced, because deleterious reactions can be avoided for considerable lengths of time. With the more refractory matrix compositions, respectable properties have been measured at temperatures as high as 1000°C. However, under conditions at which matrix microcracking occurs, properties deteriorate rapidly due to fiber degradation and/or interfacial reactions that result in an increased fiber/matrix bond strength.

It is likely that improvements will be made in increasing the high temperature capabilities of these materials. For example, incorporation of whiskers into the matrix can increase the matrix microcrack stress. However, these materials will probably not compete with more refractory crystalline matrix materials for the highest temperature applications envisioned for ceramic matrix composites.

A limitation of this type of composite is the requirement of hot pressing to consolidate the matrix. The need for hot pressing results in restrictions in size, shape, and complexity of structures that can be fabricated by the process. In addition, hot pressing is an expensive operation, not only because of equipment cost, but because of the relatively long cycle time required to produce one structure. Despite these limitations, the attractive aspects of these materials justify further development, and commercial applications seem likely.

The limitations with regard to hot pressing also apply to crystalline matrix composites. In addition, the large diameter (~143 μm) silicon carbide fibers used in the silicon nitride matrix composites would also limit complexity. The issues of fiber diameter and the hot pressing requirement are being addressed to some extent at Textron [13] with investigation of hot isostatic pressing of parts and also by the use of somewhat smaller (75 μm diameter) SCS-9 fibers. The work cited on silicon carbide matrix composites by this technique was too limited for a realistic evaluation.

REFERENCES

1. Bacon, M., Reflections on glass, *Materials Edge*, vol. 17, 27–28 (1990).
2. Prewo, K. M., Brennan, J. J., and Layden, G. K., Fiber reinforced glasses and glass-ceramics for high performance applications, *Ceramic Bulletin*, vol. 65, no. 2, 305–313 (1986).
3. Prewo, K. M., and Brennan, J. J., High-strength silicon carbide fibre-reinforced glass-matrix composites, *Journal of Materials Science*, vol. 15, 463–468 (1980).
4. Prewo, K. M., Tension and flexural strength of silicon carbide fibre-reinforced glass ceramics, *Journal of Materials Science*, vol. 21, 3590–3600 (1986).
5. Mah, T., Mendiratta, M. G., Katz, A. P., Ruh, R., and Mazdiyasni, K. S., High-temperature mechanical behavior of fiber-reinforced glass-ceramic-matrix composites, *J. Am. Ceram. Soc.*, vol. 68, no. 9, C248–C251 (1985).
6. Allaire, R. A., Janas, V. F., Stuchly, S., and Taylor, M. P., Glass matrix composites for higher use temperature applications, *Advanced Materials Technology '87*, 32nd International SAMPE Symposium, April 6–9, 1987, 624–634.
7. Kim, R. Y., and Katz, A. P., Mechanical behavior of unidirectional SiC/BMAS ceramic composites, *Ceram. Eng. Sci. Proc.*, vol. 9, no. 7–8, 853–860 (1988).
8. Wu, C. C., Lewis, D., and McKinney, K. R., Strength and toughness measurements of ceramic fiber composites. In *Frac. Mech. Ceram.*, Proceedings of the Fourth International Symposium, Plenum Press, New York, NY, vol. 7, 1986, 53–60.
9. Wang, S.-W., and Parvizi-Majidi, A., Mechanical behavior of Nicalon™ fiber-reinforced calcium aluminosilicate matrix composites, *Ceram. Eng. Sci. Proc.*, vol. 11, no. 9–10 (1990).
10. Zawada, L. P., and Butkus, L. M., Room temperature tensile and fatigue properties of silicon carbide fiber-reinforced aluminosilicate glass, *Ceram. Eng. Sci. Proc.*, vol. 11, no. 9–10, 1592–1606 (1990).
11. Michalske, T. A., and Hellmann, J. R., Strength and toughness of continuous-alumina-fiber-reinforced glass-matrix composites, *J. Am. Ceram. Soc.*, vol. 71, no. 9, 725–731 (1988).
12. Gadkaree, K. P., Chyung, K. C., and Taylor, M. P., Hybrid ceramic matrix composites, *Journal of Materials Science*, vol. 23, 3711–3720 (1988).
13. Thomson, B., and LeCostaouec, J.-F., Recent developments in SiC monofilament reinforced Si_3N_4 composites, *SAMPE Quarterly*, vol. 22, 46–51 (1991).
14. Anon., *Adv. Mater.*, vol. 12, no. 10, 3–4 (1990).
15. Miyoshi, T., Kodama, H., Sakamoto, H., Gotoh, A., and Iijima, S., Characteristics of hot-pressed fiber-reinforced ceramics with SiC matrix, *Metallurgical Transactions A*, vol. 20A, 1989–2419 (1989).

12

Ceramic Matrix Composites Produced by Reaction Bonding

I. BACKGROUND

Reaction bonding provides a method of consolidating and densifying crystalline ceramic matrices derived from slurry infiltration of continuous fiber tows or two- or three-dimensional preforms. With reaction bonding, the matrix of a slurry-infiltrated preform is formed by chemical reaction, rather than by joining of like particles as in sintering or hot pressing. Three general variations are possible. The slurry-infiltrated preform containing a particulate material can react with an infiltrating gas. Alternatively, the slurry-infiltrated preform containing a particulate material can be impregnated with a second, molten material, causing reaction. Finally, two (or more) different types of particulate materials can be included in the slurry and these can be heated to cause reaction with each other.

A favorable condition for reaction bonding is a product density lower than the density of the particulate material, so that a volume increase occurs on reaction, reducing porosity of the composite. For example, conversion of silicon to silicon nitride involves a 23% increase in volume, which generally results in a porosity decrease with no change in external dimensions of the specimen that is nitrided. Also, a system requiring a reaction temperature below that at which the fibers used in the composite degrade appreciably is obviously desirable.

Although not quite the same as reaction bonding, a somewhat related concept, cement bonding, will be included in this chapter. The term "cement" is used here to mean any material included in a ceramic slurry that can undergo reactions to form a binding phase to join matrix particles (or, conceivably, to form the entire matrix material). Relatively little has been reported on the concept of cement bonding for high performance ceramic matrix composites.

II. REACTION BONDED COMPOSITE MATERIALS

A. Silicon Nitride Matrix Composites

The most widely reported work on a reaction-bonded composite system has been carried out by R. T. Bhatt and coworkers at NASA-Lewis Research Center on an SCS-6 fiber, silicon nitride matrix system. Figure 1 [1] schematically illustrates one variation of the process. The SCS-6 fibers were first wound onto an axially translating cylindrical drum to provide a desired fiber spacing and coated with a polymeric binder, so that fiber mats could be cut from the drum. High purity silicon powder obtained from Union Carbide Corp. was used. Since the as-received powder contained large agglomerates, attrition milling was used to reduce the particle size and improve its reactivity to nitrogen during the nitridation step. The average particle size of the attrition milled powder was 0.3 μm. This powder was mixed with a polymeric binder and an organic solvent and rolled into silicon "cloth." The thickness of the cloth determined the volume fractions of the matrix and fibers in the composite.

The two types of material were alternately stacked to form a multilayer preform and, after binder removal, hot pressed to consolidate. The resulting preforms were transferred to a nitridation furnace to convert the silicon to silicon nitride. Both unidirectional and 0/90 cross-plied composites have been produced. A typical cross section of a unidirectional composite is shown in Figure 2.

Tensile testing was conducted at room temperature, both on as-fabricated specimens and on specimens that had been subjected to heat treatment in air and in nitrogen for 100 h at temperatures up to 1400°C. Interfacial shear stress at room temperature was calculated using a relationship based on the average matrix crack spacing after fracture. Tensile tests were also carried out at room temperature using notched specimens. Four-point flexural tests were conducted from room temperature to 1400°C in air. Interlaminar shear strength was determined from compression testing of specimens with notches machined from the opposite faces, using calculations including the maximum failure load and the shearing area. Fracture toughness for unidirectional composites was measured using the single edge notched beam method in 3-point bending. Thermal shock resistance was indicated by comparing room temperature tensile and flexural strengths of as-produced specimens with those of specimens water quenched from temperatures up to 1100°C.

Typical room temperature tensile stress-strain curves for unidirectional and cross-plied specimens and typical fractured specimens are shown in Figures 3 and 4, respectively. Stress-strain curves showed three regions, an initial linear elastic region, a nonlinear region, and a second linear region. Curves for unidirectional composites tested normal to the fiber orientation direction and for unreinforced reaction bonded silicon nitride are shown also. Interfacial shear strength was calculated to be 18 MPa. Average interlaminar shear strength for unidirectional composites was 40 MPa.

During fracture toughness testing, cracks were initially formed normal to the crack tip, then grew along the fiber/matrix interface. With continued load, cracks formed parallel to the notch tip. Even at the maximum load, cracks

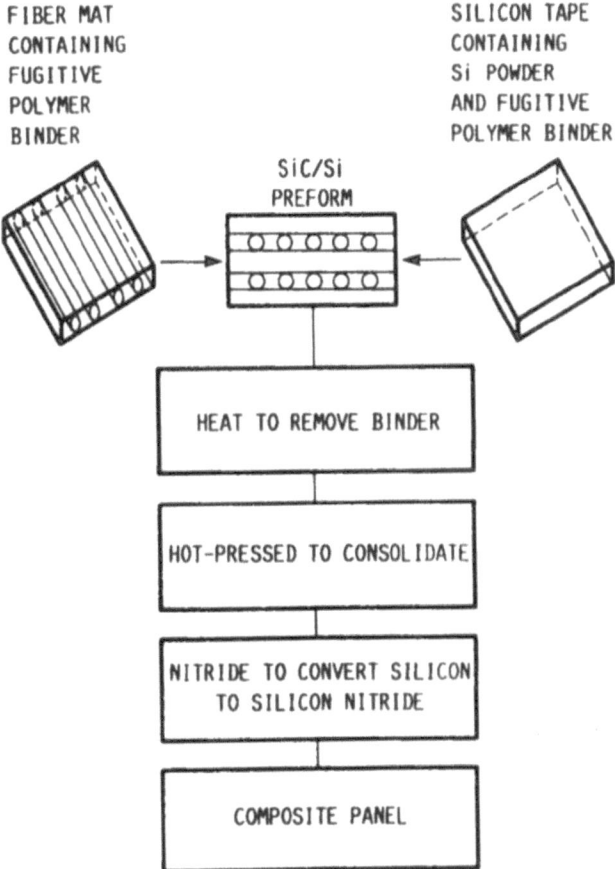

FIG. 1 Schematic flow diagram for formation of SCS-6 fiber reinforced silicon nitride by reaction bonding. (From Ref. 1.)

propagating from the notch tip did not reach the compression side, and the specimen did not break, but bent into a U-shape. The measured K_{Ic} value for unidirectional material was 13 MPa·m$^{1/2}$, over three times a typical value for unreinforced reaction bonded silicon nitride. As another indication of increased toughness, tensile strength was not affected by the presence of notches.

Water quenching from temperatures as high as 1100°C did not appreciably affect room temperature tensile strength of unidirectional composites. Flexural strength was not affected by a quench from 500°C, but it was reduced by about 50% by quenches from 650° to 1100°C, presumably due to matrix microcracks that did not affect tensile strength. In comparison, flexural strength of unreinforced reaction bonded silicon nitride was not affected by a water quench from 300°C, but was decreased by about 75% by quenching from 500° or 800°C.

Flexural strength at either room temperature or at 1200°C was about 750 MPa. Strength at 1400°C was still about 600 MPa. In comparison, unreinforced hot pressed silicon nitride had a higher room temperature strength, about 800 MPa, but strength decreased to less than 600 MPa at 1200°C and to less than 400 MPa at 1400°C. Overall flexural strengths of unreinforced reaction bonded

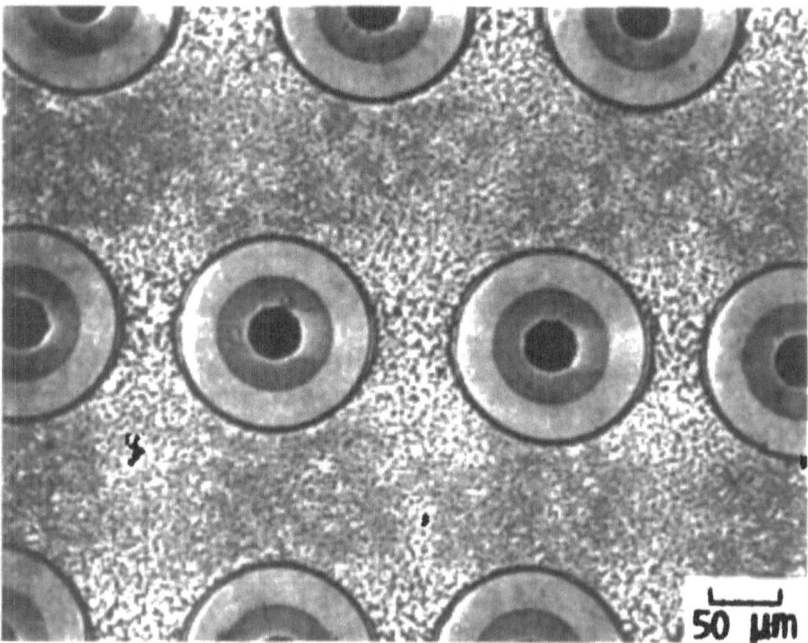

Fig. 2 Typical cross section of an SCS-6-fiber, silicon nitride matrix composite produced by reaction bonding. (From Ref. 1.)

silicon nitride were considerably lower. Room temperature flexural strength was less than 300 MPa. Strength increased to about 400 MPa at 400°C and about 350 MPa at 1400°C.

Ultimate tensile strength values and first matrix microcrack stress values after the 100 h heat treatments are shown in Figure 5. Neither ultimate tensile strength nor first matrix microcrack stress was affected by heat treatment in nitrogen. However, there was a significant decrease in both of these properties after heat treatments at 600° and 800°C in air. After heat treatment at higher temperatures, ultimate tensile strength was reduced relative to as-fabricated strength, but to a lesser extent than after heat treatment at the intermediate temperatures, and first matrix microcrack stress was about the same as for the as-fabricated material.

It was proposed that the large loss in strength after heat treatment at the intermediate temperatures in air was due to oxygen penetration of the porous matrix and oxidation of the carbon coating at the fiber/matrix interfaces. At

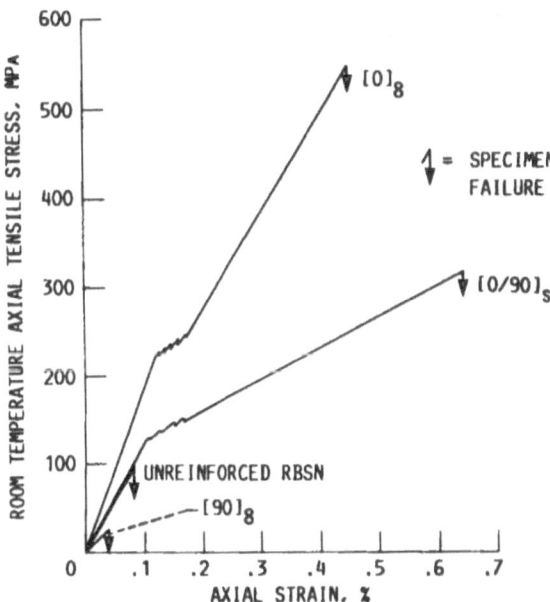

FIG. 3 Typical room temperature tensile stress-strain curves for SCS-6 fiber, silicon nitride matrix composites produced by reaction bonding. (Includes 8-ply unidirectional composites tested in the fiber direction, unidirectional composites tested 90° to the fiber direction, and cross-plied composites tested along one of the fiber directions.) (From Ref. 1.)

higher temperatures, the oxygen might react with some of the silicon nitride to form a protective silica coating on the composite surface. However, instability of this coating as well as fiber degradation were possible reasons why some strength reduction occurred at these higher temperatures.

Work carried out to reduce porosity of the silicon carbide fiber reinforced reaction bonded silicon nitride was carried out by Bhatt and Kiser [2] to attempt to avoid the loss in strength after heat treatment in air at intermediate temper-

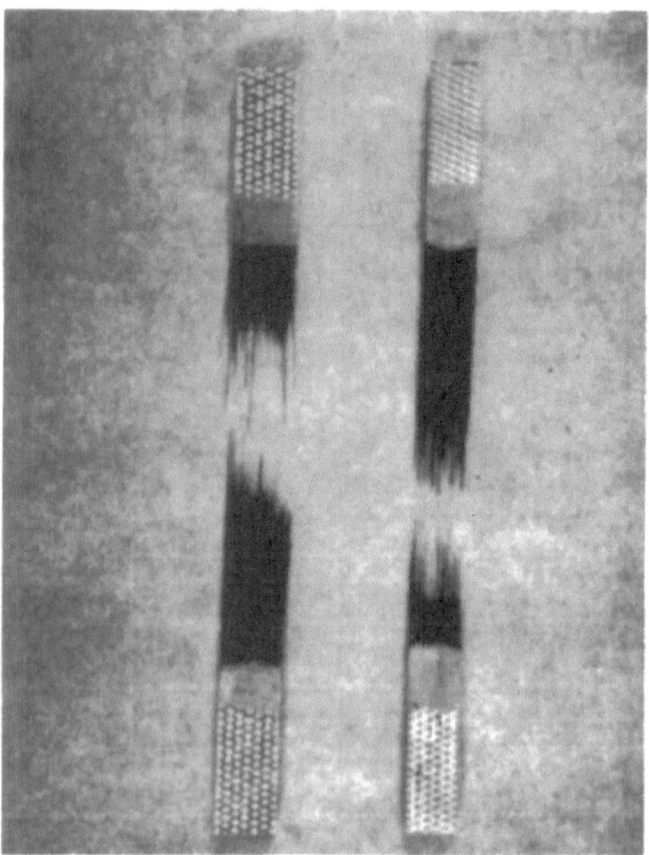

FIG. 4   Fractured tensile test specimens of SCS-6 fiber, silicon nitride matrix composites showing extensive fiber pullout. (From Ref. 1.)

atures. Composite fabrication was similar to that described above. For these experiments, 2.5 wt% nickel oxide was added to the silicon powder to enhance nitridation. For experiments in which hot isostatic pressing was to be investigated, 5 wt% magnesia (in addition to the nickel oxide) was added as a sintering aid. Hot pressing was carried out at pressures ranging from 27 to 138 MPa for up to 1 h from 600° to 1000°C. The consolidated preforms were heat treated in high purity nitrogen for up to 100 h at 1000° to 1400°C.

Some of these composites were hot isostatically pressed. For this process, the specimens were coated with a boron nitride slurry, wrapped with Grafoil (Union Carbide Corp.), and then placed over a Grafoil wrapped silicon carbide

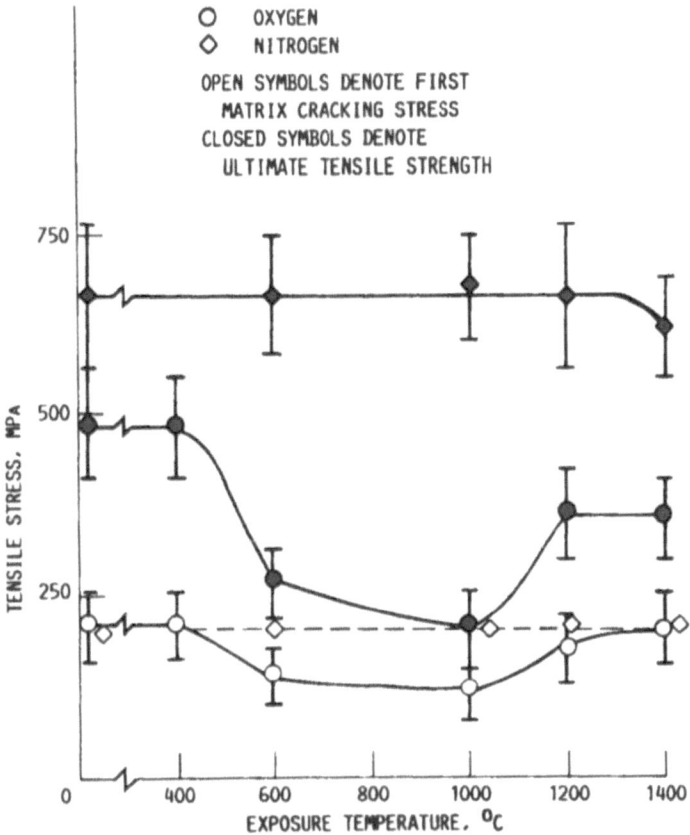

FIG. 5 Ultimate tensile strength values and first matrix microcrack stress values for SCS-6 fiber, silicon nitride matrix composites after 100 h treatments in air and in nitrogen at various temperatures. (From Ref. 1.)

plate, which prevented warping of specimens during HIPing. The composite and silicon carbide plate were sealed into a tantalum can having inside surfaces coated with boron nitride to prevent reaction with the Grafoil.

The sealed cans were hot isostatically pressed for 1 h in argon under 138 MPa pressure. After HIPing, the cans were cut open to remove the specimens. In addition to mechanical property testing, density and pore size measurements were made and microstructural characterization tests were conducted.

Composite density increased with increasing hot pressing temperature. It increased from 2.07 to 2.45 g/cm^3 as pressure was increased from 27 to 138 MPa. There was a decrease in mean pore diameter from 3.5 to 2.0 μm over the same range. However, there was little effect of density on ultimate tensile strength, Young's modulus, or first matrix microcrack stress. Microstructural analyses indicated that the largest cracks and voids between the fibers resulted from the lowest consolidation pressure. At the two higher pressures, there were few cracks, but there was nonuniform distribution of porosity.

The hot isostatically pressed specimens were fully dense, with uniform microstructure and little matrix porosity. The carbon-rich coating on the fibers was intact after the HIPing process. Room temperature tensile stress-strain curves for unidirectional as-nitrided and hot isostatically pressed specimens are shown in Figure 6. The HIPed composite had a higher Young's modulus and a higher first matrix microcrack stress, but lower strain to first matrix microcracking and a much lower ultimate tensile strength. The fracture surfaces of the HIPed material had much less fiber pullout. It was postulated that fiber strength degradation during HIPing was responsible for the lower ultimate strength. This was supported by experiments in which fibers were tensile tested as-received and after heat treatment simulating the hot isostatic pressing conditions. Average fiber tensile strength was 3.8 GPa as-received and 0.8 GPa after heat treatment.

In another investigation [3], effects of nitridation conditions on composite properties were determined. Nitridation was carried out in either high purity nitrogen, or in nitrogen containing 4 vol% hydrogen. Temperature was varied from 1000° to 1400°C. Times up to 100 h were used. For one nitridation condition, specimens were subsequently heat treated in oxygen at 600°C for 100 h. Properties similar to those outlined above were determined.

Table 1 shows mechanical properties as a function of nitridation conditions and subsequent heat treatment. Nitridation at 1200°C in nitrogen for 40 h resulted in best composite properties, including an ultimate tensile strength of near 700 MPa. Properties were poorer with a hydrogen addition to the nitriding atmosphere and, especially, with use of a higher temperature, 1350°C. Heat treatment of specimens produced using the best nitridation conditions degraded properties considerably.

Based on the interfacial shear strength measurements and microstructural examinations, it was concluded that the effects of the various conditions on the carbon-rich outer layers of the SCS-6 fibers dominated composite properties. Fabrication of composites in hydrogen-containing nitrogen or oxidation of composites at 600°C in air caused degradation of the carbon-rich surface coating, which, in turn, caused a decrease in interfacial strength. An interfacial shear strength of greater than 10 MPa was considered desirable.

Other work on this type of composite involved testing of unidirectional composite specimens at several orientations with respect to the fiber direction and testing of cross-plied composites at 0/90 and +45/−45 orientations.[4] As would be expected, there was a large anisotropy in unidirectional composites depending on specimen orientation, and in-plane off-axis properties were improved with cross-plied composites, at the expense of properties along the tensile axis.

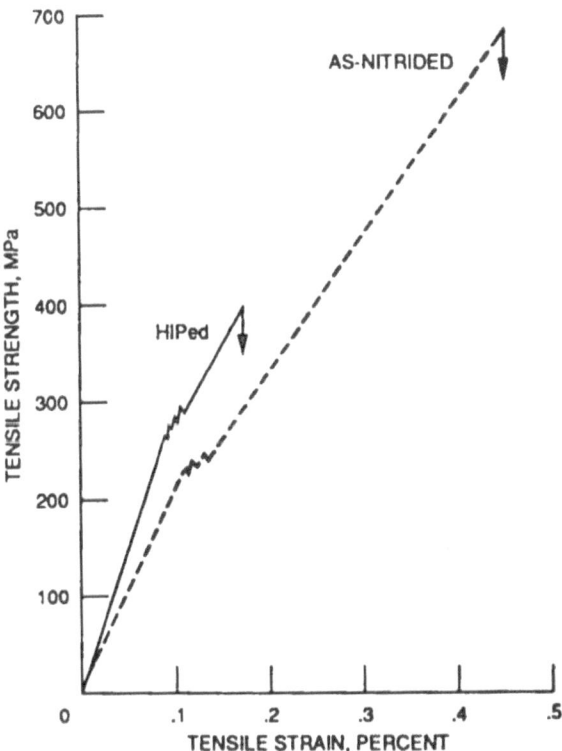

FIG. 6 Room temperature tensile stress-strain curves for unidirectional as-nitrided and hot isostatically pressed specimens of SCS-6 fiber, silicon nitride matrix composites produced by reaction bonding. (From Ref. 2.)

Reaction bonded silicon nitride matrix composites have also been produced at Oak Ridge National Laboratory.[5] This work differed from the NASA work in a number of ways. Nicalon fibers were used. Powder was suspended in an aqueous solution of organic monomers and after casting was polymerized into a gel. Drying yielded parts with only about 3 wt% polymer. Microwave heating was utilized for the nitridation process. This permitted the use of lower temperatures and shorter sintering times than conventional heating sources.

High purity silicon powder (Elkem Inc. grade HQ) with a mean particle size of 4.2 μm was used. The gelation chemicals were an acrylamide monomer, a cross linking agent (N,N'-methylene bisacrylamide), and a free radical initiator (ammonium persulfate). Unidirectional Nicalon fiber tows were cut to length and hand laid into a mold.

In order to eliminate a problem of premature gelation during blending of the silicon powder and the organic additives, the surface activity of the powder was reduced by heat treatment in nitrogen at 800°C for 4 h. Slurries made with the heat-treated powder did not prematurely polymerize. Attempts to remove air from the slurries by stirring under vacuum also resulted in premature geling, so the slurries were not deaired but simply poured into the mold, which had been chilled with ice water. The initiator was then mixed into the slurry. To improve wetting of the fibers, part of the slurry was brushed onto the fiber tows, which were then manually arranged in the mold. A block was placed on the mold to

TABLE 1. Effects of nitridation conditions and heat treatment on mechanical properties of SCS-6 fiber, silicon nitride matrix composites produced by reaction bonding.

Fabrication conditions	Primary elastic modulus (GPa)	First matrix cracking stress (MPa)	First matrix cracking strain (%)	Average matrix crack spacing (mm)	Interfacial shear strength (MPa)	Ultimate tensile strength (MPa)
1200°C in N_2 for 40 h	193 ± 7	227 ± 40	0.12	0.8 ± 0.2	18 ± 4	692 ± 150
1200°C in N_2 + 4% H_2 for 40 h	190 ± 8	217 ± 30	0.11	1.0 ± 0.4	14 ± 3	440 ± 110
1350°C in N_2 + 4% H_2 for 40 h	150 ± 6	176 ± 35	0.12	7.0 ± 0.5	1.8 ± 0.5	320 ± 56
1350°C in N_2 + 4% H_2 for 72 h	90 ± 8	46 ± 10	0.05	—	—	46 ± 10
1200°C in N_2 for 40 h (treated in O_2 at 600°C for 100 h)	140 ± 10	154 ± 30	0.11	12 ± 2	0.8 ± 0.4	230 ± 50

Source: Ref. 3.

compress the fibers and squeeze excess slurry to the side of the mold. Geling was accomplished by heating to 50°C. The green composite was dried at room temperature and binder removed by heating in air at 600°C.

Reaction bonding was carried out in a 2.45-GHz microwave furnace in a nitrogen atmosphere of about 0.1 MPa. Heating rate was 5°C/min to 800°C, 2°C/min to 1100°C, and 0.5°C/min to 1250°C or 1350°C. Heating was continued for 8 h at the maximum temperature. Four-point flexural strengths were measured and fracture surfaces examined by SEM.

Compared with the properties of the NASA-Lewis reaction bonded silicon nitride matrix composites, flexural strengths were very low. Strength averaged 85 MPa for composites nitrided at 1250°C and 70 MPa for composites nitrided at 1350°C. There was relatively little fiber pullout, and failure was brittle for the composite nitrided at 1350°C, suggesting fiber degradation and/or strong interfacial bonding. Obviously, more work would need to be done using these fabrication techniques before a material with properties competitive with other successful composites can be demonstrated.

Reaction bonded silicon nitride as a composite matrix has also been investigated at Massachusetts Institute of Technology (MIT) by J. S. Haggerty and coworkers.[6] Although the emphasis of this section is on continuous fiber reinforcement and the MIT work emphasized whisker reinforcement, it will be cited here to provide more information on reaction bonded silicon nitride as a matrix material.

Silicon powder was produced from gaseous precursors using a laser as a heat source. These submicron powders were reported to be extremely pure, nonagglomerated, and producible at commercially acceptable costs. It was reported that powder handling, dispersion, shaping, and drying techniques permitting parts with random close packing and no introduction of contaminants have been developed. Nitridation was at 1250°C for 1 h. Density was about 75% of theoretical. Under these nitridation conditions, there was no degradation of silicon carbide whisker surfaces (whisker source not specified).

The process for the silicon powder production and monolithic silicon nitride production at MIT has been reported in numerous articles over the last ten years. Recent information on this process has been reported by workers at MIT and the University of Massachusetts.[7] This work is referenced to provide a source for background information on processing conditions, but will not be described here since it did not involve whisker or fiber reinforcement.

Work on reaction bonded silicon nitride as a composite matrix has also been conducted at Georgia Tech Research Institute by Starr and coworkers.[8] This work involved short fiber reinforcement. Nicalon fibers and silicon carbide whiskers (Advanced Composite Materials Co.) were used as reinforcements. With the Nicalon fibers, silicon nitride powder having a mean particle diameter of 5 μm was used. With the whiskers, a very fine powder (mean diameter of

0.03 μm) from Tokyo Tekko Co. was used. Since previous studies had indicated that a desirable fiber aspect ratio was about 20:1, the chopped Nicalon fibers, which had aspect ratios of about 100:1, were reduced by crushing and the distribution of fiber lengths was narrowed by a combination of screening and washing operations.

Blending of aqueous fiber and powder dispersions was carried out using low shear dispersion techniques. Deairing techniques were used to avoid formation of large flaws in the composites. Target density was about 75% of theoretical, since it was reported that a higher density results in incomplete nitridation, while a lower density results in higher porosity and reduced strength. Fiber contents of up to 23 vol% were attained. Nitridation required temperatures of up to 1400°C in nitrogen, which resulted in severe disintegration and pitting of Nicalon fiber surfaces and formation of silicon nitride whiskers near the fiber surfaces. Reaction was less severe with ceramic grade Nicalon fibers than with standard grade Nicalon fibers.

Flexural strengths (4-point) for composites with 25 vol% Nicalon fibers ranged from 150 to 190 MPa and fracture toughness (single edge notched beam) ranged from 1.8 to 2.7 MPa·m$^{1/2}$. Apparently fiber degradation limited the strength and toughness values. Whisker reinforced materials were lower in density and no mechanical testing was performed. Obviously, improvements would have to be made using these processing techniques to provide successful composite materials.

B. Silicon Carbide Matrix Composites

Silicon carbide formation by reaction bonding to produce a composite matrix has also been investigated, but to a lesser extent than silicon nitride. In reaction bonding of silicon carbide, a carbon preform is usually used and reaction is with infiltrated molten silicon. Work on fiber reinforced reaction bonded silicon carbide has been reported by Fitzer and Gadow [9] at the University of Karlsruhe (Germany).

Carbon fibers and silicon carbide fibers and whiskers were used as the reinforcement. Carbon fibers were coated with silicon carbide or with titanium carbide by chemical vapor deposition. Fibers were wet-wound with phenolic, epoxy, or polyphenylacetylene resin and dried to form specimens with unidirectional reinforcement. Production of carbon bonded composites was analogous to that used in forming carbon/carbon composites, with pressing, hardening, and solid-state carbonization of the precursor. Pore size distribution and microstructure could be modified by addition of dispersed carbon fillers and variation in the chemical composition of the precursor. Resins producing carbon residues ranging from 28 to 85% were used. For reinforcement with silicon carbide whiskers, dry mixing of resin powder, filler, and whiskers, with subsequent melting and pressure-hardening was reported as a suitable process. The preforms were heat treated to form silicon carbide whisker/carbon composites and then reacted with molten silicon.

Siliconization was performed at temperatures ranging from about 1425° to 1475°C to minimize fiber reactions or at about 1925°C to attempt to obtain faster or more complete conversion of the carbon matrix into silicon carbide. The selected temperature also depended upon the starting porosity and pore size distribution of the preform.

With siliconization at the lower temperatures, flexural strength (3-point) was improved compared with monolithic materials, particularly when ''high modulus CVD-deposited silicon carbide fibers'' (perhaps SCS-6, although this was not reported) or high modulus silicon carbide coated carbon fibers were used (Figure 7). When the carbon matrix was completely converted to silicon carbide, relatively brittle fracture was obtained. With silicon carbide fiber reinforcement, failure was abrupt, whereas the composite containing the silicon carbide coated carbon fibers showed a stepwise decrease in residual strength, to about 50% at double the strain to failure. Titanium carbide coated carbon fibers behaved in a manner similar to silicon carbide coated carbon fibers, although the coating process resulted in more degradation of the fibers.

Failure was nonbrittle when dense multiple resin impregnated carbon/carbon composites were siliconized at about 1925°C and the carbon was sufficiently protected by the original matrix against attack by the silicon melt. A precursor with an extremely high carbon residue (H-A43, Hercules Inc.) could not be quantitatively converted to reaction bonded silicon carbide. However, the partially siliconized composite had a high strain to failure, about 1%.

Alumina fiber was reported to be unsuitable for reinforcement of reaction bonded silicon carbide because of incomplete wetting of the pure fiber by liquid silicon and because of inadequate stability at the temperatures required for siliconization. Reaction bonded silicon carbide reinforced with Nicalon fibers had brittle fracture behavior and low flexural strength. There was apparently fiber degradation by reaction with liquid silicon. In general, results were best when conditions were such that silicon mass transfer occurred only by grain boundary or surface diffusion, not by bulk diffusion of molten silicon.

Additional information on composites having a silicon carbide matrix produced by reaction bonding was provided by Krenkel and Hald.[10] A number of precursor resins were examined. A phenolic resin resulting in a 63 wt% yield on carbonization was used in most of the tests. The resin was considered suitable for conventional prepreg technology as well as for injection molding with two- or three-dimensional reinforcement. Carbon fibers were used.

During carbonization, unidirectional preforms had a shrinkage of about 10% perpendicular to the fibers and near zero shrinkage parallel to the fibers. Open porosity was about 8%. Infiltration with molten silicon resulted in only a

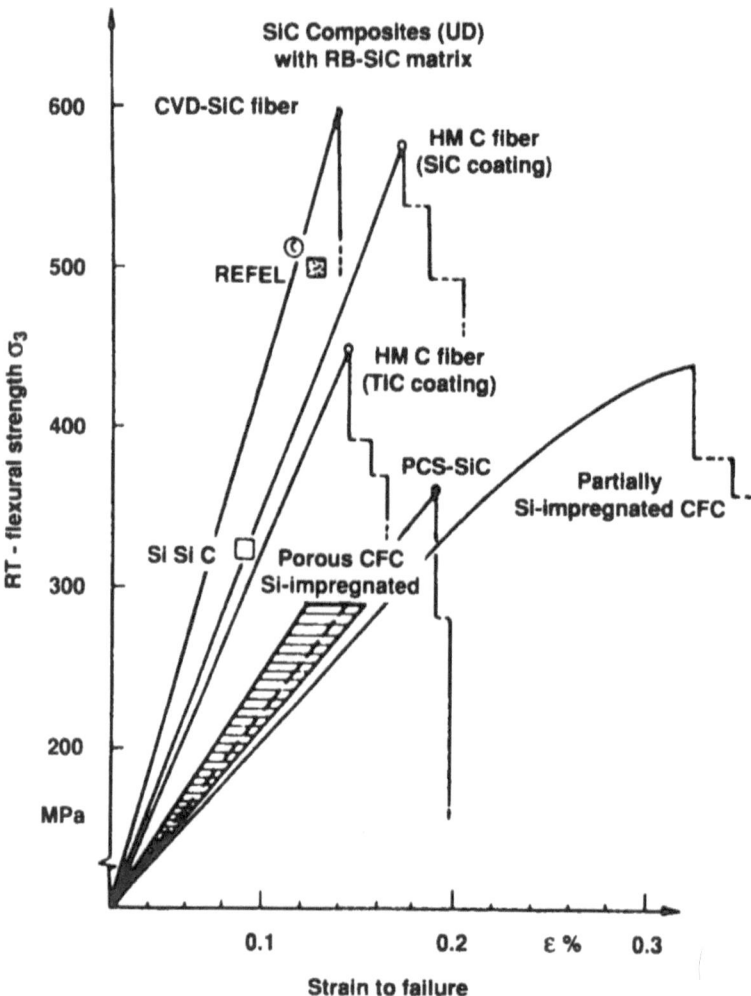

FIG. 7 Room temperature flexural (3-point test) stress-strain curves for unidirectional composites with reaction bonded silicon carbide matrices siliconized at 1525°C. (CVD-SiC fiber type is not specified; PCS indicates polycarbosilane derived fiber; CFC indicates a carbon bonded/carbon fiber preform; Refel is a commercial monolithic reaction bonded silicon carbide.) (From Ref. 9.)

thin reaction zone at the surface (about 15 μm deep). Use of woven fabrics instead of a unidirectional fiber arrangement resulted in nearly zero shrinkage for both directions. However, microscopically there was shrinkage perpendicular to the fibers. Cracks occurred around each fiber bundle due to the restriction of shrinkage caused by the transversely oriented neighboring fiber bundles.

Siliconization of the preforms containing woven fibers involved reaction mostly in the microcracks, with the fibers relatively protected from attack. Each fiber tow was divided into up to ten fibers, each of them surrounded by silicon carbide and a certain amount of free silicon. (In reality, then, this type of material might be better characterized as oxidation inhibited carbon/carbon than a silicon carbide matrix composite.)

Strengths depended upon the resin used, and whether or not a second cycle of resin impregnation and carbonization was conducted. Highest flexural or tensile strength for a unidirectional composite was about 200 MPa. A maximum strength of only about 50 MPa was obtained with fabric reinforcement. Although these strengths are not very high, the composites did perform relatively well in high temperature erosion tests. Although the composites were not considered suitable for long time high temperature applications in air, they were superior to externally coated carbon/carbon in space craft reentry conditions.

C. Aluminum Borate Matrix Composites

An example of the third type of reaction bonded composite system, one in which the matrix is formed from reaction sintering of blended solid components, has been described by Ray.[11] The matrix is based on aluminum borate, $Al_{18}B_4O_{33}$ (or, $9Al_2O_3 \cdot 2B_2O_3$), produced by reaction of alumina and boric oxide (B_2O_3). This reaction is accompanied by a 14% volume increase, which is useful in reducing porosity in the fabricated composite relative to the green composite.

In a typical application of the process, alumina having an average particle size of about 0.4 μm (Alcoa A16 super ground) and −325 mesh boric oxide (Fisher A-76) were blended in the ratio required to produce aluminum borate, along with a 1 wt% addition of calcium oxide as a sintering aid, and milled. The carrier/binder was a variation of a nonaqueous polymerizable binder system described earlier. Slurry solids loading was 57 vol%. Plain weave Nicalon cloth with the size thermally removed was used as the reinforcement.

The slurry was poured over one layer of cloth and allowed to infiltrate the fiber tows. Air bubbles were removed using a steel roller. Successive cloth layers (total of nine) were stacked and infiltrated. Excess liquid was squeezed from the preform and binder was polymerized in a mold at a pressure of 3.5 MPa and a temperature of 65°C for 1 h. After removal from the mold, the preform was heated in argon using a slow heat-up rate to a maximum temperature of 400°C. Reaction sintering was carried out at 1080°C for 2 h.

The composite contained about 30 vol% fibers and the matrix was 60% dense, relative to the theoretical density of aluminum borate. Flexural strength was 83 MPa and tensile strength was 97 MPa. Strain to failure in the tensile test was 0.7%, and failure mode was nonbrittle.

In another variation of the process, 8-harness satin weave Nicalon cloth that had been coated by CVD with a layer of boron nitride over a layer of carbon was used as the reinforcement. Fiber volume fraction of the completed composite was about 29% and matrix density was 58% of theoretical. This composite specimen had a flexural strength of 193 MPa after heat treatment for 1 h at 815°C, indicating some high temperature capability.

III. CEMENT BONDING

Before describing work on advanced continuous fiber reinforced ceramics by cement bonding, it seems appropriate to discuss another type of material that has relatively low mechanical properties and limited elevated temperature capabilities relative to most of the composites described in this book, but is certainly a composite utilizing cement bonding. This material is carbon fiber-reinforced portland cement.

Portland cement powder has four major components, tricalcium silicate, dicalcium silicate, tricalcium aluminate, and calcium aluminoferrite solid solution.[12] When mixed with water, cement undergoes an exothermic hydration-hydrolysis reaction. Setting and hardening are the result of a complex sequence of processes. Major binding component of hardened portland cement paste is calcium silicate hydrate. Fibers and particles of this material grow and link together larger crystallites and residual anhydrous cement grains, producing a microporous material with minimal interconnected capillaries. Although recent advances in cement processing and in particulate reinforcement covered in the cited reference are noteworthy, they will not be described here.

Improvement of cement properties by addition of chopped carbon fibers has been clearly demonstrated, and this area is reported to be very active in Japan. Inagaki [13] reviewed progress in Japan with this material, including commercial applications. In the work cited, GP-grade pitch-based fibers were utilized. These had an average length of 10 mm, diameter of 14.5 μm, and a specific gravity of 1.63. Mechanical properties included a tensile strength of 8 GPa, a Young's modulus of 37 GPa, and an elongation of 2.1%. To reduce the weight of the material in the application cited, hollow aluminosilicate spheres with a specific gravity of 1.0 and an average particle diameter of less than 150 μm were added.

Cement processing conditions were not given in detail, although two different curing conditions were described: curing at 20°C and 65% relative humidity for seven days and curing in an autoclave at 180°C under a pressure of 10 atm for 5 h and then holding in a room at 20°C and 65% relative humidity.

Incorporation of only about 2 vol% fibers into the cement increased ultimate tensile strength from less than 2.5 MPa to about 4.0 MPa. More significantly, strain to failure increased from less than 1×10^{-3} to about 5×10^{-3}. With 4 vol% fibers, maximum tensile stress increased to about 5.0 MPa. Properties were further enhanced by autoclave curing. With 4 vol% fibers, ultimate tensile strength was about 40% higher than with room temperature curing. Increase in initial crack strength was about 2.5 times greater with autoclave curing than with room temperature curing. Durability tests were performed also. After treatment in hot water at 75°C or after freeze/thaw cycles at 10°C and -18°C, no changes in dimensions or mechanical properties were detected. More technical details on carbon fiber reinforced cement are available from Nishioka et al.[14]

Work on cement bonded continuous fiber composite systems with the more general use of the word "cement" has been reported relatively recently. An alternative term for this type of matrix formation is "chemical bonding."

Continuous fiber reinforced material in which matrix particles are bonded together by aluminum phosphate (Al_3PO_4) has been developed at Alcoa Laboratories.[15] A schematic flow diagram is given in Figure 8. A particulate material such as alumina was slurried with a soluble phosphate material. The slurry was applied to a fibrous preform. Compaction and curing were carried out. Compaction procedures analogous to those used in polymer matrix composites could be used, permitting fabrication of large, complex components. A final curing temperature of less than 400°C was required to provide the phosphate binding phase. With alumina as the particulate material, the berlinite form of aluminum phosphate was found to be the major phase.

Properties included a density of 2.32 g/cm^3 and void volume of 15% as-fabricated, or 11.5% after a post-infiltration cycle. Weight loss on heating was less than 1% to 1650°C. Mechanical properties at room temperature were compared for composites containing uncoated and carbon coated fibers (type, volume fraction, and architecture unspecified). Flexural strength decreased from 254 to 166 MPa with the carbon coating, while tensile strength increased from 142 to 312 MPa. Fracture toughness varied from 3 to 16 MPa·m$^{1/2}$, depending on the interface. Although maximum use temperatures of 1400°C with non-oxide fiber reinforcement and 1500° to 1600°C with oxide fibers were estimated, no test results were cited. Development work at Alcoa Laboratories on this composite system, as well as the one described earlier, was discontinued in 1991.

IV. SUMMARY

Ceramic matrix composites fabricated using the reaction bonding technique are promising for eventual commercialization. In particular, the work on silicon nitride matrix composites by Bhatt and associates at NASA-Lewis has shown that very attractive room temperature properties can be obtained, and property

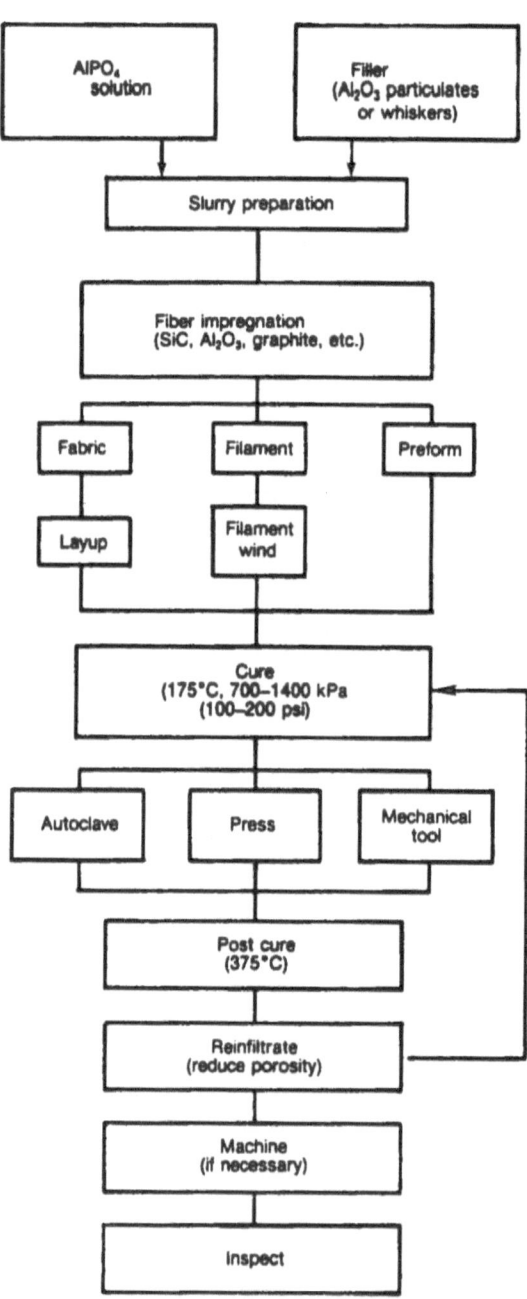

FIG. 8   Schematic flow diagram for fabrication of composites having a matrix of alumina chemically bonded by aluminum phosphate. (From Ref. 15.)

retention at elevated temperatures is quite good (although there is an intermediate temperature range in which oxidation is a problem).

However, the method is not without shortcomings. For example, use of the large diameter SCS-6 fibers limits structural shapes. In addition, although hot pressing (with its limitations with regard to part size, complexity, and cost) is not used for the actual matrix formation step, it is used to form the fiber/silicon powder preforms. Variations in reaction bonded silicon nitride matrix composites using smaller diameter fibers and without hot pressing reported from other organizations have not yet demonstrated impressive properties.

Although reaction bonded silicon carbide matrix composites show some promise, they are at a very early stage of development. Fitzer and Gadow demonstrated reasonably good strengths, but many of the stress-strain curves indicated relatively brittle failure. In addition, elevated temperature mechanical properties have not been adequately evaluated for reaction bonded silicon carbide composites. Reported properties for reaction bonded composite materials having an aluminum borate matrix are only moderate, but this system was at a relatively early stage of development.

Cement bonding appears to have potential for producing ceramic matrix composites, but relatively little work in this area has been reported. Although not in a class with most of the matrix materials discussed in this book in terms of strength and high temperature capability, portland cement fits into this general category, and impressive property gains from reinforcement of this material with chopped carbon fibers have been achieved. A composite system based on bonding of matrix particles with aluminum phosphate has also been described. The method has a number of processing attributes, and reasonably good room temperature mechanical properties have been reported. However, high temperature capabilities of these composite materials have not been given.

REFERENCES

1. Bhatt, R. T., Properties of silicon carbide fiber-reinforced silicon nitride matrix composites, NASA Technical Memorandum 101356, NASA-Lewis Research Center, Cleveland, OH (1988).
2. Bhatt, R. T., and Kiser, J. D., Matrix density effects on the mechanical properties of SiC/RBSN composites, NASA Technical Memorandum 103098, NASA-Lewis Research Center, Cleveland, OH (1990).
3. Bhatt, R. T., Influence of interfacial shear strength on the mechanical properties of SiC fiber reinforced reaction-bonded silicon nitride matrix composites, NASA Technical Memorandum 102462, NASA-Lewis Research Center, Cleveland, OH (1990).
4. Bhatt, R. T., Thermal effects on the mechanical properties of SiC fibre reinforced reaction-bonded silicon nitride matrix composites, *Journal of Materials Science*, vol. 25, 3401–3407 (1990).

5. Omatete, O. O., Tiegs, T. N., and Young, A. C., Gelcast reaction-bonded silicon nitride composites, CONF-910162, December 1991, Oak Ridge National Laboratory, Oak Ridge, TN.
6. Anon., Reaction bonding strengthens ceramic matrix composites, *Materials and Processing Report*, vol. 1, no. 11, 3 (February 1987).
7. Lightfoot, A., Ker, H. L., Haggerty, J. S., and Ritter, J. E., Properties of RBSN and RBSN-SiC composites, *Ceram. Eng. Sci. Proc.*, vol. 11, no. 7–8, 842–856 (1990).
8. Starr, T. L., Harris, J. N., and Mohr, D. L., Reaction sintered silicon nitride composites with short fiber reinforcement. Paper presented at the 11th Annual Conference on Composites and Advanced Ceramic Materials, Cocoa Beach, FL, January 18–23, 1987.
9. Fitzer, E., and Gadow, R., Fiber-reinforced silicon carbide, *Ceramic Bulletin*, vol. 65, no. 2, 326–335 (1986).
10. Krenkel, W., and Held, H., Liquid infiltrated C/SiC—an alternate material for hot space structures. In *Spacecraft Structures and Mechanical Testing*, (Proc. International Conference, Noordwijk, The Netherlands, October 19–21, 1988 (ESA SP-289, January 1989).
11. Ray, S. P., Aluminum borate ceramic matrix composite, US Patent 5,053,364, October 1, 1991.
12. Roy, D. M., New strong cement materials: chemically bonded ceramics, *Science*, vol. 235, no. 4789, 651–658 (1987).
13. Inagaki, Research and development on carbon/ceramic composites in Japan, *Carbon*, vol. 29, no. 3, 287–295 (1991).
14. Nishioka, K., Yamakawa, S., and Shirakawa, K., Properties and applications of carbon fiber reinforced cement composites. In *Developments in Fibre Reinforced Cement and Concrete*, vol. 1, Proc. Third International Symposium, R. N. Swamy, R. L. Wagstaffe, and D. R. Oakley (Ed.), RILEM Technical Committee 49-FTR, 1986.
15. Sheppard, L. M., Cost-effective manufacturing of advanced ceramics, *Ceramic Bulletin*, vol. 70, no. 4, 692–701 (1991).

13

Matrix Formation by Macromolecule Decomposition

I. BACKGROUND

An alternative to slurry processing to form ceramic matrices is the use of a liquid to impregnate a fiber network, with subsequent thermal decomposition of the liquid to produce the matrix. There are a number of different types of liquid precursors, but all involve relatively large organometallic or inorganic molecules, which can be termed macromolecules. They are discussed together in this chapter, since there are overall similarities in processing despite the differences in the composite systems.

It should be noted that for some of the systems to be described, a distinction from slurry processing is somewhat arbitrary. For example, in one of the types of macromolecule processing to be described, use of sol-gel chemistry to produce oxide matrices, the liquid can contain particulate material having an extremely small (colloidal) size. In the other general type of macromolecular processing to be described, use of polymeric precursors to form non-oxide matrices, filler particles have been tried to increase matrix density and decrease shrinkage cracks.

II. SOL-GEL PROCESSING

Sol-gel processing for formation of monolithic ceramics is a rather active area of research and development, but use of this method to produce ceramic matrix composites is as yet somewhat limited. The general principles of this method will be summarized prior to describing some of the work on composite matrix formation.

Most work on the sol-gel process has been on formation of silica.[1] Typically, an organometallic compound such as tetraethoxyorthosilicate (TEOS), $Si(OCH_2CH_3)_4$, is partially hydrolyzed to form $(CH_3CH_2O)_3SiOH$ and ethyl alcohol. By use of a catalyst, polycondensation reactions (joining of these molecules with the elimination of water molecules) occur, forming organometallic macromolecules with Si-O-Si bonds. As molecular weight increases, the viscosity of the system increases, changing from a relatively low viscosity liquid (sol) to a gel. After drying to remove the bulk of the liquid phase, the gel is converted to silica by heat treatment. The condensation catalyst is usually either an acid or base, and microstructure, especially pore size distribution, of the product differs depending upon which was used. The water-to-TEOS ratio also has a considerable effect on the product. Figure 1 [2] shows a reaction sequence for production of silica by a sol-gel process.

Drying of gels is accompanied by considerable shrinkage. A problem with sol-gel processing has been cracking due to the stresses induced during drying. This stress is a function of the pore size and the rate of evaporation of the pore liquid, which, in turn, is a function of the vapor pressure. A breakthrough in this area was the discovery of the benefit of drying control chemical agents, which include formamide, glycerol, and oxalic acid. When used with alkoxide precursors, they permit the reliable production of a wide range of sizes and shapes of transparent silica and other materials.

After drying, the gel is porous and amorphous.[3] At temperatures between 300° and 450°C, residual organic material and more water is removed. There is additional shrinkage and there may be further chemical reactions. At this stage, sol-gel derived "glasslike solids" have similarities to melt derived glasses without having undergone the high temperature melting process, as illustrated in Figure 2. Formation of crystalline matrices using sol-gel technology can be accomplished at temperatures lower than the sintering temperatures required with conventional processing. In general, the sol-gel route to composite formation offers promise for lower temperature processing, with a reduction of deleterious reactions between the matrix and the reinforcing fibers or whiskers.

Another attribute of sol-gel processing is control of homogeneity of materials superior to that obtained with conventional processes using ceramic powders.[1] An example of interest for ceramic matrix composites is formation of mullite ($3Al_2O_3 \cdot 2SiO_2$) from TEOS and alumina monohydrate. Using the sol-gel technique, it has been shown that the molecular mixture of alumina and silica forms mullite directly without the formation of other extraneous phases. In conventional processing an amorphous silicate phase is retained at the mullite/mullite grain boundaries, but this is not observed in mullite formed by sol-gel processing. There is promise that mullite with glass-free grain boundaries may ultimately be superior to silicon nitride or silicon carbide at temperatures up to 1400°C.

In principle, at least, formation of both whisker reinforced and continuous fiber reinforced composites using sol-gel techniques should offer advantages over slurry processing. In whisker composite processing, better dispersion of whiskers would appear likely in a sol, relative to that in a powder dispersion. In addition, casting into near net shapes might be possible, whereas the hot pressing usually required with slurry processing is expensive, is limited to relatively simple geometries, and can result in whisker damage. For continuous fiber reinforcement, sols might more easily penetrate fiber tows than do slurries. A serious disadvantage, however, is the large shrinkage that occurs between the sol stage and the final product. A linear shrinkage of over 30% is not uncommon. This is significantly greater than shrinkage during pressureless sintering of ceramics, which results in problems in the presence of non-shrinking reinforcing fibers (cf. Chapter 10, Figure 5).

Despite the attractive potential of sol-gel processing of ceramic matrix composites, success has as yet been limited. Fitzer and Gadow investigated the sol-gel process for composite fabrication in considerable detail.[4] Their work emphasized a silica matrix formed using TEOS. They were unable to produce satisfactory composites using the sol-gel method without hot pressing, which was probably a consequence of the large shrinkage mentioned above. It was concluded that because a hot pressing temperature of about 1425°C was re-

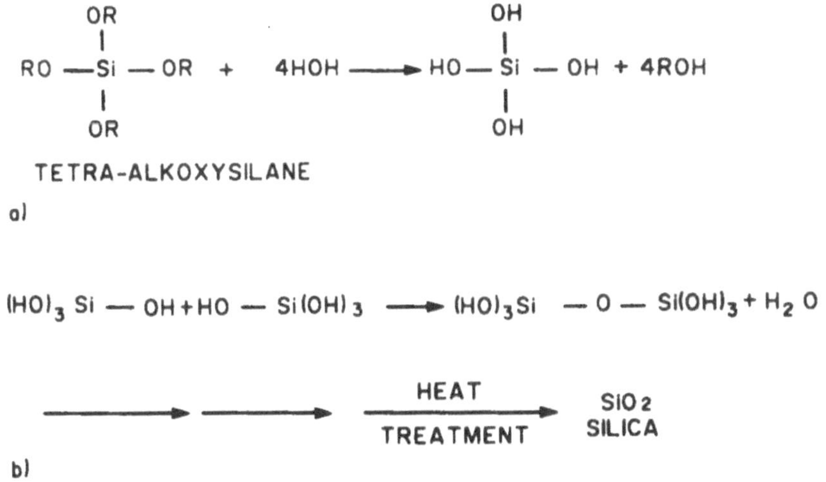

$$RO - \underset{\underset{OR}{|}}{\overset{\overset{OR}{|}}{Si}} - OR \ + \ 4HOH \ \longrightarrow \ HO - \underset{\underset{OH}{|}}{\overset{\overset{OH}{|}}{Si}} - OH \ + \ 4ROH$$

TETRA-ALKOXYSILANE

a)

$$(HO)_3 Si - OH + HO - Si(OH)_3 \ \longrightarrow \ (HO)_3 Si \ - O - Si(OH)_3 + H_2O$$

$$\longrightarrow \qquad \longrightarrow \qquad \underset{\text{TREATMENT}}{\overset{\text{HEAT}}{\longrightarrow}} \qquad \underset{\text{SILICA}}{\overset{SiO_2}{}}$$

b)

FIG. 1 Schematic representation of chemical reactions occurring during formation of silica by a sol-gel process. (From Ref. 2.)

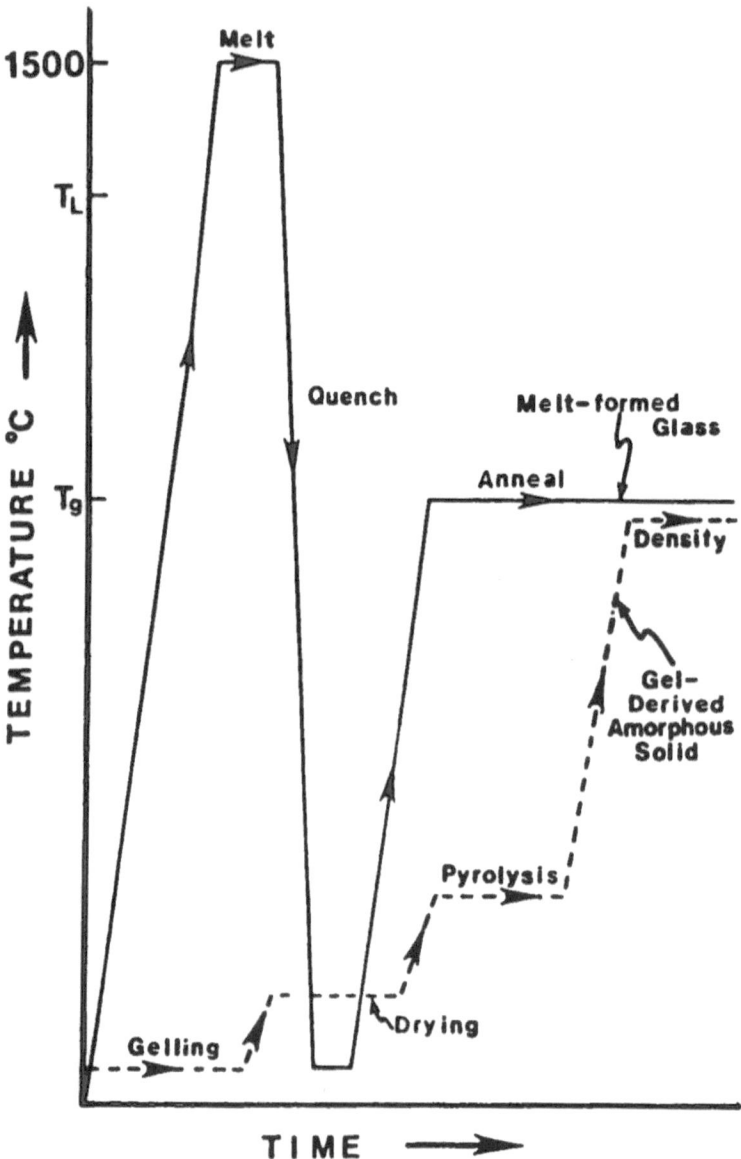

FIG. 2 Comparison of the processing steps for melt-derived glasses and gel-derived "glasslike solids." (From Ref. 3.)

quired, carbon fibers should be used with this matrix because alumina fibers or silicon carbide base fibers produced from polymeric precursors would not have adequate thermal stability. They did conduct experiments with other kinds of fibers by lowering the softening point of the silica through the use of other oxide additives.

Since uncoated carbon fibers would react with pure silica, fiber coatings were used. CVD was used to apply titanium carbide, titanium nitride, and silicon carbide coatings to Thornel 400 fibers (Amoco Performance Products, Inc.). It was found that a titanium carbide coating resulted in greatest pullout during composite fracture and silicon carbide resulted in the least. Composite strengths up to those predicted from fiber strengths were obtained in some cases. With 48 vol% of unidirectionally aligned silicon carbide coated fibers, room temperature flexural strength was close to 900 MPa and strain to failure was over 0.6%. Heat treatment at elevated temperatures in air for 50 h destroyed the carbon fibers. However, a 0.5-mm external coating of unreinforced silica formed by the same process resulted in maintenance of as-fabricated room temperature strength after a 50 h treatment in air at 1000°C. Nonetheless, as with other carbon containing composites, it was recognized that this is only a partial solution, since any cracks in the coating will permit oxygen ingress.

Fitzer and Gadow also examined silica fiber reinforced silica produced by sol-gel processing, but this will not be discussed here since silica fibers are not generally considered promising for use in high performance ceramic composites. Work with alumina (Fiber FP) and silicon carbide base (Nicalon) fibers involved use of germania (GeO_2) to lower the softening point of the matrix and, therefore, reduce the hot pressing temperature required. With 10 mol% germania added, a hot pressing temperature of 1125°C could be used. Figure 3 shows flexural stress-strain curves with 40 vol% unidirectional alignment. Strengths were improved manyfold over the matrix strength and strain to failure increased somewhat. Fracture toughness (K_{Ic}) values were about 1.2 MPa·m$^{1/2}$ for the matrix material, 2.5 MPa·m$^{1/2}$ with silicon carbide base fibers, and 4.0 MPa·m$^{1/2}$ with alumina fibers. The strength values corresponded to only about 50% utilization of original fiber strength. This was attributed to internal stresses built up during cooling and/or degradation of fiber strength during composite processing.

Silicon carbide whiskers were also used to reinforce germania modified silica produced by the sol-gel technique. Flexural strength was about 300 MPa and strain to failure was 0.7%, with 60 vol% whiskers (type unspecified). Fracture toughness value was not given.

Reinforcement of alumina and mullite produced using sol-gel processing was investigated also. Alumina was formed using aluminum sec-butylate as a precursor. Hot pressing at about 1725°C was required to achieve a matrix porosity below 5%. Mullite was formed from TEOS and aluminum sec-butylate. For mullite, hot pressing temperature could be reduced by about 100°C.

Uncoated carbon fibers as well as fibers coated with silicon carbide were used for reinforcing alumina. With 40 vol% unidirectionally aligned fibers, flexural strength was 500 MPa. (It is not clear whether this was with coated or uncoated fibers.) With fiber contents below 10 vol%, strength was lower than the matrix strength. With 50 vol% fraction of coated fibers, fracture toughness was about 10 MPa·m$^{1/2}$ with coated fibers, but there was no improvement over the fracture toughness of the matrix (about 4 MPa·m$^{1/2}$) with uncoated fibers.

Because of a lower hot pressing temperature and better densification of mullite matrix composites, higher flexural strengths (800 to 900 MPa) were attained. Highest fracture toughness was about 6 MPa·m$^{1/2}$. Values were not influenced by the silicon carbide coating. It was postulated that with the mullite matrix a silicon carbide layer formed during processing.

Using alumina fibers, flexural strengths were not increased with either alumina or mullite matrices. This was attributed to fiber degradation with the hot pressing temperatures required. However, substantial increases in fracture toughness (up to about 14 MPa·m$^{1/2}$ with 60 vol% fibers) were measured on the mullite matrix material. No explanation could be postulated for this toughening effect, especially since fracture surfaces indicated no fiber pullout.

The sol-gel technique for matrix formation for ceramic composites was also investigated by Pierre et al.[5] Matrices were formed from aluminum sec-

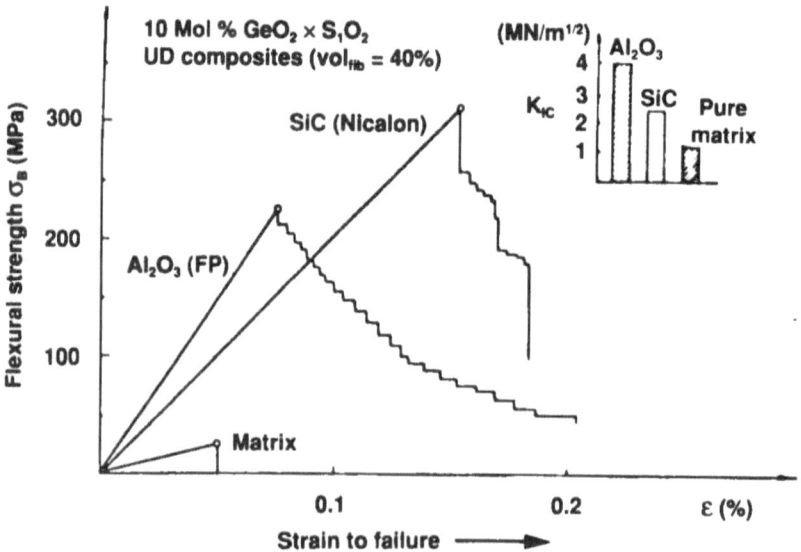

Fɪɢ. 3   Room temperature flexural stress-strain curves for fiber reinforced and unreinforced silica containing 10 mol% germania. (From Ref. 4.)

butoxide and zirconium sec-butoxide precursors. Fiber FP, SCS-6, and Nicalon fibers were used. Nitric acid was used for catalysis in the sol-gel processing. Geling was carried out at 90°C in most cases. Nitric acid-to-alkoxide ratio was varied. Results were best with the lowest ratio used (0.035, presumably by mass). For alumina matrix composites, transformation of the gel into α-alumina was carried out at 1150°C in air. Zirconia matrix composites were not formed by direct sol-gel processing. Instead, zirconia powder was produced by this route, and hot pressing was carried out at 1500°C at about 10 MPa.

With alumina matrices and SCS-6 fibers, there was extensive cracking of gel transversely to the fibers. The matrix shrunk by sliding along the fibers, leaving large gaps. With Nicalon fibers, maximum flexural strength (3-point) was only 25 MPa, with 16 wt% unidirectionally aligned fibers. Results were somewhat better using alumina fibers. With 40 wt% unidirectional reinforcement, flexural strength was 136 MPa, and strain to failure was 1.25%. Flexural strength of a zirconia matrix composite (with an unspecified fiber fraction) was only 57 MPa and strain to failure only 0.1%.

Better results were reported by Chen et al [6] on fiber reinforced mullite by sol-gel processing. They used a hydrated alumina powder containing 0.2 wt% titania and a colloidal silica sol as precursors. Alpha-alumina particles were used as a filler. The alumina filler was deflocculated and dispersed in dilute nitric acid having a pH of 3.5. The hydrated alumina powder was added and pH readjusted to 3.5. The materials were chosen so that 80% of the alumina would result from the filler and the remainder from the hydrated alumina sol. The appropriate amount of silica in the form of a sol was added and the pH was adjusted to 2.5 with nitric acid.

For composite fabrication, unsized, surface treated, high strength carbon fibers (Grafil XAS, from Hysol Grafil Ltd.) were used. These were coated with a thin layer of alkoxide sol of a sodium borosilicate (Pyrex) composition, and heated at 500°C for 0.5 h. This coating was reported to improve handleability of the fibers. Fibers were aligned in a mold and vacuum impregnated with the sol. The viscosity of the sol was sufficiently low that good infiltration resulted. Geling occurred in about 25 min. The good infiltration combined with a relatively high mass yield from this gel/particulate system permitted only one infiltration stage to be used.

The unsintered composites were dried at 50°C for 20 h, then heated in nitrogen using an unspecified slow heat-up rate to 500°C, a rate of 5°C/min to 1000°C, and 10°C/min to the maximum temperature, which was varied in different experiments. A number of characterization tests were conducted and flexural strengths in 3-point bending were measured.

It was determined that relatively little densification of the matrix occurred until about 1100°C. After firing for 2 h at 1000°C, density was only 61% of theoretical. After firing at 1250°C, it was about 85%, and after firing at 1400°C, it was about 95% of theoretical. It was postulated that a viscous phase aided sintering of this alumina filled gel, as had been previously found for a similar material without the filler particles. Differential thermal analysis indicated that mullite forms at 1350°C in the alumina filled system, and this was confirmed by X-ray diffraction.

For composites fired at 1200°C, shrinkage in the fiber direction was negligible, but there was 5.6% shrinkage perpendicular to the fibers. This can be compared with about 12% linear shrinkage with unreinforced material. The presence of the fibers inhibited densification of the matrix, as indicated by higher densities for unreinforced matrix specimens.

For composites sintered at 1200°C, flexural strength averaged 300 MPa with 21 vol% fibers, 277 MPa with 27 vol% fibers, and 289 MPa with 35 vol% fibers. Young's moduli averaged 30 GPa, 50 GPa, and 60 GPa, respectively. With 21 vol% fibers, the composites failed in a brittle manner, while fracture was more gradual with 35 vol% fibers.

An industrial firm, Pratt and Whitney, is also known to have conducted research and development on sol-gel processing for ceramic matrix composites.

III. POLYMERIC PRECURSOR PROCESSING

A method of forming a non-oxide ceramic such as silicon carbide is thermal decomposition of a polymeric precursor. Nicalon fibers and a number of other fibers described earlier are produced by this route. The matrix of a reinforced ceramic can also be produced from a polymeric precursor, although like sol-gel processing, experimental work has been limited as yet.

The attributes, as well as shortcomings, of this method are similar to those described for sol-gel processing. Composition can be controlled by altering the precursor composition. Reduction in temperature required for production of a consolidated ceramic matrix, relative to sintering or hot pressing, is even more dramatic than for sol-gel processing. For example, a temperature of over 2000°C is required for pressureless sintering of silicon carbide, whereas polymeric precursors that form a material rich in silicon carbide can be decomposed at temperatures closer to 1000°C. However, in common with sol-gel processing, shrinkage and mass loss can be considerable. In addition, although composition can readily be altered by changing the composition of the precursor, a single phase ceramic is not usually obtained, as indicated previously for Nicalon and related fibers.

A schematic flow diagram for Nicalon fibers, which is representative for polymer precursor processing, was given previously (Chapter 5, Figure 5). Figure 4 [7] is a more general representation, showing a variety of starting materials and the range of products that can be made from these materials depending on the atmosphere used during polymerization and/or during thermal decomposition. The ceramic product of polymer precursor decomposition is usually non-crystalline when formed at the minimum temperature required to produce conversion to a ceramic, but a crystalline material is usually formed at a sufficiently high temperature. For example, Figure 5 shows X-ray diffraction patterns for a silicon nitride base material produced by nitridation of polycarbosilane. At temperatures up to about 1300°C the material is amorphous, but at 1400°C the pattern for α-silicon nitride is seen.

Among the first researchers to investigate this method of ceramic matrix composite fabrication were Jamet et al [8] of the Naval Research Laboratory.

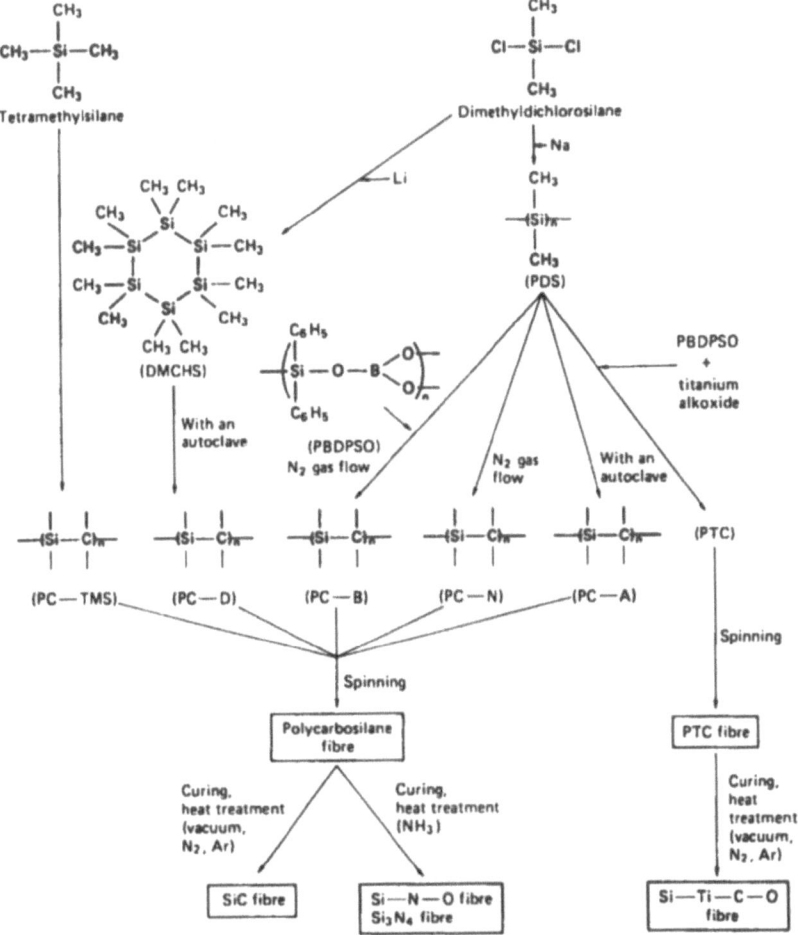

FIG. 4 Schematic illustration of polymer precursor routes to ceramic materials, showing effects of both starting materials and processing conditions. (From Ref. 7.)

They considered a number of silicon carbide polymeric precursors for investigation, and selected polysilanes (mostly, vinyl polysilanes), based on the fact that these have similarities to precursors (such as polyfurfural alcohols) successfully used in producing carbon matrices for carbon/carbon composites. Although there are large shrinkages of these materials during thermal decomposition, fillers were used to attempt to minimize cracking caused by shrinkage.

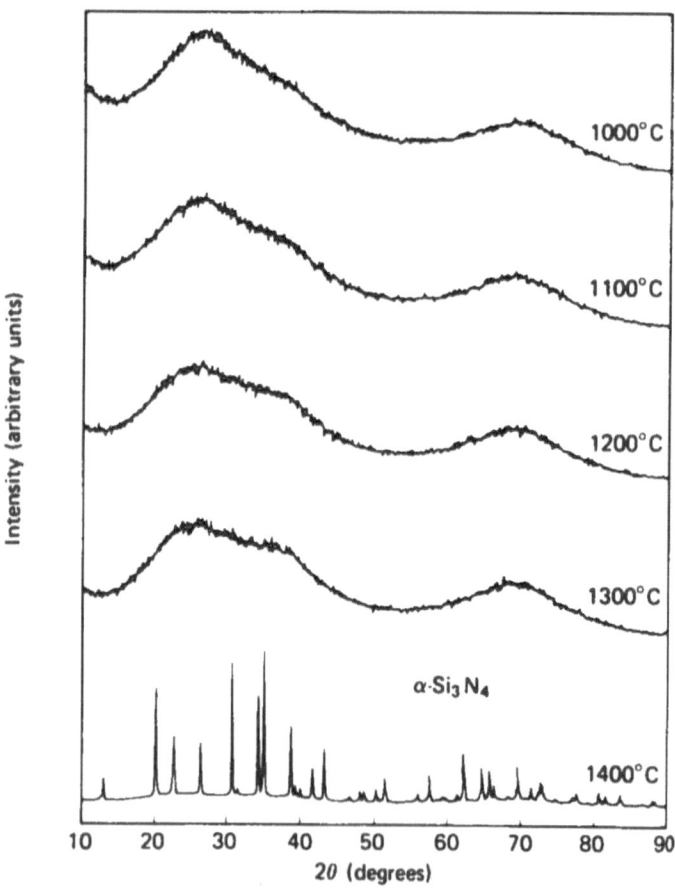

Fɪɢ. 5   X-ray diffraction patterns of silicon nitride produced by the nitridation of polycarbosilane, after heat treatments at various temperatures. (From Ref. 7.)

They reported that colloidal graphite fillers are used to overcome cracking problems in carbon/carbon processing.

Decomposition studies on a vinyl polysilane material alone indicated a mass loss of 36% between 270° and 1000°C. Up to 400°C, there was no appreciable cracking of the material with a heat-up rate of up to 2°C/min. However, extensive cracking developed before 1000°C, which reinforced their conclusion that fillers were needed for successful composites using this material. A number of fillers, including silica, silicon carbide, silicon nitride, and boron nitride, were considered. High viscosity slurries with filler concentrations of 50 to 67% were molded and cured after slow solvent extraction. Curing was carried out by heating the material to 270°C at a heat-up rate of 5°C/min. Heating these samples in nitrogen to 1000°C, using a relatively rapid heat-up rate of 10°C/min, indicated that the fillers did allow pyrolysis without cracks. Also of importance, some of the fillers increased pyrolysis yield (Table 1), indicating favorable chemical interactions. Yield increased by 10% with boron nitride filler. Additional tests in which the boron nitride content was varied indicated an influence of the filler concentration on the reduction in porosity and cracking.

Based on the preliminary work, composites were fabricated using a polysilane with 50 wt% boron nitride as the matrix precursor material. Nicalon silicon carbide base fibers and Sumitomo alumina fibers were used. After removal of fiber size at 700°C in air, tows were impregnated by passing them through a bath containing boron nitride particles suspended in a solution of polysilane in a solvent. Fiber tows were unidirectionally aligned using a rotating drum. The impregnated tows were dried at 68°C for 2 h. The unfired composites were cured in a mold under 30 MPa pressure at a heat-up rate of 5°C/min to 270°C. After removal from the mold, the matrix material was pyrolyzed under nitrogen at a heat-up rate of 15°C/h to 1100°C. For comparison, a composite without filler material was produced using similar processing. Other composites were fabri-

TABLE 1. Pyrolysis yields of a vinyl polysilane with various fillers.

	Yield (wt%)	
Filler	Theoretical[a]	Measured
None	—	62
50 wt% BN	81	91
59 wt% SiO$_2$	84	79
67 wt% Si$_3$N$_4$	87.5	90
67 wt% SiC	87.5	88

[a]Theoretical yield, i.e., calculated using filler wt plus observed 62% yield without filler.
Source: Ref. 8.

cated using boron nitride coated fibers and using a higher heat-up rate during pyrolysis (100°C/min). Flexural strength (3-point) was determined.

Table 2 summarizes results using Nicalon fibers. As anticipated from the preliminary studies, a composite made using a boron nitride filler was far superior to one without the filler. Density was considerably higher, porosity was lower, and flexural strength was about six times greater with the filler. In addition, fracture was brittle without the filler but nonbrittle with the filler. The boron nitride coating was not beneficial to flexural strength. However, it was shown in stress-strain curves that the coating resulted in greater load carrying capability at higher strains and shown from SEM micrographs that fiber pullout at fracture surfaces was greater with the coating. The higher heat-up rate was detrimental to composite properties. A composite made using uncoated alumina fibers had a flexural strength of 120 MPa and nonbrittle fracture.

Fitzer and Gadow [9] were also among the first researchers to investigate polymer precursor processing for ceramic matrix composite fabrication. They reported on the fabrication of continuous fiber reinforced composites using polycarbosilane as a matrix precursor. Porous fiber preforms were first prepared using a small amount of binder. Preforms were placed in an autoclave, which was then evacuated. Liquid impregnations of the preforms were carried out at temperatures of up to 525°C and pressures of up to 40 MPa. Polymerization of the precursor occurred under these conditions. After removal of the fiber/polymer composites from the autoclave, the matrix material was pyrolyzed in an inert atmosphere at an elevated pressure at temperatures ranging from 525° to 1025°C. An unspecified number of reimpregnation and pyrolysis cycles were carried out. Finally, the composites were heat treated at temperatures ranging from 1025° to 1525°C. The silicon carbide yield for an optimum precursor was about 60%. Matrix density was 3.0 to 3.3 g/cm^3. Properties of these composites were not reported.

Other work on matrix fabrication from preceramic polymers involving a class of precursors called silsesquioxanes has been carried out at NASA Lewis Research Laboratory by F. I. Hurwitz and coworkers. As an example [10] of work conducted on these materials, polymer synthesis was carried out using phenyl- and methyl-trimethoxysilanes (from Petrarch Chemical Systems) as starting materials. The monomers were added to an acetic acid solution, which was stirred until the solution became clear, then kept at 30°C for 72 h. After this time, a viscous liquid polymer had settled to the bottom of the flask. The polymer was refrigerated to increase its viscosity and decanted to remove excess water.

As-received ("P-size") Nicalon fibers were used in composite fabrication. Fiber tows were wound onto a mandrel. The polymer was dissolved in acetonitrile. Fillers were used in some composites. In those cases, filler particles were suspended in the polymer solution by vigorous stirring, and the weight of the

TABLE 2. Characteristics of silicon carbide fiber composites having matrices produced from vinyl polysilane plus 50 wt% boron nitride.

SiC Nicalon fibers	Matrix	Before pyrolysis		After pyrolysis				
		Density (g/cm³)	Porosity (Vol%)	Density (g/cm³)	Porosity (Vol%)	Fiber Content (Vol%)	Flexural strength (MPa)	Fracture Character
Without coating	VPS alone	1.53	0.0	1.48	25.0	53	56	Very brittle
	VPS +50 wt% BN	2.0	0.5	1.91	7.1	67	326	Large fracture surface area
With coating	VPS +50 wt% BN	1.85	1.4	1.84	15.8	57	221	Fiber pullout
	High heating rate	1.86	1.1	1.76	22.0	57	164	Large strain

Source: Ref. 8.

liquid polymer was adjusted to maintain a total polymer plus filler weight of 40 g. The fiber tows were coated with the polymer solution and the solution allowed to evaporate under ambient conditions. The resulting prepreg was removed from the mandrel, cut into plies, and stored in a freezer.

Twelve-ply composites were formed by pressing stacked prepregs in a mold at 180°C under a pressure of 689 Pa for 2 h. The composites were pyrolyzed at ambient pressure in flowing argon purified to have a low oxygen content. After holding periods at intermediate temperatures to allow full evolution of volatiles, final pyrolysis temperature was varied from 1200° to 1400°C, with a 30-min holding time. Flexural (4-point) and tensile strengths were determined. Tensile specimens were straight sided. Strain was measured using strain gages bonded to the specimens. Several techniques for specimen characterization were used.

To assess high temperature stability in oxygen, unreinforced polymer was pyrolyzed in flowing argon to 1400°C using the same conditions as for composites. These specimens were then heated in flowing oxygen for 100 h at temperatures of 1150° and 1250°C and weight monitored.

Various conditions of pH and water-to-monomer ratios produced primarily low molecular weight oligomers. Molecular weight ranged from about 900–1600. There was some variation in polymer characteristics with pH. Material produced using a pH of 3 was chosen for composite fabrication because it appeared to produce the lowest viscosity liquid best suited to the prepreg techniques used and for potential reinfiltration.

Polymer viscosity decreased with increasing water-to-monomer ratio over the range investigated. Two ratios were used in composite fabrication, 14.5 and 3.6. The polymer synthesized at the higher ratio was more viscous and seemed better suited to filament winding. The less viscous polymer was judged to be better for reimpregnation. Composites were fabricated using both. Polymers prepared using three phenyl-trimethoxysilane/methyl-trimethoxysilane ratios, 30/70, 50/50, and 70/30, were selected for composite fabrication. The composition of the siliconoxycarbide pyrolysis products from the silsesquioxane copolymers had been found in earlier work to vary with polymer composition, such that carbon increased nonlinearly with phenyl content of the copolymer, producing amorphous glasses with carbon contents of 37, 44, and 52 at%. Silicon content decreased linearly from 26 to 23 at% with increasing phenyl-trimethoxysilane content, and oxygen accounted for the rest of the composition.

Highest flexural strength for composites from the three copolymers, with pyrolysis at 1200°C, was 179 MPa. Failure mode was brittle. Flexural strength was higher and composite-like fracture behavior was observed when processing was done at 1400°C. However, in tensile testing, failure was brittle even with processing at 1400°C. Failure was by fiber bundles, with shear of the matrix, rather than by debonding at the fiber/matrix interfaces. Hence, the composite-like appearance with flexural testing was considered to be misleading. Similar kinds of results were found when other processing conditions were varied.

For studies on fillers, the filler particles were prepared from the same material used in the matrices. Two filler levels, 10 wt% and 20 wt% of the filler/polymer blend were used. Strength was reduced with the filler additions, which was postulated to be due to an increase in viscosity, producing matrix-rich regions and decreasing the fiber volume fraction.

The oxidative stability tests indicated that the matrix material was stable to oxygen at 1150°C (0.4% weight loss after 100 h). At 1250°C, material was stable for several hours, but had a 12% weight loss in 10–100 h. In all cases, a layer of silica could be detected on the surface.

Dow Corning has produced composites using polymer impregnation and pyrolysis, although few details have been given.[11] Polymers are generalized as having a silicon carboxide or silicon carbonitride composition. Vacuum bagging and pressing or autoclaving is used for densification and curing of the matrix precursor, and the precursor is thermally decomposed to produce the matrix. Several iterations of impregnation, densification/curing, and pyrolysis are used. With two-dimensional fiber orientation, flexural strengths have ranged from about 275–415 MPa, and tensile strengths from about 275–310 MPa. Properties are reduced by less than 20% after exposure to air at 1100°C for 50 h.

Work by Mohr et al [12] did not entail using a polymeric precursor to produce a matrix, but, rather, using it as a binder for silicon powder prior to using the reaction bonding technique for silicon nitride production. A copolymer, 1,2-dimethyl-silazane(1-methyl-silazane), from Petrarch Corp. was used. This polymer has a predicted char yield of 50–55%. Silicon powder (Keminord Industries, distributed by Superior Graphite Co.) that had a median particle diameter of 7 μm was reduced to about 0.5 μm by dispersing in toluene and grinding in an attritor mill for 24 h. Silicon carbide whiskers (American Matrix Corp.) were 1–3 μm in diameter and 30–60 μm long. Whisker bundles and small particles were removed. For whisker reinforcement, components were slurried in toluene under an argon atmosphere. The slurry was then pressed between microporous filters to remove excess solvent.

Composites using 8-harness satin weave Nicalon fiber cloth were also produced. The cloth was desized and coated with silicon powder. The coated fabric layers were impregnated with the polysilazane precursor, stacked, and compacted using a hydraulic press. Each specimen was clamped in a silicon nitride fixture, prior to pyrolysis and nitridation, to minimize separation of the fiber layers during these operations.

Preliminary results indicated that a prestabilization process was required. This was carried out in a mold at 1.4 MPa pressure at 200°C under nitrogen. Pyrolysis was carried out at a heat-up rate of 1°C/min to 800°C. Nitridation was performed using a heat-up rate of 0.5°C/min to 1100°–1200°C.

Some retardation of nitriding was noted for specimens containing the precursor material binder, compared with those without the binder. This might have been due to the increased green density, which could have decreased the rate of transport of nitrogen into the interior. Fracture surfaces showed no pullout, and failure was brittle. Mechanical properties were not reported.

Work at Ube Industries by Yamamura et al [13] involved preparation of materials using Tyranno fibers. In one aspect of the work, composites having the matrix produced from the polymer used for Tyranno fiber production, polytitanocarbosilane (PTC), were fabricated. Powder was produced by pyrolysis of PTC at 1300°C in nitrogen, followed by crushing and screening. Average particle size was 2 μm and maximum size was 5 μm. Composites were fabricated using alternate layers of plain weave Tyranno fiber cloth and PTC powder. Hot pressing was carried out at 1600°–2100°C under a pressure of 30–70 MPa for 1 h in argon. Flexural strength (3-point) was reported as 400 MPa for a specimen that had been hot pressed at 1950°C under a pressure of 70 MPa.

Greater emphasis was placed on materials fabricated by hot pressing Tyranno fiber cloths alone (no matrix material). Although these are described as composites in the cited reference, they are, of course, not composites, since there is no separate matrix phase. Nonetheless, they have mechanical performance similar to some of the composite materials described in this book, and will be described here.

In hot pressing cloths of Tyranno fibers, little distortion of fibers occurred using a pressure of 70 MPa and a temperature up to 1800°C, as indicated by scanning electron microscopy. However, on hot pressing at 1950°C, the fibers distorted to a hexagonal cross section. On hot pressing at 2100°C, no fibrous shape was observed, and only traces of edges based on the hexagonal shape could be observed. Chemical analyses for silicon, carbon, titanium, and oxygen showed a reduction in oxygen content from 17 wt% in the original fibers to below 0.5 wt% during hot pressing at 1600°–1900°C, and to near zero during hot pressing at 2100°C. Although the original fibers contained excess carbon, the atomic ratio of carbon to silicon and titanium decreased to the stoichiometric value on hot pressing at 1600°–1800°C, and the ratio did not change during hot pressing at 1800°–2100°C. Thus, it was considered that material hot pressed at the higher temperatures consisted of silicon carbide and titanium carbide, partially in solid solution.

Flexural strength of material produced by hot pressing Tyranno fibers at 1950°C under 70 MPa pressure was about 240 MPa. This did not vary appreciably after heat treatments at temperatures up to 1600°C for 1 h. Neither was flexural strength reduced significantly when the test was conducted at 1600°C. The area under the stress-strain curve was large, indicating high toughness. Preliminary fracture toughness values obtained using an indentation method were 16–20 $MPa \cdot m^{1/2}$.

Other industrial firms that have experimented with matrix formation using polymer precursors include Allied-Signal Inc. A brochure on the subject [14] reports excellent properties using a matrix material termed "Black Glass," but details will not be given here since it is indicated that the information is subject to US Government export control laws. Strife et al [15] of UTRC and GA Technologies have also investigated the polymer precursor route to ceramic composite fabrication under a DARPA contract. In addition, firms such as Dow Corning, Atochem, and Toa Nenryo Kogyo offer, or have offered, polymers of this type. Work on polymeric precursors for ceramics has also been carried out in a number of university laboratories. A notable example is development by D. Seyferth of the Massachusetts Institute of Technology of a number of promising precursor materials, some of which have been licensed commercially.

IV. SUMMARY

Sol-gel processing is an attractive method of producing specialized monolithic ceramic materials. Advantages over conventional ceramic slurry processing include more homogeneous mixing of components for multiphase materials and reduced sintering temperatures. It can be expected that these attributes will be useful for ceramic matrix composites as well. In particular, any reduction in the temperature required for matrix densification is more important for ceramic composites, with the possibility of deleterious reactions between the fibers or fiber coatings and matrix, than for monolithic ceramics.

However, results to date range from poor to only moderately good. The major problem with processing by the sol-gel technique has been matrix cracking and limited densification due to the large shrinkage accompanying the transformation from sol to gel and from gel to a ceramic material. Linear shrinkages of 30% are not uncommon. Although such large shrinkage can be acceptable in a monolithic ceramic specimen if processing conditions are carefully controlled, the nonshrinking fibers in a sol impregnated composite restrict the possible change in external dimensions and cracking is inevitable.

Steps taken to minimize the shrinkage problem have included hot pressing of the composite specimens. However, this entails the inherent disadvantages described previously with regard to specimen size, shape, and complexity, as well as cost. Another approach was to incorporate filler particles into the sol material to restrict shrinkage during processing, and this was reasonably successful. Another approach, which was not stressed in the cited works, is to perform multiple impregnation/densification cycles. This would seem to be a reasonable approach, although cost could increase appreciably. It seems likely that an increased effort in this area will result in more promising results.

A number of research groups have investigated the fabrication of ceramic matrix composites by impregnation of fiber networks with polymeric materials that decompose to non-oxide ceramic materials on pyrolysis. For the most part, mechanical properties of these materials have been only moderately good. Since in most cases the fibers and matrix materials have related chemical compositions, suitable fiber coatings will probably be required to prevent undesirably strong bonds between fibers and matrix from developing.

Much of the work has involved either the use of filler particles or multiple impregnation/pyrolysis cycles, both of which address the large mass loss and shrinkage inherent with most of the precursors used. It can be anticipated that polymers with greater mass yields and lower shrinkages upon thermal decomposition will be developed, resulting in improved composite properties.

Hot pressing Tyranno fiber networks alone (no matrix material) resulted in fiber deformation into a close packed hexagonal array and a material with reasonably good room temperature strength and excellent retention of strength at high temperatures or after heat treatment at high temperatures. However, hot pressing has several limiting aspects, as described earlier.

REFERENCES

1. Ulrich, D. R., Sol-gel processing, *Chemtech*, vol. 18, no. 4, 242–248 (1988).
2. Hyde, A. R., Fibre reinforced glass and ceramic composites, *GEC Journal of Research*, vol. 6, no. 1, 44–49 (1988).
3. Mackenzie, J. D., Amorphous oxides from gels. In *Ultrastructure Processing of Advanced Ceramics*, J. D. Mackenzie and D. R. Ulrich (Ed.), John Wiley & Sons, New York, NY, 1988, 589–601.
4. Fitzer, E., and Gadow, R., Fibre reinforced composites via the sol/gel process. In *Tailoring Multiphase and Composite Ceramics*, (Materials Science Research-Vol. 20), R. E. Tressler et al (Ed.), Plenum Press, New York, NY, 1986, 571–607.
5. Pierre, A. C., Uhlmann, R. R., and Hordonneau, A., Ceramic composites made by sol-gel processing, *Rev. Int. Hautes Tempér. Réfract., Fr.*, vol. 23, 29–35 (1986).
6. Chen, M., James, P. F., Jones, F. R., and Bailey, J. E., Fibre reinforced mullite composites by sol-gel processing. In *Fabrication Technology*, R. W. Davidge and D. P. Thompson (Ed.), British Ceramic Proceedings, No. 45, The Institute of Ceramics, Shelton, Stokes-on-Trent, Staffs, UK, 1990, 211–219.
7. Okamura, K., Ceramic fibres from polymer precursors, *Composites*, vol. 18, no. 2, 107–120 (1987).
8. Jamet, J., Spann, J. R., Rice, R. W., Lewis, D., and Coblenz, W. S., Ceramic-fiber composite processing via polymer-filler matrices, *Ceram. Eng. Sci. Proc.*, vol. 5, no. 7–8, 677–694 (1984).
9. Fitzer, E., and Gadow, R., Fiber-reinforced silicon carbide, *Ceramic Bulletin*, vol. 65, no. 2, 326–335 (1986).
10. Hurwitz, F. I., Gyekenyesi, J. Z., Conroy, P. J., and Rivera, A.L., Nicalon/silicon-oxycarbide ceramic composites, *Ceram. Eng. Sci. Proc.*, vol. 11, no. 7–8, 931–946 (1990).

11. Anon., Dow Corning CMC technology produces complex parts cheaply, *Materials Edge*, , no. 24, 4 (May 1991).

12. Mohr, D. L., Desai, P., and Starr, T. L., Production of silicon nitride/silicon carbide fibrous composites using polysilazanes as pre-ceramic binders, *Ceram. Eng. Sci. Proc.*, vol. 11, no. 7–8, 920–930 (1990).

13. Yamamura, T., Ishikawa, T., Shibuya, M., Tamura, M., Nagasawa, T., and Okamura, K., A new type of ceramic matrix composite using Si-Ti-C-O fiber, *Ceram. Eng. Sci. Proc.*, vol. 10, no. 7–8, 736–747 (1989).

14. Brochure, "Black Glass," Allied Signal, Inc., Morristown, NJ.

15. Strife, J. R., Brunette, C. M., Pike, R. A., Streckert, H. H., and Sheehan, J. E., A study of the critical factors controlling the synthesis of ceramic matrix composites from preceramic polymers, Semiannual report under Air Force Office of Scientific Research Contract No. F49620-87-C-0093, April 15, 1988.

14

Production of Ceramic Matrix Composites by Chemical Vapor Infiltration

I. BACKGROUND

In addition to being a useful method for applying thin coatings, particularly carbon and boron nitride, onto reinforcing fibers to tailor the interface in ceramic matrix composites, chemical vapor deposition has been well developed as a method for producing the matrix of such a composite. This variation of the process is ordinarily termed chemical vapor infiltration (CVI), emphasizing the fact that deposition in the interior of fiber preforms is required. Surface deposition must be minimized, so that restriction to precursor gas ingress and by-product gas egress is minimized.

The chemical vapor infiltration method for silicon carbide matrix formation has been quite well developed and large structures can be produced. Properties of composites produced using CVI are among the best, and perhaps the best, of any of the composites described in this book. Many commercial scale structures have been produced by CVI; these will be described in a later chapter on ceramic matrix composite applications.

The deposition of ceramic materials onto a substrate by decomposition of gases or vapors is a well established process both from a fundamental and a practical standpoint. Table 1 [1] lists many materials that have been produced by this method, the precursor gases, deposition temperatures, the variation of

TABLE 1. Ceramic materials produced by chemical vapor deposition.

Coating	Chemical mixture	Deposition temp. (°C)	Method[a]	Application[b]
		Carbides		
TiC	$TiCl_4$-CH_4-H_2	900–1000	CCVD	wear
	$TiCl_4$-$CH_4(C_2H_2)$-H_2	400–600	PACVD	elec
HfC	$HfCl_x$-CH_4-H_2	900–1000	CCVD	wear, cor/ox
ZrC	$ZrCl_4$-CH_4-H_2	900–1000	CCVD	wear, cor/ox
	$ZrBr_4$-CH_4-H_2	>900	CCVD	wear, cor/ox
SiC	CH_3SiCl_3-H_2	1000–1400	CCVD	wear, cor/ox
	SiH_4-C_xH_y	200–500	PACVD	elec, cor
B_4C	BCl_3-CH_4-H_2	1200–1400	CCVD	wear
B_xC	B_2H_6-CH_4	400	PACVD	wear, elec, cor
W_2C	WF_6-CH_4-H_2	400–700	CCVD	wear
Cr_7C_3	$CrCl_2$-CH_4-H_2	1000–1200	CCVD	wear
Cr_3C_2	$Cr(CO)_6$-CH_4-H_2	1000–1200	CCVD	wear
TaC	$TaCl_5$-CH_4-H_2	1000–1200	CCVD	wear, elec
VC	VCl_2-CH_4-H_2	1000–1200	CCVD	wear
NbC	$NbCl_5$-CCl_4-H_2	1500–1900	CCVD	wear
		Nitrides		
TiN	$TiCl_4$-N_2-H_2	900–1000	CCVD	wear
	$TiCl_4$-N_2-H_2	250–1000	PACVD	elec
HfN	$HfCl_x$-N_2-H_2	900–1000	CCVD	wear, cor/ox
	Hfl_4-NH_3-H_2	>800	CCVD	wear, cor/ox
Si_3N_4	$SiCl_4$-NH_3-H_2	1000–1400	CCVD	wear, cor/ox
	SiH_4-NH_3-H_2	250–500	PACVD	elec, cor/ox
	SiH_4-N_2-H_2	300–400	PACVD	elec
BN	BCl_3-NH_3-H_2	1000–1400	CCVD	wear
	BCl_3-NH_3-H_2	25–1000	PACVD	elec
	$BH_3N(C_2H_5)_3$-Ar	25–1000	PACVD	elec
	$B_3N_3H_6$-Ar	400–700	CCVD	elec, wear
	BF_3-NH_3-H_2	1000–1300	CCVD	wear
	B_2H_6-NH_3-H_2	400–700	PACVD	elec
ZrN	$ZrCl_4$-N_2-H_2	1100–1200	CCVD	wear, cor/ox
	$ZrBr$-NH_3-H_2	>800	CCVD	wear, cor/ox
TaN	$TaCl_5$-N_2-H_2	800–1500	CCVD	wear
AlN	$AlCl_3$-NH_3-H_2	800–1200	CCVD	wear
	$AlBr_3$-NH_3-H_2	800–1200	CCVD	wear
	$AlBr_3$-NH_3-H_2	200–800	PACVD	elec, wear
	$Al(CH_3)_3$-NH_3-H_2	900–1100	CCVD	elec, wear
VN	VCl_4-N_2-H_2	900–1200	CCVD	wear
NbN	$NbCl_5$-N_2-H_2	900–1300	CCVD	wear, elec

continued

TABLE 1. *Continued*

Coating	Chemical mixture	Deposition temp. (°C)	Method[a]	Application[b]
		Oxides		
Al_2O_3	$AlCl_3$-CO_2-H_2	900–1100	CCVD	wear, cor/ox
	$Al(CH_3)_3$-O_2	300–500	CCVD	elec, cor
	$Al[OCH(CH_3)_2]_3$-O_2	300–500	CCVD	elec, cor
	$Al(OC_2H_5)_3$-O_2	300–500	CCVD	elec, cor
SiO_2	SiH_4-CO_2-H_2	200–600	PACVD	elec, cor
	SiH_4-N_2O	200–600	PACVD	elec
TiO_2	$TiCl_4$-H_2O	800–1000	CCVD	wear, cor
	$TiCl_4$-O_2	25–700	PACVD	elec
	$Ti[OCH(CH_3)_2]_4$-O_2	25–700	PACVD	elec
ZrO_2	$ZrCl_4$-CO_2-H_2	900–1200	CCVD	wear, cor/ox
Ta_2O_5	$TaCl_5$-O_2-H_2	600–1000	CCVD	wear, cor, elec
Cr_2O_3	$Cr(CO)_6$-O_2	400–600	CCVD	wear
		Borides		
TiB_2	$TiCl_4$-BCl_3-H_2	800–1000	CCVD	wear, cor, elec
MoB	$MoCl_5$-BBr_3	1400–1600	CCVD	wear, cor
WB	WCl_6-BBr_3-H_2	1400–1600	CCVD	wear, cor
NbB_2	$NbCl_5$-BCl_3-H_2	900–1200	CCVD	wear, cor
TaB_2	$TaBr_5$-BBr_3	1200–1600	CCVD	wear, cor
ZrB_2	$ZrCl_4$-BCl_3-H_2	1000–1500	CCVD	wear, cor, elec
HfB_2	$HfCl_x$-BCl_3-H_2	1000–1600	CCVD	wear, cor

[a]CCVD = conventional CVD; PACVD = plasma-assisted CVD.
[b]Wear = wear-resistant coatings; elec = electronics; cor = corrosion-resistant coatings; ox = oxidation-resistant coatings.
Source: Ref. 1.

process used, and applications. Many carbides, nitrides, oxides, and borides have been deposited. Precursor materials are either gases at ambient temperature, or vapors obtained by evaporation of a liquid or sublimation of a solid. Gases are more convenient, since for liquids or solids heating must be carefully controlled to achieve the appropriate vapor concentration in the deposition chamber as well as to avoid plugging of inlet lines with condensed material. However, for many materials of practical interest, the best precursors are solids or liquids at ambient temperature.

Conventional CVD (CCVD) is ordinarily carried out at a temperature greater than 900°C. By using electrodes and an ionizable gas in the deposition chamber to establish a plasma, deposition can be carried out at much lower temperatures, even as low as 25°C. This is termed plasma-assisted CVD

(PACVD). However, only the conventional method has been used for CVI processing of ceramic matrix composites.

Although not included in Table 1, deposition of carbon using CVD is well established, and some conditions were given in the section on fiber coatings. CVI of carbon as one of several methods of carbon/carbon composite fabrication preceded use of this method to produce ceramic matrix composites. In fact, some pioneering work in the 1970's by R. Naslain and coworkers at the University of Bordeaux in France on CVI for producing ceramic matrix composites began as an attempt to improve the mechanical properties and oxidation resistance of carbon/carbon composites.[2] This work led to a cooperative program with the French organization, Société Européene de Propulsion (SEP), that has resulted in extensive development of the method for fabrication of silicon carbide matrix composites. SEP processing technology has been licensed to E. I. du Pont de Nemours & Co., Inc. (Du Pont) in the U.S.A., and both SEP and Du Pont offer commercial scale composite structures. Many other organizations have conducted research into this process and several offer commercial scale products.

II. CVI PROCESSING FOR CERAMIC MATRIX COMPOSITES

A. Silicon Carbide Matrix Composites by Isothermal Processing

A number of precursors for silicon carbide production by CVD have been used, but most formation of silicon carbide matrices by CVI has involved methyltrichlorosilane (MTS), CH_3SiCl_3. This material, which is a liquid at room temperature and must be vaporized, has the appropriate silicon-to-carbon ratio to produce silicon carbide. A hydrogen carrier gas and a temperature of greater than 1000°C are used.

Figure 1 [1] is a schematic representation of a laboratory scale reactor for silicon carbide deposition. The MTS is placed in a vaporizer and hydrogen is swept through the container to assist in vaporization and mixing. In general, this does not result in the optimum hydrogen-to-MTS ratio, so that some pure hydrogen is blended with the hydrogen/MTS mixture flowing to the reactor. The reactor shown is a "hot-wall reactor," which means that a furnace is used to heat the substrate, a fibrous preform in the case of CVI. A process carried out in this manner is termed an isothermal process. Another variation is a "cold-wall reactor," in which the substrate is selectively heated, often by inductive heating of a carbon mandrel. This is termed a thermal gradient process. Hydrogen chloride is the major by-product of the decomposition of MTS, and it is removed in a scrubber column. Some organic and/or organosilicon compounds can also form. Although documentation was not found in the published reports reviewed, these compounds can apparently be pyrophoric and decompose rapidly when the interior of the feed lines and reactor are exposed to the atmosphere.

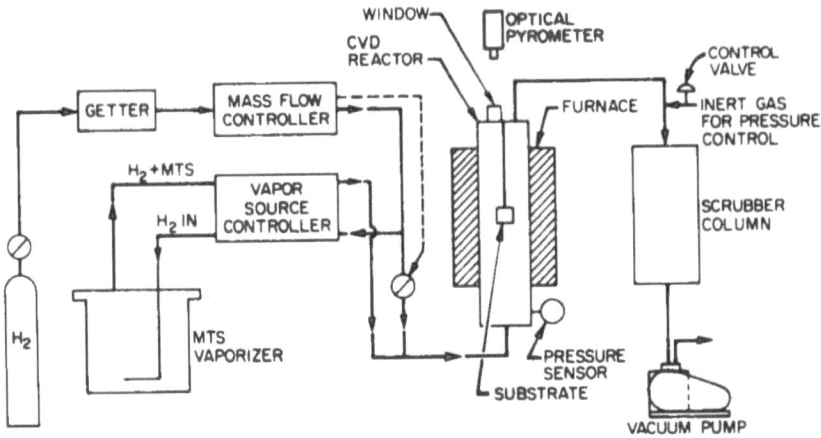

FIG. 1 Schematic illustration of a laboratory scale reactor for chemical vapor deposition of silicon carbide. (From Ref. 1.)

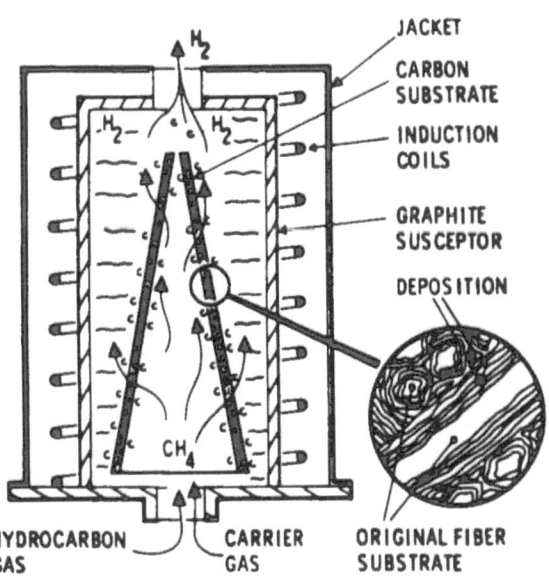

FIG. 2 Schematic representation of an isothermal system for chemical vapor infiltration of carbon. (From Ref. 3.)

Schematic representations of only the reaction chambers for larger scale CVI are given in Figures 2 and 3.[3] The diagrams illustrate fabrication of conical carbon/carbon parts by CVI, but serve to illustrate the CVI process for silicon carbide, if MTS were substituted for the hydrocarbon gas shown and hydrogen chloride as well as excess hydrogen were shown leaving the chamber. In the isothermal variation (Figure 2) gas flows through the fibrous preform held at a constant temperature. In the thermal gradient variation (Figure 3), a heated graphite susceptor also serves as a mandrel for the preform. The infiltrating mixture can be forced to flow between the mandrel and a sleeve. Deposition is most rapid at the hotter face.

Naslain et al [4] have provided some processing details. In a first step, a stack of reinforcing fiber cloth was consolidated with a small amount of carbon deposited by decomposition of methane. At this stage, the preform had an open porosity of 30–60%. These preforms could be cut to provide specimens of a desired size. In the second step, the porous preform was densified using CVI with an MTS/hydrogen mixture. Temperatures and pressures at the low side of the possible range for successful deposition were used. This favors in-depth penetration, as opposed to surface coating. Times greater than 100 h were required for satisfactory densification.

It has been reported elsewhere [1] that despite conditions favoring in-depth penetration rather than surface coating, greater deposition of matrix still takes place near the preform surfaces than in the interior, thus necessitating interruption of the process for intermediate machining using diamond tools to remove material from the surface and reestablish passageways to the interior of the specimen. This cyclic process is reported to require several weeks to several months to complete. A cross section of a composite produced by CVI is shown in Figure 4.[5] Multiple layers, as well as surface inaccessible pores, can be observed.

As with several of the other fabrication processes already discussed, Fitzer and Gadow [6] reported fundamental work on the CVI process for silicon carbide deposition. MTS was used as the precursor and temperature and pressure were varied. Infiltration depth increased but deposition rate decreased with decreasing temperature. Reduction of the pressure resulted in greater infiltration depths, but lower deposition rates. Doubling the total pressure from 5 kPa to 10 kPa doubled the deposition rate, but increasing the pressure from 5 kPa to 100 kPa did not increase deposition rate proportionally. Increasing the pressure beyond a certain point resulted in a decrease of the mean free path of the gaseous molecules to such an extent that silicon carbide powder was formed in the gas phase.

In general, deposition in the interior of a preform (CVI), relative to surface deposition (CVD), was enhanced by conditions under which deposition rate was controlled by surface reaction kinetics rather than by diffusion. When diffusion of the precursor gases into the interior of the preform and diffusion of the by-product gases out of the interior was rate controlling, surface deposition was favored. For specimens that were 4 mm × 5 mm × 50 mm in size and having a fiber content of 40–50 vol%, Fitzer and Gadow reported that optimum infiltration parameters were a temperature of about 1050°C, a pressure of 3 kPa, and a time of 4 days. However, higher temperatures can be successfully used, perhaps depending upon preform porosity and pore structure. Grateau et al [7] reported conditions consisting of a temperature of 1250°C, a pressure of 266 Pa, and a flow rate of 10 cm³/min. Deposition rate was 15 μm/h.

Depending on conditions, stoichiometric silicon carbide is not necessarily obtained. Naslain and Langais [8] indicated the importance of the starting hydrogen/MTS ratio on the composition of the product. At a constant pressure, silicon carbide, carbon, and silicon ratios varied according to the hydrogen/MTS ratio and the temperature. Maximum yields of silicon carbide were predicted to occur at a hydrogen/MTS ratio of about 100. A deposition product with excess silicon is favored by a very high hydrogen/MTS ratio and a product with excess carbon is favored by a low ratio.

As indicated in Table 1, a mixture of silane, methane, and a hydrocarbon gas can also be used to deposit silicon carbide by chemical vapor deposition. Although few details were given, Pan et al [9] reported on composites containing

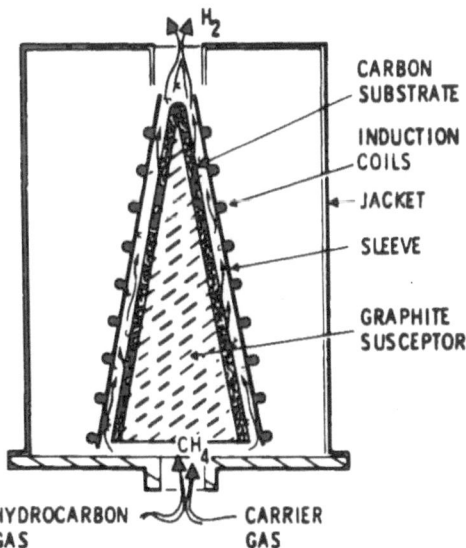

FIG. 3 Schematic representation of a thermal gradient system for chemical vapor infiltration of carbon. (From Ref. 3.)

8-harness satin weave Nicalon cloth (40 vol%) produced by CVI using a silane gas system at about 1200°C.

Carbon and silicon containing precursors other than MTS have been used also. In work cited earlier in conjunction with fiber coatings, Brennan [10] used methyldichlorosilane (MDS), CH_3SiHCl_2, and dimethyldichlorosilane (DMDS), $(CH_3)_2SiCl_2$, as well as MTS in chemical vapor infiltration studies. MDS has the advantage over MTS of a lower boiling point (41.5°, vs 66.4°C), allowing the transportation of an MDS-saturated carrier gas over long distances without heated lines. DMDS has a higher boiling point (70.5°C).

With a hydrogen/MDS ratio of over 300, essentially stoichiometric silicon carbide was formed with a substrate temperature of 1080°C. At a ratio of about 30 or lower, excess carbon was produced along with silicon carbide (about 8% more carbon than silicon). With MTS, a hydrogen-to-MTS ratio of about 5 and substrate temperature of 1120°C produced a material very high in excess carbon (about 40% more carbon than silicon) and with considerable oxygen. More

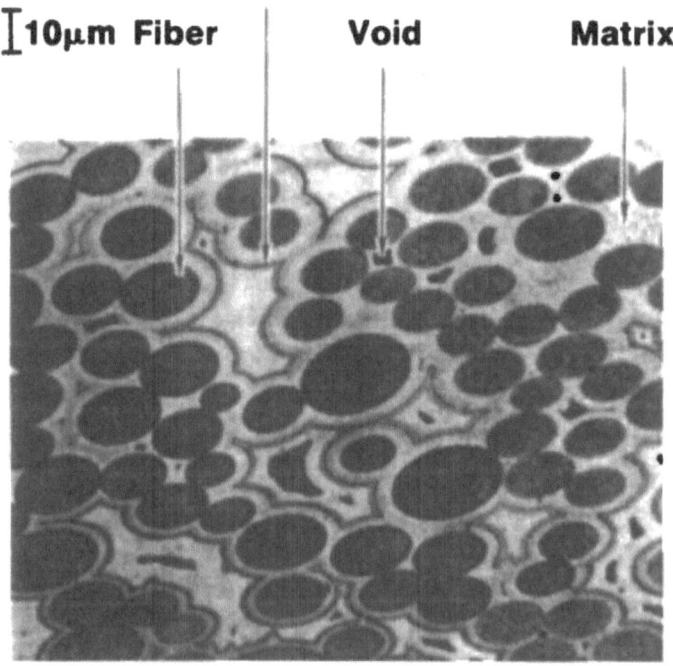

FIG. 4 Cross section of a Nicalon™ fiber, silicon carbide matrix composite produced by chemical vapor infiltration. (From Ref. 5.)

nearly equal silicon-to-carbon ratios were found with a ratio of about 50. With DMDS and a substrate temperature of 900°C, deposit compositions ranged from excess silicon at a hydrogen/DMDS ratio of about 40 to equal carbon and silicon at a ratio of 8, to almost pure carbon at a ratio 1. Many of these findings were contrary to what was anticipated from the carbon-to-silicon ratios in the precursors, indicating complex reaction mechanisms. Thus, it is possible that despite its apparent advantage from a stoichiometry standpoint, other precursors may be superior to MTS under some conditions.

Preform geometry affects densification rate of CVI-produced composites. For example, Burkland and Yang [11] have shown that for a given preform fiber volume fraction, 35%, apparent density of a three-dimensional braided Nicalon fiber preform increased by an increment about twice that with a unidirectional laminate (a 1.7 g/cm^3 increase vs a 0.9 g/cm^3 increase) after the same deposition time. A two-dimensional plain weave fabric laminate had about a 1.4 g/cm^3 increase in the same period. It was postulated that because the braided structure had a more uniform pore network, densification by CVI was enhanced. Surface grinding followed by additional densification for about 5% of the time used in the original CVI densification resulted in an increase in density of about 0.01 g/cm^3 for the three-dimensional and two-dimensional materials and about 0.025 g/cm^3 for the unidirectional material, showing that the pore network after partial densification differs with fiber architecture.

B. Silicon Carbide Deposition using Forced Flow/Thermal Gradient CVI

A variation of the CVI process that utilizes both a thermal gradient and a pressure gradient has been developed at Oak Ridge National Laboratory (ORNL) by Lackey and Caputo.[12] The system is shown in simplified schematic form in Figure 5.[1] The fibrous preform is retained in a graphite holder. By water cooling the holder, a steep temperature gradient is maintained. Typically, the hot face is kept at about 1200°C and the cold face at about 700°C. The reactant gases are forced under low pressure into the cooler side of the preform. Because of the low temperature, they do not react there, but continue into the hot region, where they deposit silicon carbide. Deposition eventually occurs throughout the preform because the hot region moves progressively from the top to the bottom of the specimen as silicon carbide deposition increases. When the top surface becomes so densified that it cannot be permeated by precursor gases and by-products, gases begin to flow radially through the preform to the annular void space and escape through vents in the retaining ring. Details of the apparatus have obviously changed over the years as experience with the process increased.

The ORNL process has been used with continuous fiber preforms having various architectures and with whiskers and platelets. Most work on continuous fiber reinforced material has been with Nicalon fibers, but Nextel, Tyranno, and other fibers have been used as well.

An advantage of the thermal gradient, forced flow method for CVI is a shorter processing time than for the isothermal method. Rather than weeks, as typically required for isothermal CVI, composites can be formed by the ORNL method in a matter of hours. However, multiple specimens can be produced simultaneously in the reaction chamber using the isothermal method, so that the processing time advantage with the ORNL method is not as significant as it might first appear. Shape making capabilities are also more limited than with the isothermal process. Most specimens produced to this point have been relatively small discs, although tubes about 12 cm in length and 4 cm in outer diameter, with a wall thickness of about 0.6 cm have been produced in a cooperative program between ORNL and Babcock & Wilcox.[13]

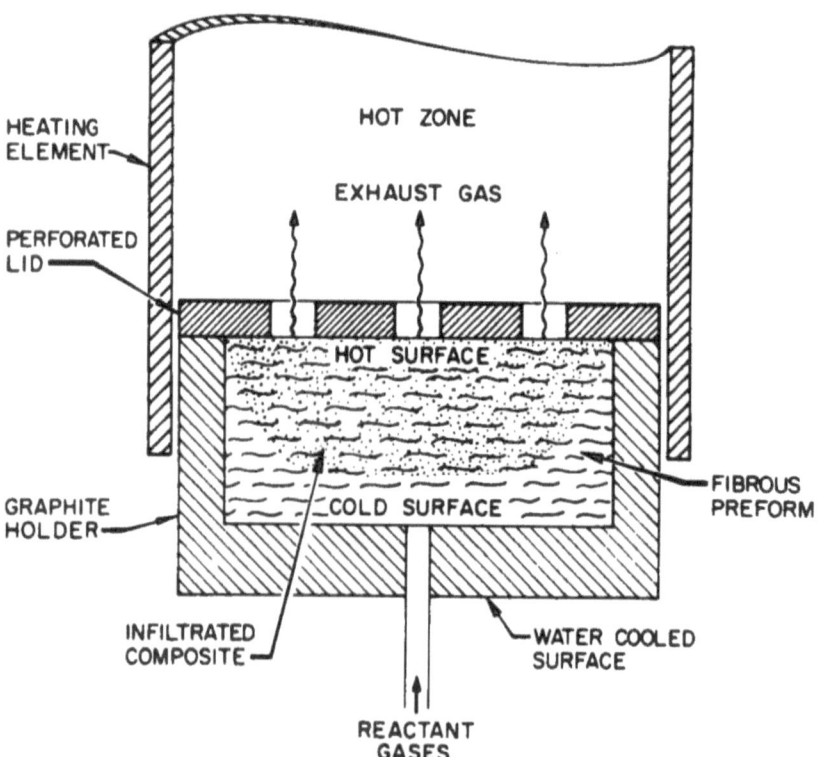

FIG. 5   Schematic illustration of the ORNL forced-flow/thermal-gradient chemical vapor infiltration system. (From Ref. 1.)

Detailed information on processing parameters for the ORNL process has been reported by Lowden et al.[14] A statistically designed experiment including effects of temperature, pressure, hydrogen/MTS ratio, and total gas flow on properties was carried out. Temperature was varied from 1000° to 1300°C, pressure from 10 to 100 kPa, total gas flow from 275 to 1100 cm^3/min, and hydrogen-to-MTS ratio from 10 to 35.

Preforms were produced from 52 layers of plain weave Nicalon fiber cloths in a 30/60/90 orientation. The layers were compressed and held into place by a perforated graphite lid pinned to the holder. Average fiber content was about 41 vol%. Diameter was 45 mm and thickness was 12.5 mm. Fiber size was removed by multiple washings with acetone. Fibers were precoated with a thin layer of carbon prior to CVI by isothermal deposition of propylene in argon at 1100°C and 5 kPa. Calculated average preform density was 2.91 g/cm^3. Infiltration was stopped in each experiment when a predetermined pressure differential across the sample, indicating complete infiltration, was reached. In addition to density, flexural strength (4-point) was used as a measure of specimen quality. Test specimens were cut from samples parallel to the 0/90 orientation of the top layer.

Processing time was most effectively decreased by increasing temperature and total gas flow rate. Temperature was the most significant variable; the majority of runs processed with a furnace temperature of 1300°C were completed in less than 20 h. The shortest time, 5.3 h, resulted from a furnace temperature of 1300°C, a pressure of 100 kPa, a total flow of 1100 cm^3/min and a hydrogen-to-MTS ratio of 10. The conditions requiring the longest time, 202.7 h, were a temperature of 1100°C, a pressure of 100 kPa, a total flow rate of 275 cm^3/min, and a hydrogen-to-MTS ratio of 35.

Density values at the tops of specimens were as high as 2.65 g/cm^3 (91% of theoretical density), but density usually decreased from top to bottom. In some cases, uninfiltrated bottom sections delaminated. Correlations of density with process variations could only be made for top layer specimens. Results suggested that density increased with decreasing temperature, decreasing total gas flow rate, and decreasing hydrogen-to-MTS ratio. Pressure had little effect. Correlation of flexural strength with processing parameters was relatively poor.

C. Pulse Infiltration

Another variation of the CVI process for ceramic matrix composite formation, "pulse CVI," has been reported by Sugiyama and Ohzawa.[15] The method consists of exposing a preform heated by radio frequency (r.f.) induction to repeated short term cycles of the precursor/carrier-gas atmosphere followed by vacuum. In principle, this should minimize diffusion restrictions that limit ingress and egress of gases into fine pores, thus enhancing internal deposition relative to surface deposition. Although the work described did not deal with continuous fiber or whisker preforms, it is cited here because of possible future use for fiber or whisker reinforced ceramics.

Six percent MTS (probably by volume, although this was not specified) in hydrogen was used as the precursor gas for depositing silicon carbide. The gas mixture was held in a reservoir, from which it instantaneously filled the reaction chamber in each pulse by use of an electromagnetic valve. Porous carbon (29% porosity) or silicon carbide bonded with a small amount of carbon produced by decomposition of a starch binder in an inert atmosphere were used as preforms. Two types of the latter preforms were used, those made from particles having an average diameter of 4 μm and those made from particles having an average diameter of 50 μm. Porosities of these preforms were 45–50% with the smaller particles and 42–48% with the larger particles.

Temperature was varied from 1000° to 1300°C, as measured with an optical pyrometer. Time in the precursor gas atmosphere was about 1 s and pressure was about 95 kPa. It took 2 s to evacuate to below 0.2 kPa. Up to 40,000 pulses were used. Microstructural characterizations and flexural testing (3-point) were carried out.

In general, densification increased with decreasing temperature. At higher temperatures, surface deposition increased (as typical for isothermal CVI). Increases in density and flexural strength with the number of pulses depended on the preform type and temperature.

Whether there are advantages to the pulse CVI technique in continuous fiber or whisker reinforced ceramic composite fabrication remains to be proven. Experiments that use directly comparable preforms and precursor gases with the pulse method and the other variations described above would need to be conducted. It seems obvious that pulsed CVI would be more difficult and costly to carry out on a large scale than conventional CVI.

D. Deposition of other Matrices using CVI

Although silicon carbide has been the major ceramic matrix material produced by CVI, other matrices have been investigated also.

Researchers at the University of Bordeaux and SEP have deposited a number of carbide and nitride materials. In one study [16], a stack of PAN base carbon fiber cloth layers was first consolidated with a small amount of pyrocarbon produced by CVD using methane as the precursor. Porosity of the preforms ranged from 30–60 vol%. The preforms could be machined at this stage. In addition to silicon carbide deposition using MTS/hydrogen, titanium tetrachloride/methane/hydrogen was used for depositing titanium carbide (TiC), boron trichloride/methane/hydrogen was used for depositing boron carbide (B_4C), and boron trifluoride/ammonia was used for depositing boron nitride (BN). Isothermal CVI was used. Maximum densification usually required a duration of over 100 h.

Additional fabrication details with one of the matrix materials, titanium carbide, were given by Rossignol et al.[17] Initial processing was the same as outlined above, except that rayon base as well as PAN base fibers were used. Open porosities of the pyrocarbon bonded preforms were 30–40 vol%. Titanium carbide was deposited at 950°C at a pressure of about 3 kPa using a titanium tetrachloride/methane/hydrogen mixture.

Oxide matrices have also been deposited by CVI. Colmet et al [18] reported on processes of this type. As one example, a unidirectional preform of FP alumina fibers (having about 60% porosity) was exposed to a gaseous mixture of 10% $AlCl_3$, 30% CO_2, and 60% H_2 at a temperature of 950°C, a pressure of 2.7 kPa, and a total gas flow of 100 cm^3/min. Infiltration time was 300 h. After this time, porosity was found to be 10%. The matrix was mainly α-alumina. Under otherwise identical conditions, use of a temperature of 1200°C or a pressure of about 50 kPa resulted in mostly surface deposition.

In a variation of this process, the preform was first partially densified by a sol-gel method based on sec-aluminum butylate. A number of impregnation, drying, and pyrolysis (1100°C) cycles reduced porosity to 50%. In this case, CVI densification to 10% porosity required only 180 h.

Zirconia matrix composites were produced using a similar method. A unidirectional arrangement of alumina fibers was used. Initial porosity was 70%, but this was reduced to about 50% by sol-gel processing based on zirconium sec-butylate. A 300-h infiltration using zirconium tetrachloride/carbon dioxide/hydrogen reduced porosity to about 10%. Composites with alumina/zirconia mixtures were produced also.

Other work on alumina matrix deposition was described by Colmet et al.[19] Thermodynamic studies indicated that alumina is deposited quantitatively from mixtures of aluminum chloride, carbon dioxide, and hydrogen in a rather narrow initial composition range. It was also calculated that carbon would be codeposited from a hydrogen-rich initial gas composition under conditions of low temperature and low pressure.

Several types of oxide fibers were used. The fiber preforms were prepared either by impregnation using an alumina slurry or a liquid alumina precursor followed by high temperature pyrolysis, or simply by laying up the fibers in alumina molds. Fiber mats and unidirectional and two-dimensional fiber networks were used. Optimum CVI parameter ranges were found to be a temperature of 900°–1000°C, a pressure of 2–3 kPa, a total gas flow of 100 cm^3/min, and an initial hydrogen-to-carbon dioxide ratio of about 1. Densification kinetics were dependent upon the nature of the preform. There were density gradients from the specimen surfaces to the interior, but it was pointed out that similar gradients exist with carbon and silicon carbide matrices produced using CVI. In most of the composites, the fibers and matrix were strongly bonded, so that failure was brittle.

Additional information on zirconia matrix composites was given also.[20] Both oxide fiber and carbon fiber preforms were used. These included mats and two-dimensional arrangements. Detailed experimental work on reaction parameters was carried out. The variation of residual porosity with densification time followed an exponential law. The behavior of different preforms under the same CVI conditions was strongly dependent on the features of the network pore structure. A low porosity and a low mean pore size were conducive to good densification.

An example of fabrication of silicon nitride matrix composites by CVI has been given by Hoyt and Yang.[21] Trichlorosilane ($SiCl_3H$) and ammonia were used as the precursor gases. A three-dimensional braided Nextel 480 preform was used. The fibers were heat cleaned and a carbon coating applied by CVD. The preform was then placed in a graphite mold in order to maintain the proper shape and fiber volume fraction during infiltration. After partial infiltration, the mold was removed and infiltration was continued until there was no open porosity remaining. The impervious surface layer was then ground away to reestablish channels to the interior of the specimen, and infiltration conducted again. Total CVI time was 100 h. Preliminary work to establish infiltration conditions was conducted using coarse carbon particles and carbon fibers held in a carbon boat.

Results using the carbon preform indicated that at a temperature of 800°C deposition rate was independent of the distance from the surface, while at higher temperatures deposition rates were greater near the surface, indicating that premature plugging would be a problem. Presumably the Nextel fiber reinforced materials were produced using this temperature, although this was not indicated.

A plot of density as a function of processing time, with a maximum of 50 h, was approximately linear. Density increased from about 1.2 g/cm³ for the uninfiltrated preform to 2.2 g/cm³ after 50 h. Based on the fiber density and volume fraction and assuming a 2.7 g/cm³ density for silicon nitride, porosity was 23%. Fracture behavior of the as-processed composite was nonbrittle.

III. MODELING OF THE CVI PROCESS

Mathematical modeling of the CVI process has received considerable attention. An in-depth discussion of this subject will not be given here. However, brief comments on selected studies will be cited to provide references for those readers interested in this subject.

The fibrous preform used for CVI can have a rather complex pore size distribution. As an example, a number of types of porosity quite different in nature are inherent with a preform consisting of stacked layers of cloths woven using multifilament tows, as schematically illustrated in Figure 6.[22] In addition to the spaces among the individual fibers in a tow, there are openings among the individual tows in the cloth and openings between the layers. Precise modeling of the pore structure as well as other factors would obviously be extremely difficult, so many simplifications are ordinarily made.

Fitzer and Gadow [6] calculated theoretical penetration depths in cylindrical pores of various diameters as a function of temperature, making the assumption that the kinetic aspect is similar to that of heterogeneous catalysis. In that case, the combined effects of chemical reaction and diffusion are characterized by several dimensionless numbers. Results agreed reasonably well with their experiments on silicon carbide deposition by CVI.

A model by Rossignol et al [23] for two-dimensional fiber preforms, which built on an earlier model by Van den Breckel et al [24], was based on deposition of a solid from gaseous precursors in capillary tubes 0.1–1 mm in diameter. This model gave the thickness of the deposition as a function of the distance from the tube entrance, taking into account diffusion and reaction kinetics. It was assumed that a given elemental volume of a two-dimensional porous preform can be represented by an equivalent cylindrical pore. Both ordinary diffusion and Knudsen diffusion (diffusion through a very small pore via wall collisions) were taken into account. Calculations from this model indicated that deposition is reasonably homogeneous at low temperatures and pressures, but that it becomes more concentrated at the pore openings as temperature and total pressure are increased, which is in good qualitative agreement with experimental results.

Tai and Chou [25] provided a model for deposition of alumina and titanium carbide within a ceramic fiber bundle. The model considered vapor diffusion, chemical reaction on the inner surface of the capillary, deposition film growth, porosity, and effects of reactant composition at various temperatures and pressures. Binary, multicomponent diffusion and Knudsen diffusion were taken into account for different stages of CVI deposition. Both diffusion controlled and reaction controlled processes were considered. Good agreement with experimental data was indicated. Currier and Valone [26] provided a time-dependent solution to the Tai/Chou model. A number of heat transfer models that might be useful for CVI modeling were also cited. Currier [27] extended models that consider CVI to be occurring in cylindrical pores by providing for overlap of pores as densification proceeds. This model indicates a surface area maximum as deposition proceeds and predicts spatial variations within fiber tows.

Gupte and Tsamopoulos [28] assumed a pore structure represented by a Bethe lattice (a treelike network with an arbitrary, uniform number of branches at each intersection). The model predicted significant diffusion limitations even for well connected pore networks. Use of thermal gradients was predicted to improve densification only to a limited extent. Broad pore size distributions were predicted to increase under-utilized pore space, resulting in reduced densification. The model predictions compared qualitatively with results from the ORNL CVI process.

Middleman [29] provided a model which included the role of the interaction of gas-phase and surface kinetics and which coupled the events occurring within the densifying porous structure and those in the rest of the reactor. It was predicted that if a chemical scheme can be found that produces the CVI precursor through a gas-phase reaction, and if the reactor can be operated at a very short residence time, densification can proceed from the inside out, minimizing the trapping of pores within the interior of the composite as densification proceeds. Middleman and coworkers [30] subsequently provided a model indicating that it is possible to select an initial preform geometry that leads to a uniform densification profile. The compromise between improved properties and increased processing time can be assessed using the model.

Starr [31] specifically modeled the ORNL forced-flow/thermal-gradient process. Deposition rate was calculated for one-dimensional flow and temperature gradients. Density gradients as a function of infiltration conditions were predicted. Gupta and Evans [32] provided a mathematical model for CVI using microwave heating and external cooling. They predicted that microwave heating can lead to more rapid infiltration and lower residual porosity than isothermal heating and that by adjustment of microwave power and external cooling, the temperature profile in the preform can be altered to optimize CVI.

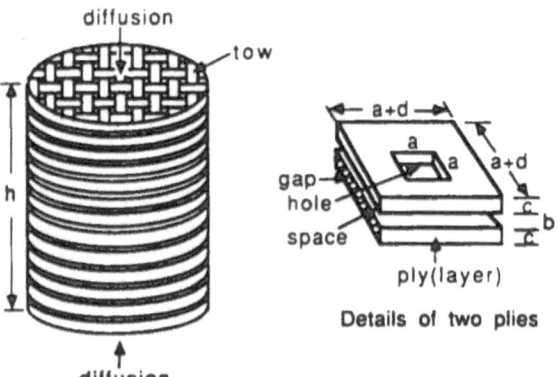

FIG. 6 Schematic drawing of a woven fabric preform showing the sides of a square hole in the weave (a), the distance between plies (b), the thickness of one ply (c), the width of a tow (d), and the height of the fibrous preform (h). (From Ref. 22.)

IV. MECHANICAL PROPERTIES OF CVI-PRODUCED COMPOSITES

A. Properties of CVI-Produced Silicon Carbide Matrix Composites

Ceramic matrix composites produced by the chemical vapor infiltration process have exhibited excellent properties. Naturally, there are variations that depend on the reinforcing fibers and architecture, as well as the nature of the fiber/matrix interfaces. Because of the lack of standardization of ceramic matrix composite mechanical test methods, it is difficult to distinguish material variations from test variations. Therefore, no attempt will be made to critically assess composites produced by different organizations. Selected results on materials produced by a number of organizations will be described here.

Fitzer and Gadow [6] showed flexural (3-point) stress-strain behavior for silicon carbide matrix composites (Figure 7) produced using a variety of fibers. Highest strength was about 950 MPa with uniaxially aligned large diameter CVD-produced silicon carbide fibers (probably SCS-6, although not identified). However, this composite failed in a brittle manner. A composite containing uniaxially aligned high modulus carbon fibers had a lower strength, about 750 MPa, but some indication of nonbrittle failure. Composites having two-dimensional fiber alignment had considerably lower strengths, but nonbrittle failure. A Nicalon fiber (termed "PCS-SiC" in the figure) cloth reinforced material had a flexural strength of about 350 MPa and very nonbrittle failure. Since the matrices were apparently produced in the same manner, strength and stress-strain behavior were obviously affected by fiber properties and orientations and by the interfacial details of the composites. Also shown in the figure are the result for an SEP material (having the trade name Cerasep, and described below) and a value for advanced carbon/carbon.

The most complete listing of properties of silicon carbide matrix composites produced using CVI has been on material produced by SEP and by Du Pont under the SEP license. One type of material, trade named Sepcarbinox, contains carbon fibers. A second material, having the trade name Cerasep, contains Nicalon fibers. Cerasep composites are more stable toward oxidation, but Sepcarbinox composites have some mechanical property advantages.

Typical tensile strength data for Nicalon reinforced composites reported by Du Pont are shown in Figure 8.[33] This is for material having a 0/90 fiber orientation, about 40 vol% fibers, a density of 2.5 g/cm^3, and a porosity of 10 vol%. At room temperature, ultimate strength was 190 MPa. The stress-strain curve is typical for a continuous fiber reinforced ceramic having an appropriately weak interface. That is, the curve is linear initially but deviates from linearity at the point at which matrix microcracking initiates (termed the proportional limit). After this point, load transfer depends upon fiber bridging and, eventually, pullout of individual fibers as they break. Strain to failure was 0.22%. At

1000°C, the stress-strain curve is very similar, although the composite strains more until failure, resulting in a higher ultimate strength.

Although many of the composite property data given in this book are for a very limited number of test specimens, a rather extensive database was reported for properties of the Du Pont composites. Various tests were conducted on many specimens. Results are given as the mean value plus/minus one standard deviation. Ultimate tensile strength was 190 ± 15 MPa (248 tests), elongation during tensile testing was 0.22 ± 0.04% (242 tests), tensile modulus was 215 ± 25 GPa (238 tests), compressive strength was 800 ± 76 MPa (72 tests),

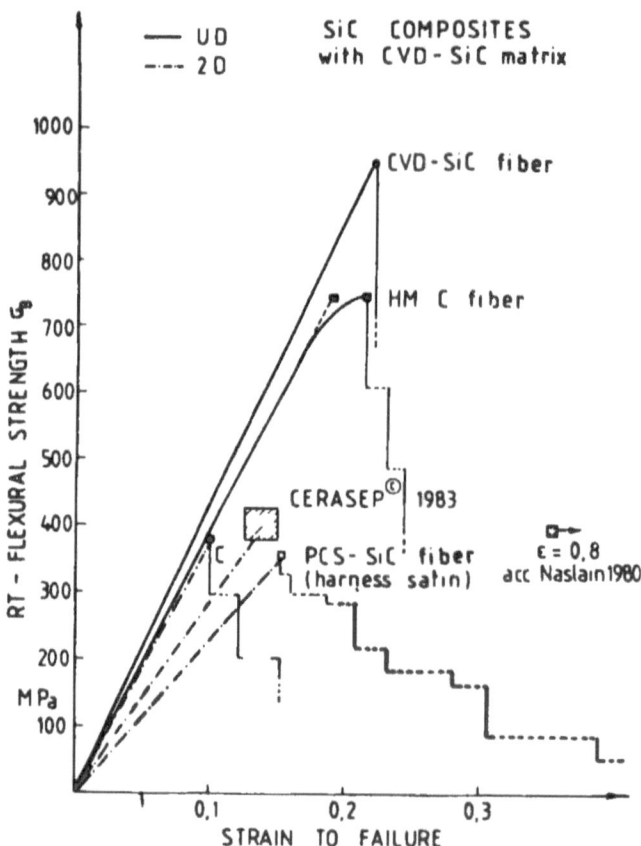

FIG. 7 Room temperature flexural stress-strain curves for fiber reinforced silicon carbide matrix composites produced by CVI. (CVD-SiC fiber is not identified; PCS-SiC stands for polycarbosilane derived silicon carbide fiber; Cerasep is a product of SEP.) (From Ref. 6.)

and interlaminar shear strength was 32 ± 10 MPa. Most of the test data were derived from specimens removed from the edges of many different plates over a long time period. Other typical properties include a room temperature flexural strength of 259 MPa, with an elongation of 0.27%, and a flexural modulus of 100 GPa. Fracture toughness was 24 MPa·m$^{1/2}$ (unspecified method).

Typical tensile stress-strain curves for carbon fiber reinforced silicon carbide produced by Du Pont are shown in Figure 9. This was for material having about 45 vol% of fibers having a 0/90 orientation, a density of 2.1 g/cm^3, and porosity of 10 vol%. Tensile strength was about 530 MPa and strain-to-failure over 1%. At 1200°C, strength was slightly higher and strain-to-failure slightly lower. The differences in strengths and stress-strain curve shapes between Nicalon fiber reinforced and carbon fiber reinforced material are due to differences in fiber mechanical properties, fiber volume fraction, and the nature of the fiber/matrix interfaces.

Database results with carbon fiber reinforcement include a tensile strength of 425 ± 95 MPa (134 tests), a tensile elongation of 0.93 ± 0.15% (134 tests), a tensile modulus of 76 ± 6 GPa (133 tests), a compressive strength of 520 ± 45 MPa (41 tests), and an interlaminar shear strength of 26 ± 3 MPa. Other typical properties include a flexural strength of 454 MPa, with an elongation of 0.72%, and a flexural modulus of 68 GPa. Fracture toughness was 40

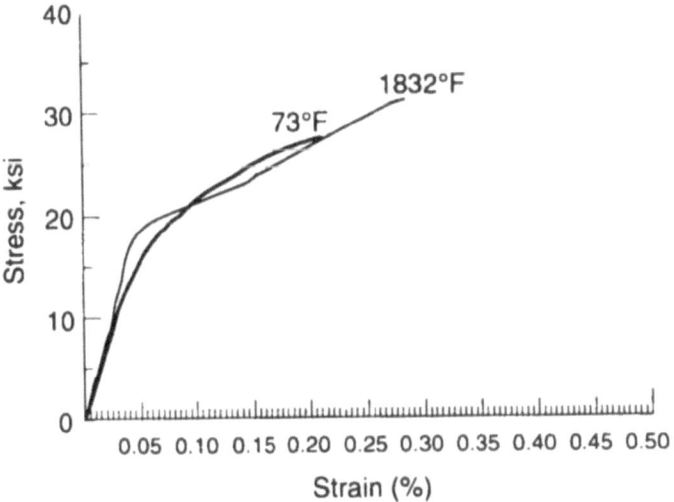

FIG. 8 Tensile stress-strain curves for CVI-produced carbon fiber, silicon carbide matrix composites at room temperature and at 1000°C (1832°F). (From Ref. 33.)

MPa·m$^{1/2}$ (unspecified method), which is certainly among the highest measured for a ceramic material.

Du Pont also reported that some mechanical properties of composites containing Nicalon fibers are relatively unaffected by fiber orientation, because matrix properties tend to dominate these properties. In contrast, properties of composites containing carbon fibers are more affected by orientation, indicating that fiber strength is the dominant factor in these composites. As examples, flexural strengths of composites with carbon cloth layers having 0/90, 0/±45/90, and ±45 orientations were about 430, 365, and 170 MPa, respectively. Corresponding strengths with Nicalon fibers were about 190, 180, and 165 MPa.

Fatigue properties were also reported. Specimens subjected in tension to 50 Hz fatigue cycles at about 315 MPa showed no signs of failure after 5 × 10^6 cycles. The fatigue stress-strain curves showed that the material was permanently strained during the initial loading, then continued to cycle along the same line, with no further permanent strain.

Oxidation resistance and tensile strength retention at elevated temperatures for Nicalon fiber containing composites were reported also. Specimens were exposed, in air, to one or more typical turbine engine cycles representing full power operation at about 800°C, temperature spikes to about 1250°C due to after-burner operation, and extended cruise operation at 650°C (Figure 10). They were then tensile tested at room temperature. After a few cycles, measured strength retention, compared with as-fabricated material, was over 100%, and even after 16 cycles, over 95% of the original strength was retained. It should

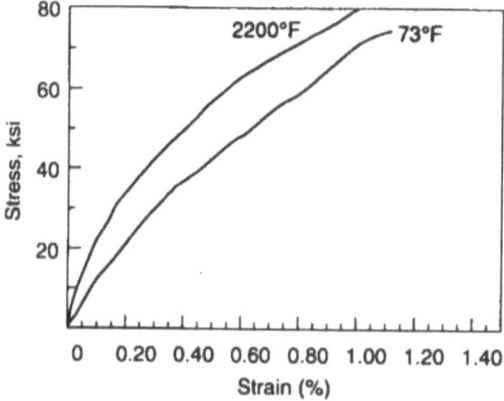

FIG. 9 Tensile stress-strain curves for CVI-produced Nicalon™ fiber, silicon carbide matrix composites at room temperature and at ~1200°C (2200°F). (From Ref. 33.)

be noted, however, that the specimens may have included a silicon carbide overcoat, preventing exposure of fiber ends. If this coating microcracks, permitting oxidation ingress at the fiber/matrix interfaces, performance may not be as good.

Thermal diffusivity, thermal expansion coefficient, emissivity, and a number of other properties have also been measured for these materials. These will not be given here; the reader interested in such measurements can consult original literature from SEP or Du Pont.

It is likely that properties of state-of-the-art material produced by SEP are similar to those reported above by Du Pont, since the Du Pont composites were based on the licensed SEP process. However, this is difficult to assess from available literature, since the time frame during which composites were produced and a detailed description of the specimens tested is usually not given in published works on these composites.

Results reported some years ago on the SEP produced material by Lamicq et al [34] for two-dimensional Nicalon fiber reinforced material indicate a flexural strength of 300 MPa, with no deterioration or even enhancement after exposure to 1100°C in air for up to 500 h, or after thermal cycling in air for up to 100 cycles. A fracture toughness value of over 25 MPa·m$^{1/2}$, with no decrease during testing at up to 1400°C, was reported.

More recent data on SEP material has been reported by Jouin et al.[35] Fiber types are specifically mentioned, Nicalon NL200 and Toray-T300 carbon fibers. Warp and fill directions of woven cloths were alternated in producing laminates. Fiber volume fractions were about 45% for carbon/silicon carbide and 40% for Nicalon/silicon carbide.

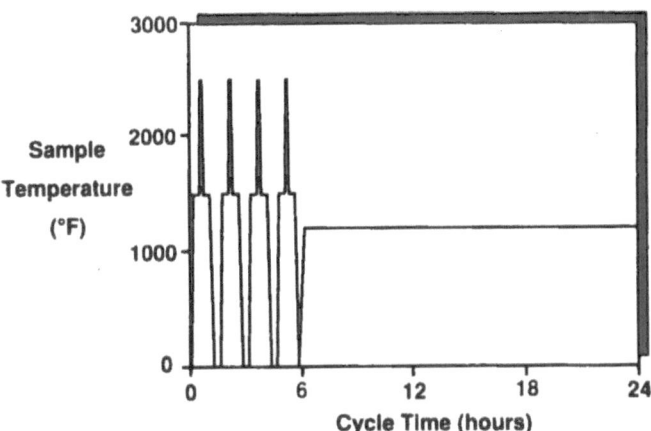

FIG. 10 Typical turbine engine temperature cycle. (From Ref. 33.)

Tensile strengths of carbon fiber reinforced silicon carbide tested at room temperature, 1000°C, and 1400°C were 350, 350, and 330 MPa, respectively. Corresponding Young's moduli were 90, 100, and 100 GPa. Elongations were 0.9% at the lower two temperatures. Flexural strengths at the three temperatures were 500, 700, and 700 MPa. Compressive strengths along one of the two fiber orientation directions were 580, 600, and 700 MPa at the three test temperatures, given in the same order as above. Compressive strengths in the thickness direction were 420, 450, and 500 MPa.

For Nicalon fiber reinforced silicon carbide, tensile strengths at room temperature, 1000°C, and 1400°C were 200, 200, and 150 MPa, respectively. Corresponding Young's moduli were 230, 200, and 170 GPa. Elongations were 0.3, 0.4, and 0.5%. Flexural strengths at the three temperatures, in increasing order, were 300, 400, and 280 MPa. Compressive strengths along one of the two fiber orientation directions were 580, 480, and 380 MPa. Compressive strengths in the thickness direction were 420, 380, and 250 MPa. A number of other properties were also reported in this work.

Several other studies on mechanical properties of SEP-produced material have been reported. Examples include the work of Bouquet et al [36] and Gomina et al.[37] Publications cited earlier in the section of CVI fabrication also include mechanical property data.

A particularly interesting study on high temperature properties was conducted by Frety and Boussuge.[38] They tested two types of SEP-produced Nicalon fiber reinforced materials. One material contained Nicalon fibers designated NLP101 while the second material contained fibers designated MLN202. In the first material, the matrix was directly deposited on the fibers, while the fibers were coated with a 0.5 μm thick carbon coating in the second material.

Three-point flexural strength tests were conducted on each material, both as-received and after various heat treatments at temperatures of 800° and 1400°C in air for periods of 1, 10, 100, and 1000 h. Other tests were conducted on "material 1" for intermediate temperatures of 1000° and 1200°C and on both materials under vacuum at 800°C for 1 h. During loading, specimens were monitored acoustically to detect first damage.

At 800°C in air, strength decreased nearly linearly with heat treatment time, from values of about 300 MPa to values of less than 50 MPa after 1000 h. Along with the decrease in strength, there was a progression toward more brittle fracture. Figure 11 shows fracture surfaces for material tested as-received and for material tested after treatment for 1000 h at 800°C. After heat treatment at 1400°C in air, as-fabricated strength was nearly maintained with "material 1," but strength decreased appreciably after only a 1-h treatment with "material 2." Along with the strength decrease, fracture became brittle. The improved performance of "material 1" after heat treatment at 1400°C relative to that after heat treatment at 800°C was confirmed by the treatments at 1000° and 1200°C,

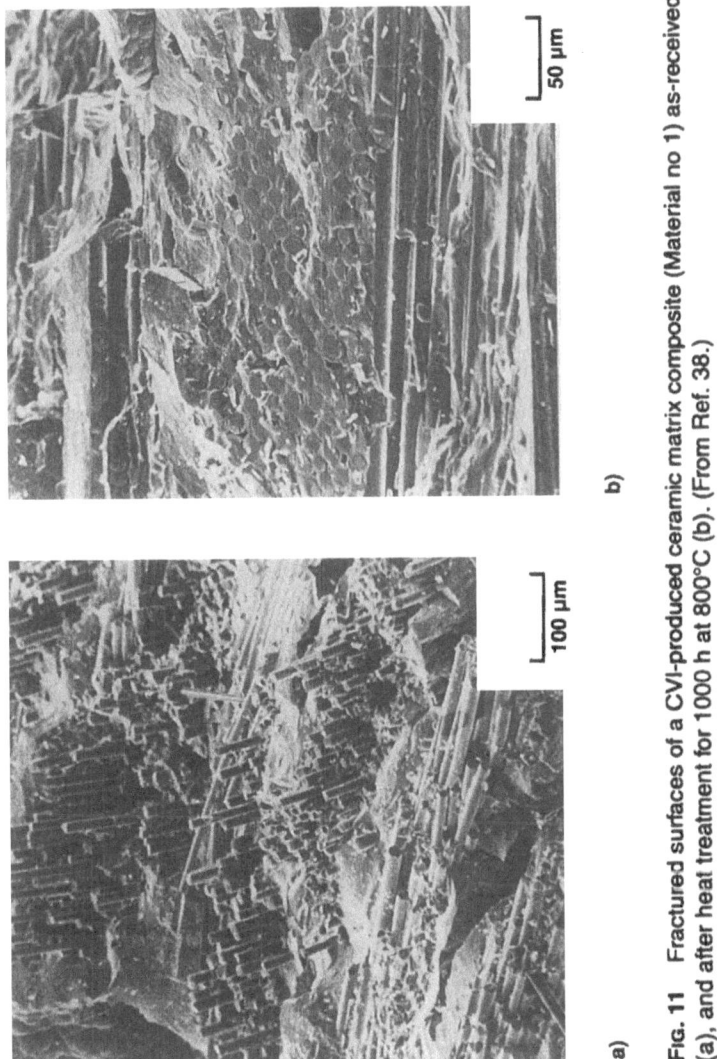

FIG. 11 Fractured surfaces of a CVI-produced ceramic matrix composite (Material no 1) as-received (a), and after heat treatment for 1000 h at 800°C (b). (From Ref. 38.)

which resulted in intermediate behavior. In vacuum, strength was maintained or enhanced after a 100 h heat treatment.

The 800°C treatment temperature effect on both materials was postulated to be due to strengthening of the interfacial bonds by oxidation. This premise was supported by the fact that strength was retained after heat treatment in vacuum. The degradation of "material 2" after heat treatment at higher temperatures for relatively short times was believed to be due to replacement of the carbon layer with a silica-rich layer that resulted in strong bonds between fibers and matrix, which was supported by scanning X-ray emission pictures. Improved performance for "material 1" with increasing heat treatment temperature must have been associated with coexistance of some bond weakening process along with the bond strengthening process.

The work cited above reemphasizes the importance of the interfacial properties on composite performance. The interfaces are subject to alteration by reactions at high temperatures, often by reactions with atmospheric oxygen. Composites produced by the same organization that apparently differed primarily by the presence or absence of a coating performed very differently after different heat treatment conditions. Whether or not fibers receive some surface treatment or are coated in the state-of-the-art SEP produced material is not reported. It is also not clear whether a relatively impervious silicon carbide overcoat is allowed to build up on the composite at the end of the CVI process, and, if so, whether mechanical test specimens have a similar coating or have been cut so that fiber containing surfaces are exposed. Undoubtedly, thermal treatment would have a different effect in each case.

Properties of silicon carbide matrix composites produced by isothermal CVI using optimized deposition conditions at other organizations are probably not appreciably different from those reported by Du Pont and SEP when similar fiber types, surface conditions, volume fractions, and architectures are used. In work reported by Yang et al [39] on composites produced at Amercom (Chatsworth, CA), two-dimensional and three-dimensional Nicalon fiber architectures were compared. The two-dimensional material was produced using plain weave fabric, with a stacking sequence of 0/30/60/90. For the braided material, the braiding axis was approximately ±20°. The fibers were coated with pyrolytic carbon in both cases.

The two-dimensional material contained 35 vol% fibers, had a density of 2.3 g/cm^3, and a porosity of 15%. The three-dimensional material contained 32 vol% fibers, had a density of 2.6 g/cm^3, and had a porosity of 11%. Optical micrographs showed that in the case of the two-dimensional material, large voids were still present between fabric layers and these voids could not be filled during infiltration. For the three-dimensional material, voids were significantly fewer and smaller, although some large voids were present where fiber tows crossed. The reduced void fraction of the three-dimensional material was postulated to be due to a more uniform pore network in the preform, assisting the diffusion of the reactant gas within the pore network.

Flexural strength (4-point) was measured on as-fabricated composites and on composites heat treated in air at 1200°C for 100 h. Fracture toughness was measured by the single edge notched beam method. The three-dimensional composite had an average flexural strength of about 690 MPa and a stress-strain curve indicating nonbrittle behavior. After the heat treatment at 1200°C in air for 100 h, average strength decreased to 350 MPa, but behavior was still nonbrittle. The two-dimensional composite had an average flexural strength of 330 MPa. After thermal treatment, strength decreased to 170 MPa. Stress-strain curves showed nonbrittle behavior in both cases, although load carrying ability dropped off faster after the maximum stress point than for the three-dimensional material.

Fracture toughness values for the three-dimensional material averaged about 30 MPa·m$^{1/2}$ or 18 MPa·m$^{1/2}$ depending on the direction of crack propagation. After thermal exposure, values decreased to 13 MPa·m$^{1/2}$ and 8 MPa·m$^{1/2}$. The two-dimensional material had values of 16 MPa·m$^{1/2}$ and 12 MPa·m$^{1/2}$, depending upon crack propagation direction, and these decreased to 8 MPa·m$^{1/2}$ and 7 MPa·m$^{1/2}$ after thermal treatment. The enhanced toughness of the three-dimensional composite was believed to be due to increased crack deflection. Micrographs showed that for the three-dimensional material, a crack propagated along the interface between fiber bundles and was then deflected near the bundle crossover due to the presence of matrix cracking and large voids.

Thus, there are several advantages of three-dimensional reinforcement. However, it is likely that tensile strength would have been lower for the three-dimensional composite than for the two-dimensional composite (when tensile stress is in the fabric plane) because of a reduced volume of fibers in the direction of tensile stress in the three-dimensional composite. Hence, optimum fiber architecture depends on the requirements of the application.

Rather extensive mechanical property evaluations have been made on the thermal gradient, forced flow ORNL process also. In general, properties for two-dimensional material are similar to those reported above for isothermal processing. A detailed study of properties as a function of specimen location relative to orientation in the thermal gradient, forced flow reactor was reported by Caputo et al.[40] Multiple test specimens were taken from four composite discs produced by the process. Average room temperature 4-point flexural strength values plus/minus one standard deviation were 288 ± 19, 312 ± 23, 314 ± 26, and 324 ± 17 MPa. Stress-strain curves indicated nonbrittle behavior (Figure 12). Although more inconsistent results, sometimes with strength values as low as 70 MPa and brittle fracture, were sometimes found during earlier work on the ORNL process, this was before carbon coatings were routinely applied to the fibers prior to CVI.

There was relatively little correlation between strength and sample position, although lower strengths were occasionally found for specimens cut from the bottom (cool side) of the discs, furthest from the center of the disc. Although a statistically significant direct relationship between strength and density was found, there was considerable scatter in the data, indicating that other factors affected strength also.

Specimens were tested at 1000°C in air also. Corresponding values for the four discs referred to above were 126 ± 51, 123 ± 14, 139 ± 18, and 173 ± 53 MPa. In contrast to the direct relationship (albeit weak) with density for room temperature flexural strength, elevated temperature strength showed no relationship with density. In fact, some of the highest elevated temperature strengths were for specimens cut from the bottoms of the discs, where density was lowest. The reduction in strength with temperature contrasts with the behavior reported by Lamicq et al [34] for similar material. However, it seems unlikely that this is related to which variation of the CVI process was used, but rather to interface factors.

Limited tensile testing was conducted. Average value was 217 ± 13 MPa, with a typical strain-to-failure value of 0.75%. Fracture toughness values were obtained for several specimens using the single edge notched beam method. Values of 8.6 MPa·m$^{1/2}$ and 10.3 MPa·m$^{1/2}$ were obtained with the load applied

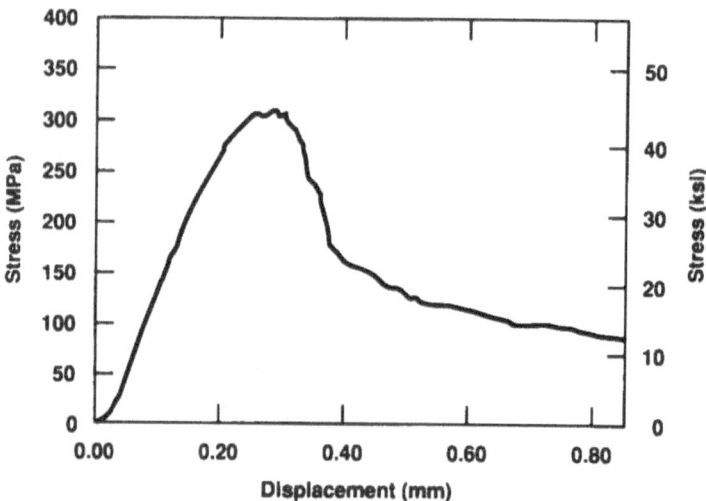

Fɪɢ. 12   Typical room temperature flexural stress-strain curve for a Nicalon™ fiber, silicon carbide matrix composite produced using the ORNL forced flow/thermal gradient process. (From Ref. 40.)

parallel to the cloth layers. Values of 6.7 MPa·m$^{1/2}$ and 8.0 MPa·m$^{1/2}$ were obtained when the load was perpendicular to the cloth layers. Although these results are somewhat lower than the best reported above using isothermal CVI, this could be due to a difference in the test details rather than in the materials.

B. Properties of CVI-Produced Composites Having Other Matrices

Mechanical properties of typical CVI-produced composites having matrices other than silicon carbide tend to be poorer than those for silicon carbide matrix composites. However, it must be recognized that most of the results are for materials in a very early state of development, so that processing conditions were undoubtedly not optimum.

Four-point flexural strength of a Nextel 480 fiber/silicon nitride matrix composite [22] determined at room temperature was about 75 MPa. Strength was relatively unaffected by a 50 h heat treatment at 800°C in air. Strength decreased to about 45 MPa after thermal treatment at 1000°C and to 15 MPa after treatment at 1300°C. Fracture proceeded to a more brittle mode as treatment temperature increased. Young's modulus as-fabricated and after thermal exposure at 800°C was about 140 GPa. Fracture toughness (single edge notched beam method) as-fabricated was less than 3 MPa·m$^{1/2}$.

For a Fiber FP, alumina matrix composite, flexural strength (3-point) was 210 MPa.[19] Strength retention at elevated temperatures was 85% for testing at 1000°C and 65% for testing at 1200°C. Strength for a zirconia matrix composite was 190 MPa.

In other work on alumina matrix composites [18], a maximum flexural strength of about 230 MPa was obtained with a uniaxial alignment of FP fibers (40 vol%). Strength decrease at 1000°C in argon was relatively small, but strength decreased appreciably at higher temperatures. Strength at 1400°C was about 50 MPa.

Compression test results of composites having a variety of matrices have been reported by Rossignol et al.[16] For all the materials, a stack of PAN-derived carbon fibers was consolidated by a small amount of carbon deposited from methane. Matrix material was then deposited by CVI as outlined earlier. In all cases, the materials had about 85% of theoretical density.

Boron nitride matrix material had a strength of 190 MPa and a Young's modulus of 37 GPa when compressive stress was applied parallel to the plane of the fiber cloths, and values of 220 MPa and 11 GPa with stress perpendicular to the fiber cloth layer planes. Corresponding values for titanium carbide matrix material were 150 MPa and 30 GPa with stress parallel to the cloth plane and 190 MPa and 8 GPa with stress perpendicular to the cloth layers. Values for boron nitride matrix material were 55 MPa and 21 GPa with stress parallel to the cloth planes and 140 MPa and 2.5 GPa with stress perpendicular to the cloth planes. Although the compressive strength values are very much lower than the value given earlier for Du Pont-produced silicon carbide matrix composites (800 MPa), this is undoubtedly due to the relative immaturity of the process when these composites were produced, rather than to inherently poor properties with other matrices. As support for this premise, compressive strength (with stress applied parallel to the cloth planes) for a silicon carbide matrix composite reported by Rossignol et al was only 180 MPa, which is close to the values for boron carbide and titanium carbide.

V. *IN SITU* CODEPOSITION OF MATRIX AND REINFORCING PHASES

Codeposition of several phases by chemical vapor deposition is well established. The resulting materials can be considered as composites, with one phase as the matrix and the other as the reinforcement (although this may be arbitrary in most cases). Quite often, the reinforcing phase is more or less spherical in nature, so that such a material could be considered as a particulate reinforced material, which is beyond the scope of this book. However, since it is possible for flake-like particles or even fibrous particles to form, this type of composite is within the purview of this book. Hence, this subject is included here, even though this type of composite is barely beyond the conjectural stage at this time.

A review of *in situ* composites produced by codeposition with the CVD process has been provided by Hirai and Goto.[41] Figure 13 indicates the types of structures possible using this process. Although over twenty references to codeposited materials were listed, the structures of only about half of these are known. Of the known structures, only one was identified as having a fibrous material as one of the phases. This material contains fibrous tin nitride in a matrix of silicon nitride (Figure 14), and is produced from tetramethylsilicon and ammonia.

The tin nitride phase was fibrous only under a specific set of fabrication conditions, which included a deposition temperature ranging from 1350° to 1450°C. Under these conditions, the matrix was β-silicon nitride and the tin nitride particles were a few tens of microns in diameter and about 2 μm in length. At lower temperatures, the silicon nitride was amorphous or found as the α-phase. The specific lattice relationships between β-silicon nitride and tin nitride were postulated to be the reason why the tin nitride was fibrous only when β-silicon nitride was deposited. Properties of the material were not given.

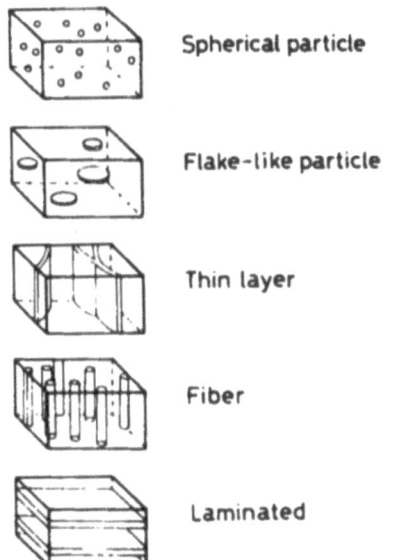

Spherical particle

Flake-like particle

Thin layer

Fiber

Laminated

FIG. 13 Schematic illustration of typical structures of CVD-produced *in situ* ceramic composites. (From Ref. 41.)

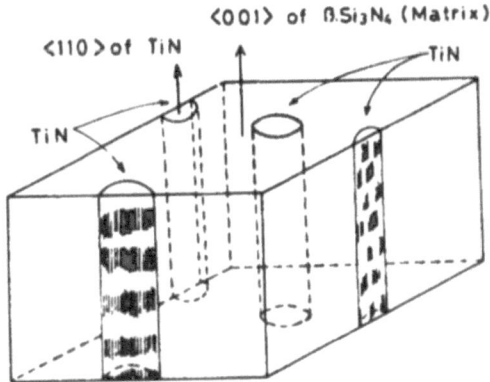

FIG. 14 Schematic microstructure of a tin nitride, β-silicon nitride CVD-produced *in situ* composite. (From Ref. 41.)

VI. SUMMARY

The chemical vapor deposition process has been successfullty adapted for depositing ceramic materials within the pores of a fibrous preform. This mode of conducting the process is termed chemical vapor infiltration (CVI). Compared with chemical vapor deposition of coatings, CVI generally requires both lower temperatures and lower pressures to cause the deposition reaction to be controlled by surface reaction kinetics rather than by diffusion. This enhances in-depth deposition relative to surface deposition. Although isothermal CVI has been most well developed, other versions involving temperature and/or pressure gradients have been used as well.

The isothermal process has the disadvantage of long processing times. In addition, restriction of infiltrating gases into the interior of the specimen due to surface deposition can necessitate one or more cycles of removal of the specimen from the reactor, surface machining, and reinfiltration. This can further add to processing time and expense. A forced flow/thermal gradient version of the process significantly reduces time, but restricts shape producing capabilities.

Large scale reactors and commercial scale structures have been produced using CVI. Major research and development efforts on the CVI process have been carried out at the University of Bordeaux and at Société Européene de Propulsion (SEP). The SEP technology has been licensed to Du Pont in the United States. Other industrial firms have also developed the process and can produce commercial scale components.

Silicon carbide deposited from methyltrichlorosilane/hydrogen has been by far the most well developed matrix material, although composites having a number of other matrices have been produced on an experimental basis by CVI. Reported mechanical properties of as-fabricated silicon carbide matrix composites are very good. Good properties at high temperatures or after exposure to high temperatures have been reported, although these properties are dependent upon maintaining the stability of the fiber/matrix interfaces and rapid deterioration of properties during thermal exposure can occur also. Mechanical properties of composites with matrices other than silicon carbide have generally been poorer than those with silicon carbide matrices, although this probably reflects the lack of maturity of these materials relative to silicon carbide.

Production of composites by codeposition of a matrix phase and a fibrous reinforcing phase by chemical vapor deposition is possible. However, there is limited experimental confirmation of fibrous phases codeposited in this way, so composite fabrication using this type of processing remains largely conjectural.

REFERENCES

1. Stinton, D. P., Besmann, T. M., and Lowden, R. A., Advanced ceramics by chemical vapor deposition techniques, *Ceramic Bulletin*, vol. 67, no. 2, 350–355 (1988).

2. Naslain, R., Ten years of research on fiber-reinforced ceramics, *Ind. Ceram.*, vol. 813, 98 (1987).

3. Buckley, J. D., Carbon-carbon, an overview, *Ceramic Bulletin*, vol. 67, no. 2, 364–368 (1988).

4. Naslain, R., Quenisset, J. M., Rossignol, J. Y., Hannache, H., Lamicq, P., Choury, J. J., Heraud, L., and Christin, F., An analysis of the properties of some ceramic-ceramic composite materials obtained by CVI-densification of 2D-C-C preforms. In *Proceedings ICCM-V*, W. C. Harrigan (Ed.), TMS, Warrendale, PA, 1985, 499–514.

5. Hopkins, G. R., and Chin, J., SiC matrix/SiC fiber composite: a high-heat flux, low activation, structural material, *J. Nucl. Mater.*, vol. 141–143, pt. A, 148–151 (1986).

6. Fitzer, E., and Gadow, R., Fiber-reinforced silicon carbide, *Ceramic Bulletin*, vol. 65, no. 2, 326–335 (1986).

7. Grateau, L., Lob, N., and Parlier, M., Microstructural studies of ceramic composites obtained by chemical vapor infiltration. In *Science of Ceramics*, D. Taylor (Ed.), Institute of Ceramics, Stokes-on-Trent, UK, 1987, 885–889.

8. Naslain, R., and Langlais, F., CVD-processing of ceramic-ceramic composite materials. In *Tailoring Multiphase and Composite Ceramics*, (Materials Science Research—Vol. 20), R. E. Tressler et al (Ed.), Plenum Press, New York, NY, 1986, 145–164.

9. Pan, Y. M., Sakai, M., Warren, J. W., and Bradt, R. C., Toughness anisotropy of a SiC/SiC laminar composite. In *Tailoring Multiphase and Composite Ceramics*, (Materials Science Research—Vol. 20), R. E. Tressler et al (Ed.), Plenum Press, New York, NY, 1986, 631–638.

10. Brennan, J. J., Interfacial studies of chemical-vapor-infiltrated ceramic matrix composites, *Materials Science and Engineering*, vol. A126, 203–223 (1990).

11. Burkland, C. V., and Yang, J.-M., Chemical vapor infiltration of fiber-reinforced SiC matrix composites, *SAMPE Journal*, vol. 25, no. 5, 29–32 (September/October 1989).

12. Lackey, W. J., and Caputo, A. J., Process for the preparation of fiber-reinforced ceramic composites by chemical vapor deposition, US Patent 4,580,524, April 8, 1986.

13. Moeller, H. H., Long, W. G., Caputo, A. J., and Lowden, R. A., SiC fiber reinforced SiC composites using chemical vapor deposition, *SAMPE Quarterly*, vol. 17, no. 3, 1–4 (April 1986).

14. Lowden, R. A., Caputo, A. J., Stinton, D. P., Besmann, T. M., and Morris, M. D., Processing and properties of SiC/Nicalon composites. CONF-9705103-11, December 1987, Oak Ridge National Laboratory, Oak Ridge,TN.

15. Sugiyama, K., and Ohzawa, Y., Pulse chemical vapour infiltration of SiC in porous carbon or SiC particulate preform using an r.f. heating system, *Journal of Materials Science*, vol. 23, 4511–4517 (1990).

16. Rossignol, J. Y., Quenisset, J. M., Hannache, H., Mallet, C., and Christin, F., Mechanical behaviour in compression loading of 2D-composite materials made of carbon fabrics and a ceramic matrix, *Journal of Materials Science*, vol. 22, 3240–3252 (1987).

17. Rossignol, J. Y., Quenisset, J. M., and Naslain, R., Mechanical behaviour of 2D-C-C/TiC composites made from a 2D-C-C preform densified with TiC by CVI, *Composites*, vol. 18, no. 2, 135–144 (1987).

18. Colmet, R., Naslain, R., Hagenmuller, R., and Lamicq, P., Method for producing a refractory composite structure, US Patent 4,576,836, March 18, 1986.

19. Colmet, R., Lhermitte-Sebire, I., and Naslain, R., Alumina fiber/alumina matrix composites prepared by a chemical vapor infiltration technique, *Advanced Ceramic Materials*, vol. 1, no. 2, 185–191 (1986).

20. Minet, J., Langlais, F., and Naslain, R., Chemical vapor infiltration of zirconia within the pore network of fibrous ceramic materials from $ZrCl_4$-H_2-CO_2 gas mixtures, *Composites Science and Technology*, vol. 37, 79–107 (1990).

21. Hoyt, J. T., and Yang, J.-M., Chemical vapor infiltration of silicon nitride matrix composites, *SAMPE Journal*, vol. 27, no. 2, 11–17 (1991).

22. Chung, G.-Y., and McCoy, B. J., Modeling of chemical vapor infiltration for ceramic composites reinforced with layered, woven fabrics, *J. Am. Ceram. Soc.*, vol. 74, no. 4, 746–751 (1991).

23. Rossignol, J. Y., Langlais, F., and Naslain, R., A tentative modelization of titanium carbide C.V.I. within the pore network of two-dimensional carbon-carbon composites, *Proc. Electrochem. Soc.*, vol. 84, no. 6, 596–614 (1984).

24. van den Breckel, C. H. J., Fonville, R. M. M., van der Straten, P. J. M., and Verspui, G., CVD of Ni, TiN, and TiC on complex shapes, *Proc. Electrochem. Soc.*, vol. 81, no. 2, 142–156 (1981).

25. Tai, N.-H., and Chou, T.-W., Theoretical analysis of chemical vapor infiltration in ceramic/ceramic composites. In *High Temperature—High Performance Composites*, Materials Research Society Synposium Proceedings, vol. 120, 1988, 185–192.

26. Currier, R. P., and Valone, S. M., Time-dependent solution to the Tai-Chou chemical vapor infiltration model, *J. Am. Ceram. Soc.*, vol. 73, no. 6, 1758–1759 (1990).

27. Currier, R. P., Overlap model for chemical vapor infiltration of fibrous yarns, *J. Am. Ceram. Soc.*, vol. 73, no. 8, 2274–2280 (1990).

28. Gupte, S. M., and Tsamopoulos, J. A., An effective medium approach for modeling chemical vapor infiltration of porous ceramic materials, *J. Electrochem. Soc.*, vol. 137, no. 5 (1990).

29. Middleman, S., The interaction of chemical kinetics and diffusion in the dynamics of chemical vapor infiltration, *J. Mater. Res.*, vol. 4, no. 6, 1515–1524 (November/ December 1989).

30. Middleman, S., Heble, B., and Cheng, H. C. T., Improved uniformity of densification of ceramic composites through control of the initial preform porosity distribution, *J. Mater. Res.*, vol. 5, no. 7, 1544–1548 (July 1989).

31. Starr, T. L., Model for rapid CVI of ceramic composites. Paper presented at the 12th Conference on Composites and Advanced Ceramics, Cocoa Beach, FL, January 17–20, 1988.

32. Gupta, D., and Evans, J. W., A mathematical model for chemical vapor infiltration with microwave heating and external cooling, *J. Mater. Res.*, vol. 6, no. 4, 810–818 (April 1991).

33. Du Pont CVI Ceramic Matrix Composites, Preliminary Engineering Data, Du Pont Composites, Newark, DE.

34. Lamicq, P. J., Bernhart, G. A., Dauchier, M. M., and Mace, J. G., SiC/SiC composite ceramics, *Ceramic Bulletin*, vol. 65, no. 2, 336–338 (1986).

35. Jouin, J. M., Heraud, L., and Malassine, B., Mechanical behavior of non-brittle ceramic C-SiC and SiC-SiC composites. In *Advances in Fracture Research*, Proceedings of the Seventh International Conference on Fracture (ICF7), vol. 4, Pergamon Press, Oxford, UK, 1989, 2437–2445.

36. Bouquet, M., Birbis, J. M., and Quenisset, J. M., Toughness assessment of ceramic matrix composites, *Composites Science and Technology*, vol. 37, 223–248 (1990).

37. Gomina, M., Chermant, J. L., Osterstock, F., Bernhart, G., and Mace, J., Applicability of fracture mechanics to fiber-reinforced CVD-ceramic composites. In *Frac. Mech. Ceram.*, Proceedings of the Fourth International Symposium, vol. 7, Plenum Press, New York, NY, 1986, 53–60.

38. Frety, N., and Boussuge, M., Relationship between high-temperature development of fibre-matrix interfaces and the mechanical behaviour of SiC-SiC composites, *Composites Science and Technology*, vol. 37, 177–189 (1990).

39. Yang, J.-M., Lin, W., Shih, C. J., Kai, W., Jeng, S. M., and Burkland, C. V., Mechanical behaviour of chemical vapour infiltration-processed two- and three-dimensional Nicalon/SiC composites, *Journal of Materials Science*, vol. 26, 2954–2960 (1991).

40. Caputo, A. J., Stinton, D. P., Lowden, R. A., and Besmann, T. M., Fiber-reinforced SiC composites with improved mechanical properties, *Ceramic Bulletin*, vol. 66, no. 2, 368–372 (1987).

41. Harai, T., and Goto, T., CVD fabrication of in-situ composites of non-oxide ceramics. In *Tailoring Multiphase and Composite Ceramics*, (Materials Science Research—Vol. 20), R. E. Tressler et al (Ed.), Plenum Press, New York, NY, 1986, 549–560.

15

Melt Processing of Ceramic Matrix Composites

I. BACKGROUND

Despite extensive work on melt infiltration of fibrous preforms for fabricating metal matrix composites, analogous work on ceramic matrix composites has been limited for several reasons. Nevertheless, ceramic melt infiltration might have promise for some systems.

A totally different use of a melt in the production of a ceramic matrix composite is growth of an oxidation product of a molten metal alloy into a particulate or fibrous preform placed atop the melt. This type of process has been extensively developed by the Lanxide Corporation. This method is commonly called the Lanxide process, although the term Dimox™ (for "directed metal oxidation") process is sometimes used by the Lanxide Corporation to distinguish it from other processes of the company.

The Dimox process was conceived by M. S. Newkirk, who founded the Lanxide Corporation (Newark, DE) in 1983 to develop and commercialize the process.[1] In the mid-1980s, the company also formed joint ventures with Du Pont (Du Pont Lanxide Composites Inc. and Lanxide Armor Products Inc.) and one with Alcan Aluminum Co. (Alanx Aluminum Co.).

II. COMPOSITE FORMATION USING CERAMIC MELTS

A number of considerations limit the usefulness of production of ceramic matrix composites by impregnation of preforms by molten ceramics, followed by solidification of the melt. Melting points of many ceramic materials of promise for high temperature applications are higher than the temperatures that the available ceramic reinforcing fibers can withstand without severe degradation. Other ceramic materials of interest decompose rather than melt at high temperatures. Also, ceramic melts have much higher viscosities than metal melts, making infiltration of preforms more difficult. Finally, shrinkage can occur during solidification of ceramic melts, leading to cracking problems.

Most of the recent work on ceramic melt infiltration has been conducted by W. B. Hillig and coworkers at the GE Research and Development Center. An example [2] involved the use of SCS-6 fibers and a strontium aluminosilicate ($SrAl_2Si_2O_8$) melt in which silicon carbide particles and whiskers were dispersed. Good infiltration was obtained. Nonbrittle fracture was observed (Figure 1) during flexural testing at both room temperature and at 1450°C. However, it was found that fiber strength was degraded during processing and that batch-to-batch variation was large.

III. THE DIMOX™ PROCESS

A. Process Description

As one example of the process, a ceramic material, aluminum oxide, is produced by the oxidation of a molten aluminum alloy. The oxide increases in thickness as molten metal is wicked through the structure. After attaining a desired thickness (and shape, if a retainer is used to limit growth in different directions) the oxide is removed from the molten metal pool. Since it still contains metal stringers, it is a kind of ceramic/metal composite. Although this is not a composite of the type of interest in this book, the placement of ceramic fibers, whiskers, or particles atop the melt surface prior to oxidation results in a reinforced ceramic/metal matrix material. If the residual metal is subsequently converted to additional oxide, a true ceramic/ceramic composite is produced.

Figure 2 [3] schematically represents the Dimox process without any reinforcing phase, thus resulting in a ceramic/metal composite. Figure 3 is a schematic representation of the process when reinforcing material is placed on the surface of the metal melt, which results in a ceramic reinforced ceramic/metal.

A typical feature of the Dimox process is the alloying with one or more "dopants" in the parent metal to stimulate oxidation.[4] Concentrations range from less than 1% to several percent by weight. Also, it is often found that the desired oxidation properties occur only within a limited temperature range. In general, these temperatures are low compared to temperatures required to produce ceramic bodies by sintering. Once the appropriate conditions have been selected, the process generally grows a composite of uniform microstructure as long as the molten metal and oxidant are available and the temperature is main-

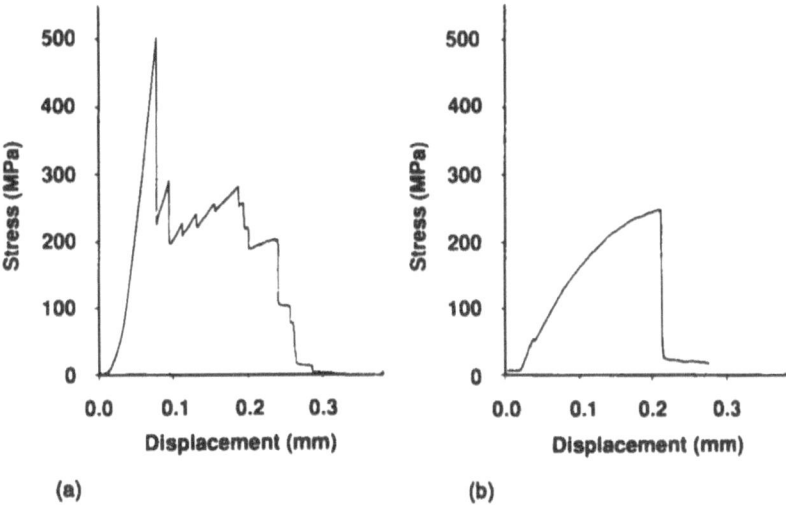

FIG. 1 Flexural stress-strain curves at room temperature (a), and at 1450°C (b), for SCS-6 fiber, strontium aluminosilicate matrix composites produced by melt infiltration. (From Ref. 2.)

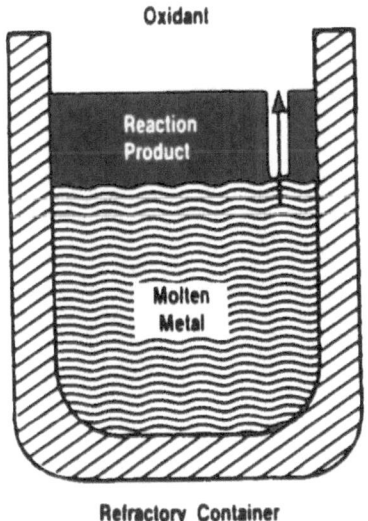

FIG. 2 Schematic representation of formation of an oxide/metal material with oriented oxide growth from the surface of a molten metal upon exposure to air. (From Ref. 3.)

tained. Composites over 10 cm thick have been grown with no indication of a decrease in rate. It appears that one-piece ceramic bodies of virtually any size can be formed.

The volume fractions of metal and "grown" ceramic, the microstructure, the porosity, and the degree of interconnectivity of the metal phase all depend upon processing conditions, such as temperature, time, and the dopants utilized. The ceramic phase is generally interconnected in all three directions. If the oxide is removed prior to depletion of the metal pool, a dense ceramic/metal material results. If the pool is depleted, the product is more porous.

In a typical series of experiments used to develop the process, melts of aluminum containing additions of magnesium, silicon, and other metals were used. Oxidation was carried out at various temperatures, with a typical exposure cycle involving about 5 h heating to maximum temperature, 24 h at the maximum temperature, 14 h cooling to 600°C, and rapid cooling to ambient temperature. Various experiments demonstrated that in the aluminum-magnesium-silicon system, the presence of both magnesium and silicon were required for accelerated oxide growth. In addition, growth was largely confined to temperatures in the range 1100° to 1400°C with these dopants. Germanium, tin, and

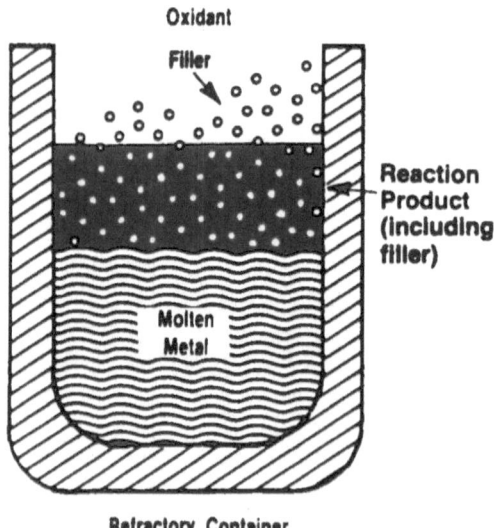

FiG. 3 Schematic representation of formation of a reinforced oxide/metal material by placing reinforcing material above the molten metal prior to oxidation. (From Ref. 3.)

lead were also found to be effective in combination with magnesium as dopants to promote accelerated oxidation.

Visual examination of specimens formed at 1125°C indicated that oxide growth occurred on the exposed surface of the metal. The material was of a columnar nature. X-ray diffraction indicated that corundum was the form of alumina produced and that there was a thin layer of magnesium aluminate ($MgAl_2O_4$) adjacent to the original oxide/metal interface. The columns contained fully interconnected networks of alumina and aluminum. Typically, they were several millimeters wide and had indistinct boundaries. Through each column, the oxide phase had nearly the same crystallographic orientation. Material grown by continuing heating after depletion of the molten metal pool resulted in removal of molten aluminum from the microstructure to form additional oxide at the external surface. The result was a porous alumina structure with interconnected voids replacing nearly all of the metal.

Material grown at 1325°C also had a columnar microstructure, but the columns were narrower and the boundaries between them defined a hexagonal cell structure. Relative to material grown at the lower temperature, this material had a lower metal content, less interconnectivity of the metal phase, a finer alumina grain structure, and a greater interconnectivity of the oxide.

Many other systems and processing parameters, have been investigated since the conception of the Dimox process. Processing temperatures as low as 900°C have been investigated for aluminum base melts. Processes for metals other than aluminum and oxidants other than the oxygen in air have been developed. Other metals include titanium and zirconium, and nitrogen has been used as the oxidant (resulting in nitride formation). In addition to the basic process development, composites containing fibers, whiskers, and particles have been developed. In all, by 1989 the company had filed about 1500 patent applications worldwide and had already been granted many patents.[1] A subsidiary company, Lanxide Technology Company, has been formed specifically for patent processing.

A number of variations on the basic process outlined above can be used. Many of these variations have been outlined.[5] Rather than alloying the oxidation-enhancing dopants into the molten metal, a dopant can be applied to the surface of the metal. In addition to the basic idea of growing the ceramic material within the pores of a filler material, a structure of a desired shape can be provided by using a cover having the inverse geometry of the desired structure, but which is permeable to the oxidant gas (Figure 4).

Other modifications include variation of the partial pressure of the oxidizing gas by either increasing or decreasing the ambient pressure or by dilution with an inert gas. Wetting can be enhanced by pretreatment of the filler material to remove oxygen and/or occluded moisture. If desired to optimize properties that are enhanced by the presence of the metal phase in the product, alloying additives favorable for the properties can be added to the metal and, in addition, the infiltration and oxidation rates can be controlled so that the oxidation of the parent metal is only partially complete. In general, the metal content of the product can range from 1 vol% to greater than 40 vol%.

B. Mechanical Properties of Fiber Reinforced Dimox™ Composites

Properties of Nicalon fiber reinforced aluminum oxide and aluminum nitride produced by the Dimox process has been reported by Fareed et al.[6] Twelve-harness satin weave cloths were stacked in a 0/90 orientation. An undisclosed fiber coating was applied using CVD. The coating was reported to provide protection from the aluminum alloy during matrix growth, as well as to provide a relatively weak fiber/matrix interface. Growth of matrix material was carried out either in air to form aluminum oxide or in nitrogen to form aluminum nitride. An aluminum metal removal process was used to reduce the residual aluminum content from 6–7 wt% to 1–2 wt%. Fiber volume fraction was about 35%.

Four-point flexural tests and fracture toughness (Chevron notch) tests were conducted. Aluminum oxide specimens were all cut from a single 15-cm × 10-cm plate. Specimens cut from various areas of the plate were reasonably uniform. Average strength of 18 specimens was 527 ± 52 MPa. Strength of 31 comparable specimens which had the residual metal content lowered averaged 477 ± 81 MPa. Therefore, reduction of the residual metal phase had only

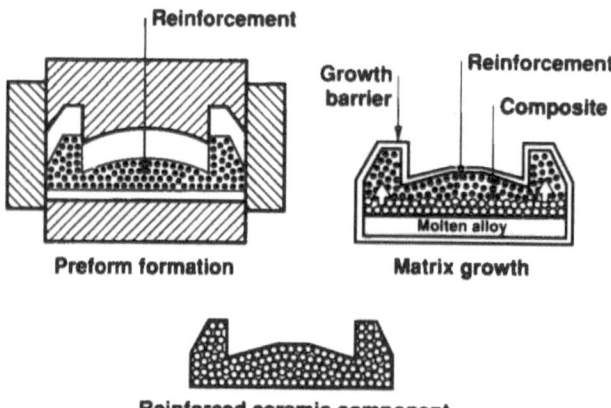

Fig. 4 Schematic illustration of formation of a net shape composite by the Dimox™ process by use of a gas permeable cover having the inverse geometry of the desired structure. (From Ref. 1.)

a small effect on flexural strength. When tested at 1000°C, strength of seven specimens averaged 398 ± 23 MPa. Strength of 14 specimens tested at 1200°C averaged 351 ± 41 MPa. A typical stress-strain curve is shown in Figure 5. As with other composite systems discussed herein, the deviation from linearity coincided with the onset of matrix microcracking, and the residual strength represented fiber pullout.

Fracture toughness for four as-grown specimens averaged 19 ± 1 MPa·m$^{1/2}$. Toughness was not significantly affected by a reduction in the residual metal phase. Average toughness after residual metal reduction was 21 ± 1 MPa·m$^{1/2}$ for four specimens. At 1000°C, toughness of five specimens averaged 23 ± 3 MPa·m$^{1/2}$ and at 1200°C, toughness of five specimens averaged 18 ± 1 MPa·m$^{1/2}$.

Thermal shock resistance was determined also. Specimens having the residual metal phase removed were subjected to elevated temperatures for 10 min, followed by quenching into water at room temperature and determination of flexural strength. For specimens similar to those that had an average flexural strength of 385 MPa prior to the thermal shock procedure, strength averaged 357 ± 11 MPa after one cycle from 1000°C and 321 ± 7 MPa after five cycles from 1000°C. Strength was 322 ± 24 MPa after one cycle from 1200°C.

Interlaminar shear strength was also measured using the short-beam flexural method. Strength as-grown for five specimens averaged 83 ± 17 MPa. Strength of five specimens in which the metal phase had been reduced averaged 62 ± 6 MPa. Flexural strengths following thermal exposure in air at 1200°C

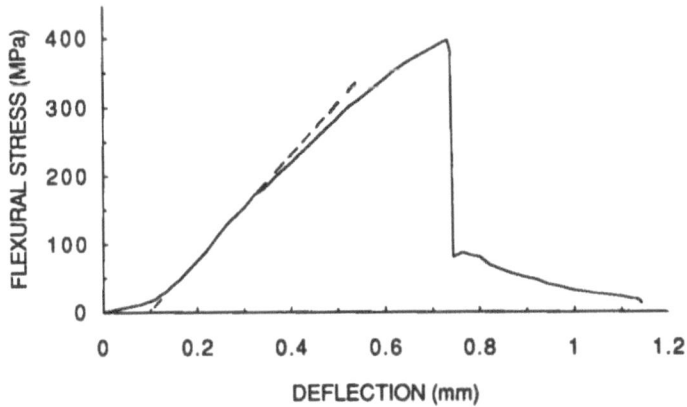

FIG. 5 Flexural stress-strain curve at 1200°C for a unidirectional Nicalon™ fiber, alumina matrix composite (metal removed) produced by the Dimox™ process. (From Ref. 6.)

for 24 h were also determined. Average strength of four specimens having the residual metal content removed prior to treatment was 310 ± 46 MPa (compared to the original strength of 477 MPa). Strength reduction was attributed to degradation of the Nicalon fibers.

Properties of alumina oxide matrix composites having a uniaxial Nicalon fiber alignment were also reported.[7] This material had a fiber content of 15-20 vol%, and this was compared with material containing about 35 vol% 12-harness satin weave. Presumably, the latter material was similar to that reported on above. Average flexural strength for the uniaxial material was 720 ± 150 MPa, compared with 540 ± 50 MPa for the two-dimensional material. Average fracture toughness values were 27 ± 3 MPa·m$^{1/2}$ and 15 ± 1 MPa·m$^{1/2}$, respectively, for the unidirectional and two-dimensional composites.

Aluminum nitride matrix composites produced by the Dimox process are in an earlier stage of development than alumina matrix composites.[6] Nonetheless, mechanical properties are good. Room temperature flexural strength for four as-grown specimens averaged 437 ± 14 MPa. Strength for three specimens at 1000°C averaged 343 ± 12 MPa. Stress-strain curves indicated nonbrittle fracture. Fracture toughness for as-grown specimens (three tested at each temperature) averaged 20 ± 1 MPa·m$^{1/2}$ at room temperature and 14 ± 1 MPa·m$^{1/2}$ at 1000°C.

IV. SUMMARY

Production of ceramic matrix composites by infiltration of a fiber or whisker preform with a molten ceramic, followed by solidification, has been limited by the generally high melting points for matrix materials of most interest, relative to maximum temperatures for fiber and whisker stabilities, and by other factors. Nonetheless, some progress in this area has been reported and this method might find some limited practical applications.

The Dimox™ process developed at the Lanxide Corporation for producing ceramic materials by growth of an oxidation product over a molten metal pool can be used for producing whisker or fiber reinforced ceramics if the reinforcing material is placed atop the molten metal pool. The Dimox process is among the most developed processes for forming ceramic matrix composites. In addition to the parent company, several large joint venture companies have been formed for pursuing specific applications for the technology.

Aluminum, doped with materials that enhance oxidation, has been the most commonly used metal and alumina, formed by reaction with the oxygen in air, has been the usual matrix material formed. As with many other composites described, Nicalon has been the most common reinforcing fiber. However, other metals have been investigated and nitrogen has been used as the oxidizing gas, with nitride formation. Mechanical properties of composites fabricated by this process compare well with those of composites produced by other well-developed methods.

Advantages of the Dimox process include the fact that shrinkage does not occur during the oxide growth process and that relatively inexpensive material is used to form the matrix. By using contoured molds that restrain oxide growth but permit permeation of the oxidizing gas, net shape processing of relatively complex structures can be carried out. Potential limitations of the process include the residual metal content, which might limit high temperature performance. However, it should be added that the residual metal content can be relatively low if appropriate processing conditions are used and/or if a post-fabrication removal process is carried out. It would appear also that some of the other processes described would be superior for very complex, thin walled structures.

REFERENCES

1. Maloney, L. D., Make way for 'engineered ceramics,' *Design News*, vol. 45, no. 5, 64–71 (March 13, 1989).
2. Brun, M. K., Hillig, W. B., and McGuigan, H. C., High temperature mechanical properties of a continuous fiber-reinforced composite made by melt infiltration, *Ceram. Eng. Sci. Proc.*, vol. 10, no. 7–8, 611–621 (1989).
3. Lasday, S. B., Unique approach to manufacture of ceramic composite components, *Industrial Heating*, vol. 55, no. 4, 14–15 (April 1988).
4. Newkirk, M. S., Urquhart, A. W., Zwicker, H. R., and Breval, E., Formation of Lanxide™ ceramic composite materials, *J. Mater. Res.*, vol. 1, no. 1, 81–88 (1986).
5. Claar, T. D., Poste, S. D., Gesing, A. J., Sobczyk, M., Raghavan, N. S., Creber, D. K., and Nagelberg, A. S., Production of ceramic and ceramic-metal composite articles incorporating filler materials, US Patent 4,824,625, April 25, 1989.
6. Fareed, A. S., Sonuparlak, B., Lee, C. T., Fortini, A. J., and Shiroky, G. H., Mechanical properties of 2-D Nicalon™ fiber-reinforced LANXIDE™ aluminum oxide and aluminum nitride matrix composites, *Ceram. Eng. Sci. Proc.*, vol. 11, no. 7–8, 782–794 (1990).
7. Andersson, C. A., Barron-Antolin, P., Schiroky, G. H., and Fareed, A. S., Properties of fiber-reinforced Lanxide™ alumina matrix composites. In *Proceedings of the International Conference on Whisker- and Fiber-Toughened Ceramics*, ASM, Metals Park, OH, 1988, 209–215.

16

Machining, Joining, Nondestructive Evaluation, and Design

I. BACKGROUND

Although a number of promising fiber reinforced ceramics have been produced and characterized by mechanical testing, several other considerations must be taken into account if these materials are to be used in practical applications.

One major consideration is machining of ceramic matrix composite materials. The term machining is used here to include production of flat faces, cutting of holes, and other types of material removal operations. Flat-face machining is required for many purposes including preparation of surfaces for joining. Holes can be required for mechanical joining as well as for other purposes such as cooling channels for high temperature aerospace applications. Other types of cuts are required for specific applications.

Another major consideration is joining. In general, a ceramic matrix composite will need to be joined to other substructures of the same material or to other materials such as metals. In some cases, such as use of whisker reinforced material as a cutting tool bit, a simple mechanical joint identical to that used for monolithic ceramics can be utilized. However, for many of the applications envisioned for fiber or whisker reinforced ceramics, more complicated joining systems will need to be developed.

Another requirement for acceptance of reinforced ceramics in many applications is nondestructive testing. Although average mechanical properties and reliability indicators such as the Weibull modulus can be used to design composite structures with an acceptably low failure probability, assurance that major flaws do not exist in a specific structure is desirable also. Nondestructive evaluation of monolithic ceramics is still a rapidly developing area, and nondestructive testing of ceramic composites is even less well developed.

As with any new material, structural designs that incorporate the desirable features of the material and minimize the effects of less desirable properties are advantageous. Designing of structural components with ceramic matrix composites presents such challenges. Documented examples of designing with ceramic matrix composites are as yet quite limited.

II. MACHINING OF CERAMIC MATRIX COMPOSITES

Conventional machining of high performance ceramic materials is relatively well developed and can be adapted for ceramic matrix composites. Because of the hardness of many of these materials, diamond is usually required as the abrading material. Machining with diamond and other "superabrasives" will not be covered here, except in comparison with other machining methods. As material hardness is increased, conventional cutting becomes increasingly more time consuming and expensive. Consequently, there has been increasing emphasis on newer machining techniques. Some of these newer techniques will be described here.

Perhaps the most promising method for machining ceramic matrix composites is laser machining. Ridealgh et al [1] have provided some fundamental information on laser cutting of a glass-ceramic matrix composite and compared laser cutting with conventional cutting using a diamond abrasive. They used a material containing about 50 vol% of uniaxially aligned Nicalon fibers. A 2 kW continuous wave carbon dioxide laser, along with a moving table, were used. Beam diameters were 1–3 mm. Power and table speed were varied to give a range of conditions. Power level was varied from 0.9 to 1.6 kW and speed from 22 to 505 mm/s. Cuts both parallel and perpendicular to the fiber orientation direction were made. After laser processing, cut widths were microscopically determined.

Conventional cutting (sawing) was carried out using two speeds, 250 and 4000 revolutions per minute, with a diamond abrasive wheel. Wheels having thicknesses of 0.3 and 1 mm were used. Topographies of laser cut surfaces were compared with those of diamond sawn surfaces using scanning electron microscopy.

The laser cutting did not cause any microcracking, so that there was no difficulty in measuring the dimensions of the cuts. With 1 kW power and a 1 mm beam diameter, full section (9 mm) cut depths were attained, either perpendicular or parallel to the fiber alignment direction, at table speeds less than about 40 mm/s. Cut depth decreased to less than 0.1 mm at a speed of 505 mm/s. Cut width was decreased from about 0.15 mm to 0.05 mm over the same range in table speed. In contrast, a full depth cut was not attained under any of the conditions used with a 3 mm beam diameter.

Use of a diamond abrasive wheel resulted in cutting of the composite by continuous multifracturing of the constituents, causing considerable damage to the glass-ceramic matrix at the cut face. At the lower sawing speed, damage varied along the length of the cut. Damage was greater, but more uniform, at the higher sawing speed. Local multiple fracture of the matrix and debonding at the fiber/matrix interface resulted in removal of matrix, leaving protruding fibers. Similar features were observed, to a lesser extent, in saw cuts made parallel to the fibers. Figure 1 shows typical surfaces cut perpendicular to the fiber direction using the lower sawing speed.

Fibers also protruded from laser cut surfaces, but to a lesser extent than the diamond sawn surface cut at the higher speed. No debonding of fibers was observed with laser cutting. A glassy appearance of the matrix indicates that laser cutting involves melting and some resolidification (Figure 2). The protrusion of the fibers might be attributable to the higher melting point and thermal conductivity of silicon carbide relative to the matrix material. However, the rounded ends of the fibers, in contrast to the jagged ends with diamond cutting, indicated that some melting of fibers had occurred. Hence, it was concluded that the main mechanism of separation of fibers was melting, and not fracture under stress as the matrix melted.

In general, the work indicated that laser cutting can be an economical process for ceramic matrix composites and that the cut surface suffers less damage than with conventional cutting using a diamond abrasive wheel.

Other newer techniques for machining advanced ceramic materials can be adapted for ceramic matrix composites, although much development work needs to be done. Machining with high pressure water jets is a well known technique that is beginning to be applied to ceramics.[2] Waterjet machining makes use of the common phenomenon of erosion due to water streams. Although the natural process is very slow, compression of water under pressures as high as 370 MPa and use of a nozzle having a diameter of about 10 to 20 μm results in water velocities up to three times the velocity of sound, and a great enhancement of cutting speed.

The cutting action of a high-pressure waterjet involves compressive shearing. Although compressive shearing of many advanced ceramic materials is made difficult because of the high hardness of these materials, the waterjet process can be made more practical by addition of solid particles to the water stream. Suspended particles in the water have the same velocity as the water and create a higher level of kinetic energy due to their higher mass. The particles also result in a very effective pulsed erosion process due to the mass difference

FIG. 1   Scanning electron micrographs showing variable damage in a glass matrix composite cut perpendicular to the fibers using a diamond saw (250 rev/min). (From Ref. 1.)

between the particles and water. Since the water pressurizing system requires very pure water for minimization of corrosion and erosion of the system, the particles are introduced at the nozzle exit. Figure 3 schematically shows nozzle configurations with and without abrasive particle additions. The waterjet cutting process is generally noisy because of the supersonic velocities attained. Equipment enclosure can provide both the necessary soundproofing and operator protection in the event of any failure in the cutting equipment.

Another technique suitable for machining of advanced ceramics, including ceramic matrix composites, is electric discharge machining, or EDM.[3] The workpiece and the shaping tool are the electrodes in EDM, and a liquid dielectric separates them so that there is no direct physical contact and, consequently, no mechanical stress on the workpiece. Machining advanced ceramics into intricate shapes, with close tolerances and a mirror finish, is possible. A significant limitation of the process for ceramic materials is that the electrical resistivity must be below about 100 $\Omega \cdot$cm.

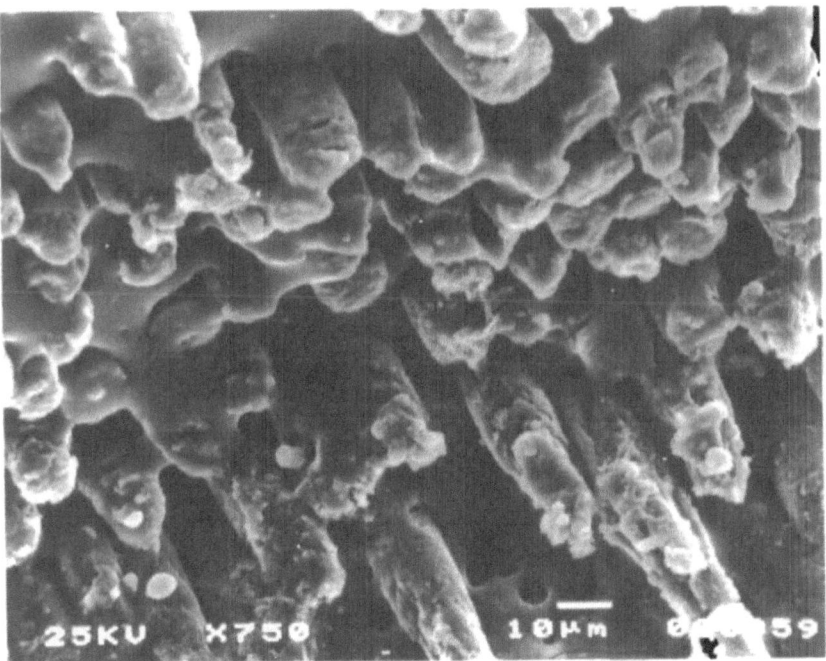

Fɪɢ. 2   Scanning electron micrograph showing the surface appearance of a cut made perpendicular to the fibers in a glass matrix composite using a laser. (1-mm diameter beam, 1.0 kW power, 252 mm/sec table speed.) (From Ref. 1.)

The EDM process is based on erosion. The dielectric fluid contains ultrafine solid particles. Applying a pulsed DC voltage of about 200 V across a gap, which is typically about 40 μm, generates a magnetic field between the anode and cathode (workpiece). Initially, the two pieces are insulated by the dielectric, so that no current flows. However, the electric field causes the particles to be suspended and bridge the gap, resulting in a breakdown of the dielectric. The voltage decreases, the current rises, and a vapor bubble forms around the channel. The input energy concentrates into a very small volume, creating a plasma and increasing temperature to tens of thousands of degrees Celsius and pressure to up to 300 MPa. Material eroded under these conditions is carried away by the dielectric fluid, which also quenches the hot spots.

There are two types of EDM machines. The die-sinking method, also called the ram-type or vertical sinking machine, can be used for tapping, cutting holes, and helical machining. A schematic illustration of this type of machining is given in Figure 4. In this type of EDM, the workpiece is the cathode and the shaping tool is the anode. Die materials are metals or graphite. Heavy hydrocarbons and kerosene are the most common dielectrics. Wire EDM uses a thin metal wire under tension to cut like a jigsaw. Wire erosion and breakage are the most common problems with this type of EDM. Water is usually used as the dielectric in wire EDM.

Although determination of the details of the EDM process is difficult because of the complexity of the process and the difficulty in obtaining accurate experimental data, the basics of the process have been determined. Three mechanisms for ceramic erosion have been identified: melting, evaporation, and thermal spalling. With low melting point ceramics, the mechanism is melting or

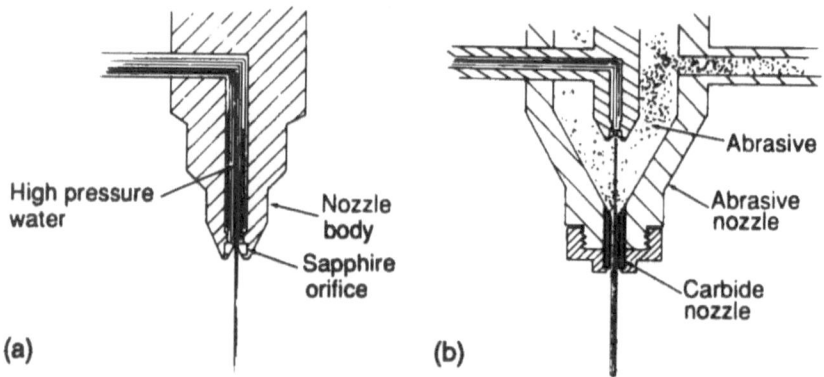

FIG. 3 Schematic illustration of waterjet cutting nozzles with pure water (a), and with incorporation of abrasive particles in the jet (b). (From Ref. 2.)

evaporation. With high melting point (<2500°C) materials having high thermal expansion and low thermal conductivity, spalling can occur.

The last method of machining to be described here is ultrasonic machining.[4] Ultrasonic machining is nonthermal, nonchemical, and nonelectrical. Materials that have been machined by this process include alumina, silicon carbide, silicon nitride, and borosilicate glass, all of which have been used as matrices in ceramic matrix composites. The process begins with the conversion of a high-frequency electrical signal into an oscillatory mechanical motion, which is acoustically transferred through a metal toolholder and tool assembly. Typical linear oscillation frequency is 20,000/s. When used with an abrasive slurry flowing around the cutting tool, the workpiece is microscopically ground. The metal toolholder and tool must be designed to transmit the acoustic energy properly and to resonate within the bandwidth of the transducer used. The maximum stroke usually required for grinding operations is 0.064 mm. Figure 5 is a schematic representation of an ultrasonic transducer and toolholder assembly.

The abrasive is a key aspect of ultrasonic machining. Boron carbide, silicon carbide, and alumina are typical abrasives. The abrasive is suspended in water, typically at a concentration of 20 to 50%. The slurry is circulated at a high rate to cool the tool and workpiece, supply fresh abrasive to the cutting location, and remove abraded particles. Abrasive size depends upon the desired speed of cut and surface finish.

It would seem likely that all of these methods, and perhaps others, will have a place in ceramic matrix composite technology, although documentation with regard to machining most of the composite systems described in this book is not yet available.

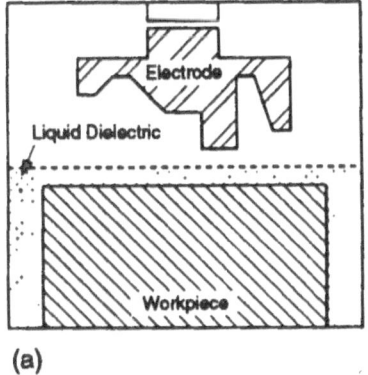
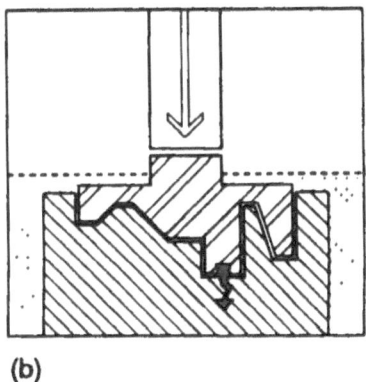

(a)   (b)

FIG. 4 Schematic drawing of the die-sinking method of electrical discharge machining, prior to machining (a), and after machining (b). (From Ref. 3.)

III. JOINING OF CERAMIC MATRIX COMPOSITES

For most applications of ceramic matrix composites, composite-to-composite or composite-to-metal joints will be required. Methods for bonding only matrix material (without any consideration of the fiber or whisker interactions) are entirely analogous to those used for bonding monolithic ceramics. Cawley [5] has reviewed bonding techniques for ceramic composites, emphasizing the fact that joining of ceramics can, in most cases, be viewed as localized variations of

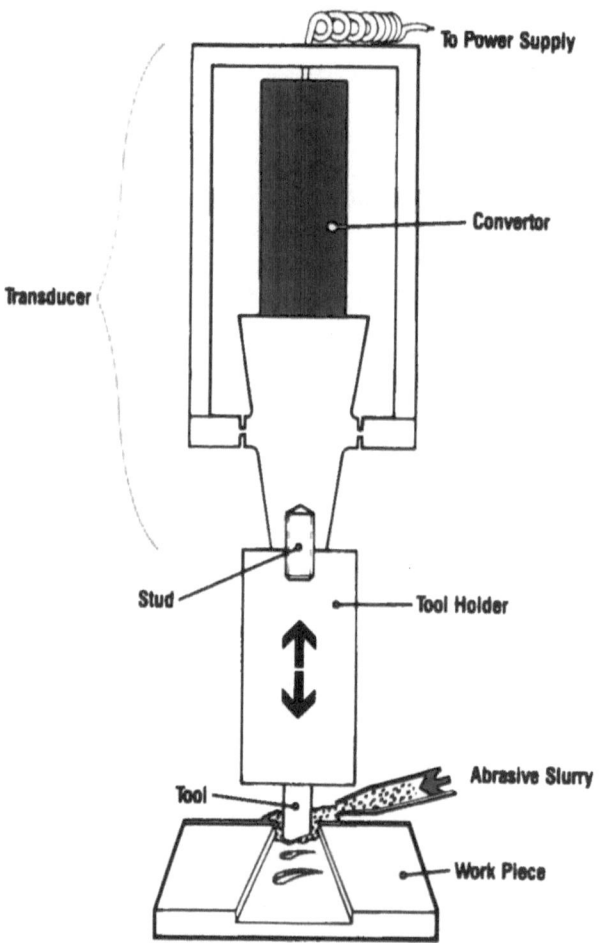

To Power Supply

Convertor

Transducer

Stud

Tool Holder

Tool

Abrasive Slurry

Work Piece

Fɪɢ. 5 Schematic representation of an ultrasonic transducer and toolholder assembly. (From Ref. 4.)

ceramic forming processes. The processes listed include metal brazing, silicate brazing (solder glasses), diffusion bonding (a variation of the sintering process for forming ceramics), cementing, and fusion welding (a variation of fusion casting). Resin bonding (adhesive bonding) can also be used, but this method will not be discussed here since the majority of potential applications for ceramic matrix composites involve temperatures above those tolerable by organic resins.

The general methods described by Cawley should be applicable for ceramic matrix composite joining, but have mostly been used or investigated for monolithic ceramic joining. Metal brazing is suitable for both ceramic-to-ceramic and ceramic-to-metal brazing. An obstacle with brazing is that most metals do not readily wet ceramic surfaces. One method of overcoming this problem is to metallize the ceramic surfaces, which can be done using several processes. The "active filler metal process" incorporates an active metal such as titanium into the filler metal. During brazing, the metal segregates to the surface of the ceramic, where it reacts to form a wettable surface. Active filler metals have been used in joining both oxide and non-oxide ceramics to themselves and to metals. It can be inferred that this process would also be suitable for joining oxide ceramics to non-oxide ceramics.

Brazing using a silicate glass is similar to metal brazing, but generally less complicated since wetting is generally not a problem. Both oxide and non-oxide ceramics based on alumina and silicon nitride have been brazed using silicates.

In producing a ceramic material such as silicon nitride, sintering aids are added to form grain boundary phases that become liquid during the sintering process. The interphase formed during silicate brazing is similar (or even identical) to this grain boundary phase. This suggests that transient liquid-phase bonding for composites is a possibility. In this case, a low-temperature-melting material would be placed between the ceramic composite faces and the temperature raised above the melting point. The molten interfacial layer could then dissolve some of the composite matrix material, forming a variable composition alloy. After a certain amount of dissolution, the alloy would become sufficiently rich in the more refractory matrix material to solidify. The initial melting should aid in the removal of porosity and capillary forces should draw the surfaces together. This process could enable reinforcing fibers or whiskers to be incorporated into the joint material, as shown schematically in Figure 6.

A number of monolithic oxide and non-oxide ceramics have been joined using the diffusion bonding process. Temperatures have ranged from 1350°C for joining alumina to alumina to 1800°C for joining silicon nitride to silicon nitride. In addition, it has been necessary to apply pressure to provide a sufficient driving force for bonding. This method is probably only of limited utility for ceramic matrix composite joining. The high temperatures required for most of the matrix systems described in this book would be detrimental to fiber or whisker properties due to thermal degradation and to reactions with the matrix material. In addition, it is difficult to apply uniform pressure to an interface in a complex component.

Cement bonding techniques might be applicable for bonding ceramic matrix composites in limited situations. For example, in the aluminum phosphate bonded system described in a previous chapter, bonding with material having a composition similar or identical to that of the matrix is a possibility. Because of the generally high temperatures required and the disruption of microstructure resulting from melting ceramic materials, it does not seem likely that fusion welding will find wide application in ceramic matrix composite joining.

Another type of ceramic to metal bonding technique that has been described [6] is friction welding. In this process, a ceramic and a metal specimen are rotated against each other under pressure. The frictional heat causes the metal to flow. The nature of the bond is not fully understood, but is believed to be part mechanical and part chemical.

Glass-to-metal bonds have also been made by electrostatic bonding. In this process, the surfaces are placed in intimate contact, and a very high voltage

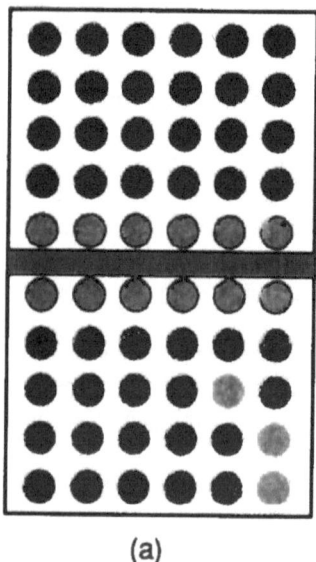
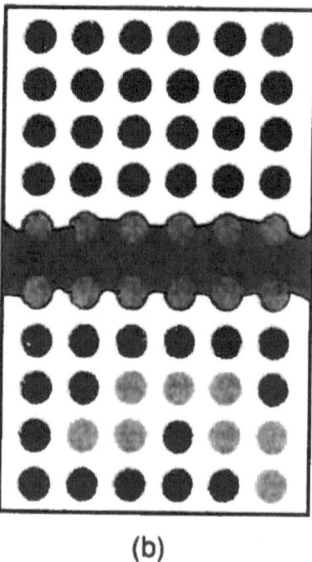

(a)   (b)

Fig. 6   Schematic depiction of transient liquid-phase bonding of a ceramic composite. The interlayer in (a) is a low temperature melting material that forms a solid solution with the matrix material. Dissolution of matrix material at a temperature in excess of the melting point of the interlayer material leads to the bonded material (b) after cooling. (From Ref. 5.)

is applied across the joint. This results in ion migration towards the joint, which produces electrostatic attraction between the two materials. The exact mechanism by which this method works is not understood.

Although all the general methods described above might be applicable to ceramic matrix composite joining, relatively little has yet been reported on non-mechanical composite joining. However, Rabin [7,8] has reported on joining work on Nicalon fiber/silicon carbide matrix material produced at Oak Ridge National Laboratory by the thermal gradient, forced flow method.

The object of the work was to provide a method that produced a joint usable to 1000°C at a processing temperature not appreciably exceeding 1400°C, to minimize fiber degradation. Two methods were used. In the first method, titanium and carbon powders, along with 15 wt% nickel powder, were reacted at the joint interface to form titanium carbide. The nickel reduced the reaction temperature from about 1600°C to 1200°C and also formed a liquid phase that aided densification. The resulting microstructure was that of a cermet of interconnected titanium carbide grains surrounded by a nickel-rich matrix. The reaction was conducted in argon under a pressure of 20 to 50 MPa at a maximum temperature of 1400°C. This method produced a joint about 125 μm thick, with some residual porosity believed to have resulted from defects in slurry application of the starting materials (Figure 7). A higher magnification scanning electron micrograph indicated that some chemical reaction occurred at the interfaces. Small particles of silicon carbide were present in the joint material.

The second method involved infiltration of carbon with molten silicon, a process described earlier. The carbon source was either fine carbon powder or woven carbon cloth. In both cases, silicon infiltration was carried out at 1480°C for 15 min in a 4 Pa vacuum. When powdered reactants were used, a thin uniform layer was applied to both joining surfaces using a slurry method. Prior to joining, the binder was removed by slow heating in air to 400°C.

A typical joint produced by this method was 200 μm thick and exhibited a uniform distribution of silicon carbide crystals in a silicon matrix. A scanning electron micrograph of the interface indicated that the molten silicon had wet the silicon carbide surface and, also, that there were a number of silicon carbide grains within the joint interlayer that were either nucleated from, or grown in contact with, the silicon carbide coating present on the original composite.

When carbon cloth was used, the silicon carbide crystals had the morphology of the original carbon fibers and there was typically several percent of unreacted carbon present. There was usually a layer of free silicon present at the interface, which is probably undesirable from a mechanical property standpoint. Figure 8 shows a joint of this type. Mechanical properties of joints produced by either of the two methods were not reported.

Unfortunately, if the matrix material of a ceramic composite is joined to other matrix material or to a metal using any of the techniques described in this section with a simple butt-type joint, the overall joined structure will not have utilized the major goals of such a composite, toughening and nonbrittle failure. That is, the joint is susceptible to the same type of failure as a monolithic ceramic because there is no bridging of the joint by the reinforcing phase. (The only exception might be a process of the type shown in Figure 6, which is somewhat speculative at this time.) Matrix joining alone can be acceptable for many applications involving whisker reinforced ceramics. Despite toughening relative to monolithic ceramics, whisker reinforced ceramics still fail brittlely, so that a joint that will have nonbrittle failure is not critical. However, for continuous fiber reinforced ceramics, where the majority of applications will require nonbrittle failure, this will not be acceptable.

Provided that joints of matrix material alone have a high shear stress, lapped or scarf joints, shown in Figure 9 [9], should provide enhanced joint performance. Further enhancement by the use of multiple laps and/or stepped laps is possible.

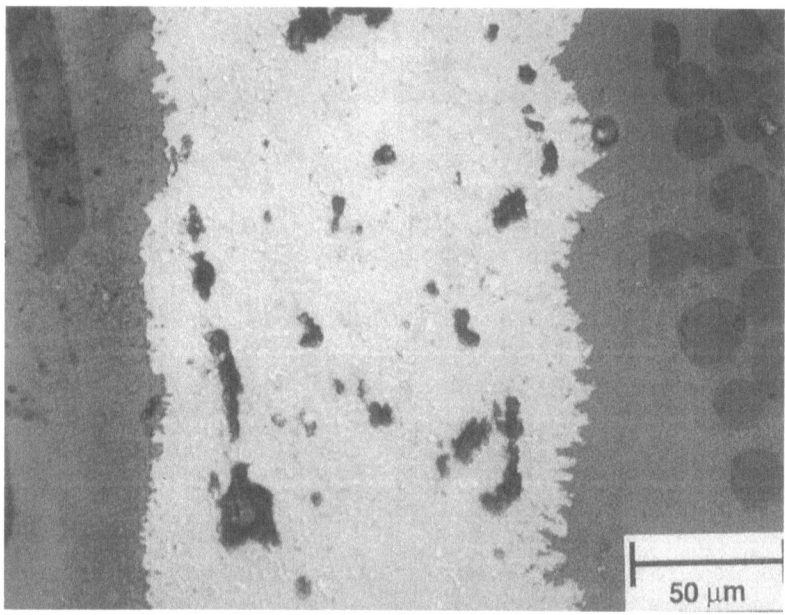

FIG. 7 Optical micrograph showing a joint between CVI-produced Nicalon™ fiber, silicon carbide matrix composite pieces that was formed by using a mixture of titanium + carbon + 15 wt% nickel. Black areas are porosity believed to have originated from the processing conditions. (From Ref. 8.)

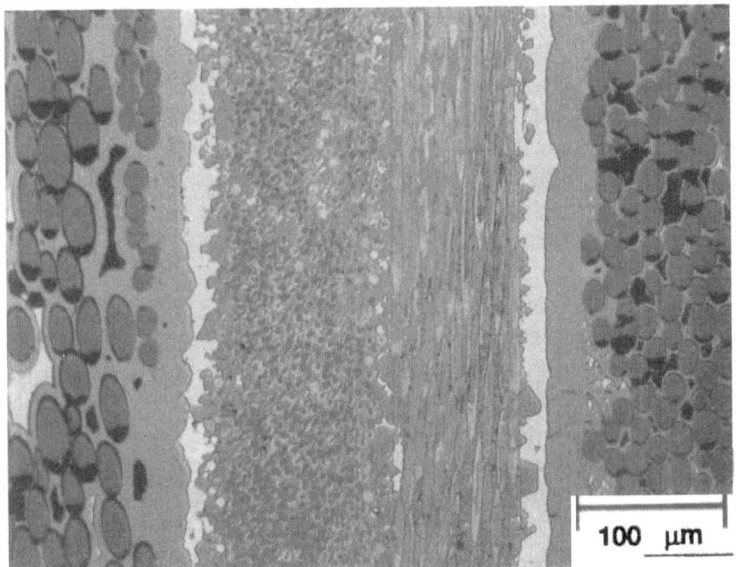

FIG. 8 Optical micrograph showing a joint between CVI-produced Nicalon™ fiber and silicon carbide matrix specimens that was formed by direct silicon infiltration of carbon cloth placed between the specimens. (From Ref. 7.)

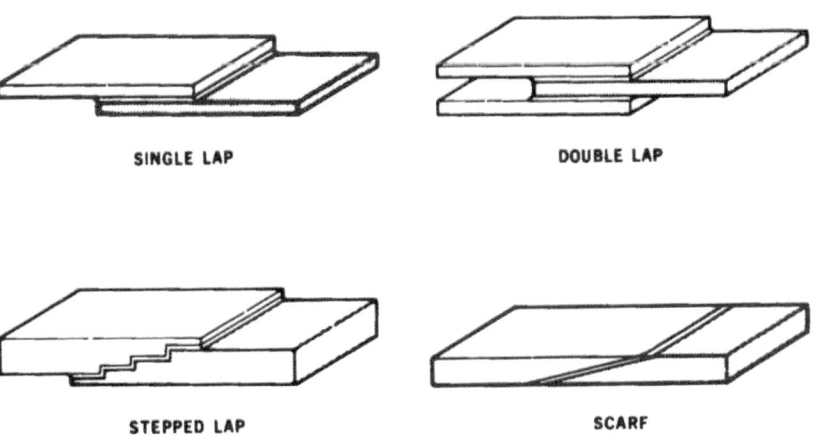

FIG. 9 Schematic examples of lap and scarf joints. (From Ref. 9.)

In addition to joining of completed ceramic composite materials, joining can be carried out at some intermediate state, as is sometimes done with monolithic ceramics. An example of this procedure, providing a lap joint with a shear resistant interface has been reported by Du Pont [10] for CVI-produced structures (Figure 10). Using a method called co-infiltration bonding, subcomponents are brought together early in the CVI process and additional CVI carried out. Bond shear strength at room temperature was about 29 MPa. Values at elevated temperatures were 21 MPa at 1000°C and 25 MPa at 1400°C. Shear strengths remained relatively constant when tested at room temperature after a 1 h or 50 h exposure in air at about 1100° or 1400°C.

Other novel methods of joining that will provide fiber reinforcement in the joined area have yet to be reported. Hence, it seems likely that mechanical joints will predominate in near-term structural applications of ceramic matrix composites in which nonreinforced joint areas are not acceptable.

Mechanical bonds can decrease the possibility that a joined assembly will fail at the joint in a catastrophic manner, as can occur with a nonmechanical bond that involves only matrix material or matrix material bonded to a metal. However, a mechanical bond of two ceramic composite materials using a metal fastener can be a problem at the high temperatures involved for many applications.

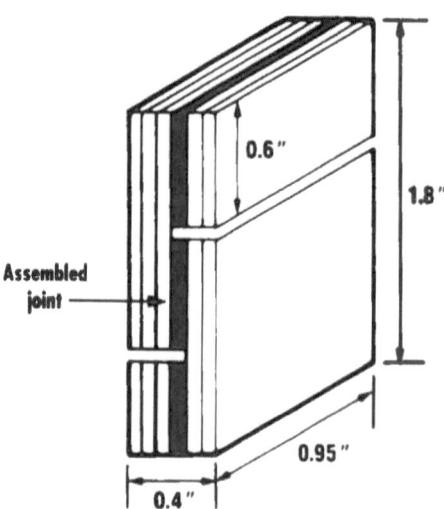

FIG. 10 Schematic illustration of a shear specimen for testing a bond produced by CVI co-infiltration. (Subcomponents are brought together early in the CVI process to provide joints with shear strengths approaching the substrate.) (From Ref. 10.)

One potential problem is that the coefficient of thermal expansion (CTE) values for the fastener metal and the composite can be appreciably different (with the metal typically having a greater expansion), resulting in loose joints or excessive stresses, depending upon the temperature regime. A snug radial fit at room temperature results in radial stresses at elevated temperatures. To achieve a snug radial fit at some elevated temperature, room temperature fit must be loose. A snug axial fit at room temperature results in a loose fit at elevated temperatures. To attain a snug axial fit at some elevated temperature, the joint must be stressed at room temperature. Another problem can be that the metal fastener is not capable of withstanding the high temperatures required for the application.

If a particular metal fastener is satisfactory from a temperature standpoint, joint configurations that can provide stress-free joints at elevated temperatures with composites have been designed. For example, Sawyer et al [11] have designed and tested elevated temperature thermal-stress-free joints for carbon/carbon composites that would be applicable to ceramic matrix composites as well. A typical thermal-stress-free joint concept suitable for materials having isotropic CTE values is shown schematically in Figure 11. The metallic components of the joint consist of a conical fastener, washer, and nut. The composite compo-

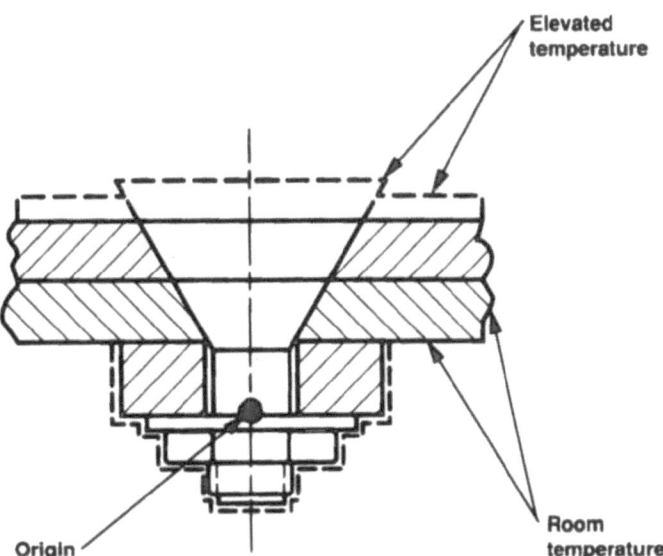

Fig. 11 Thermal stress free joint configuration with fastener and joined material having isotropic coefficients of thermal expansion. (From Ref. 11.)

nents consist of the two plates to be joined, as well as a washer of this material. The thickness of the composite material washer is determined so that the apex of the conical fastener is located at the outer plane of the washer. The inner diameter of the washer must have a clearance fit with the fastener shank if the CTE of the fastener is greater than the CTE of the material being joined (the usual case).

The clearance is determined by the difference between the CTE values and the temperature extremes to which the joint will be subjected. With this design, relative expansion or contraction between the two materials takes place on radial lines projecting from the origin of the conical fastener. Therefore, the expansion or contraction does not result in joint tightening or loosening and does not introduce thermal stresses into the material.

For materials having nonisotropic CTE values, there is an interference zone due to a reduction in the cone angle for the joined material. The change in angle is due to the inplane CTE being different from the through-the-thickness CTE. Therefore, the conical fastener configuration reduces, but does not eliminate, thermal stresses. For materials having nonisotropic CTE values, thermal-stress-free joints can still be designed but they become more complicated. The thermal-stress-free boundary for a particular fastener configuration is determined by the relative values of the CTE for the fastener and the composite materials joined together, the composite washer thickness, and the radius of the cylindrical portion of the fastener.

Changing both the radius and the washer thickness results in a variety of possible fastener and joint configurations and gives the designer a range of possible choices in joint configuration. In general, the head of the fastener is not exactly conical for materials with nonisotropic CTE values, although the deviation is often small enough that reasonably low thermal stress joints can be produced by disregarding the required curvature and using a conical head, which is easier to fabricate and use. Stresses resulting from such an approximation can, of course, be mathematically modeled to determine whether they are sufficiently low for the application.

Using the concepts summarized above, carbon/carbon specimens were joined using an OD5 superalloy fastener. The fastener had a 60° included-angle head and a no. 10 machine screw threaded shaft. Failure shear strengths of these joints were typical of those for conventional joints. The joints were exposed to a maximum temperature of about 1100°C and shear tested after cooling to room temperature. No evidence of damage or cracking due to thermal stress was observed, and the thermal cycling did not have a significant effect on joint strength or failure mode.

When metal fasteners cannot withstand the required application temperatures, fasteners made from a ceramic composite material are a possibility. In general, use of a ceramic composite fastener can also decrease the complexity of a thermal-stress-free joint, since CTE differences can be minimal. Figure 12 shows a CVI-produced silicon carbide fastener, with three-dimensional carbon fiber reinforcement (Novoltex™ process), and a Du Pont proprietary oxidation protection coating.[10] Nuts of this material have also been made. These fasteners are reported to have good shear and bearing properties. The Dimox™ process also seems suitable for producing net shape composite fasteners. Figure 13 [12] schematically illustrates the inverse process, fabrication of a structure for accepting a threaded fastener.

IV. NONDESTRUCTIVE EVALUATION

Nondestructive evaluation (NDE) is a valuable tool for quality assurance of critical structural components. In addition, NDE is useful for correlating defect generation with processing variables and for increasing understanding of phenomena occurring during mechanical or thermomechanical stressing of structural components. NDE methods are very well established for metals but are much less developed for monolithic ceramics and, especially, ceramic matrix composites. A brief summary of NDE methods for ceramics will be given, followed by a few specific examples for ceramic matrix composites.

Two of the major methods for NDE of structural metals are based on X-ray diffraction and on acoustic behavior, and these methods have been adapted

Fig. 12 CVI-produced silicon carbide fastener, with three-dimensional carbon fiber reinforcement (Novoltex™ process) and an oxidation protection coating. (From Ref. 10.)

for ceramics. X-ray diffraction techniques have been developed for ceramics to the extent that flaws as small as 2 μm can be detected using what is termed microfocus X-ray diffraction.[13] The microfocus mode is more accurate than the computer-aided-tomography (CAT) mode, but it is limited to thin parts. The CAT mode permits real-time two-dimensional sections of specimens to be examined.

A number of types of acoustic methods are under development for ceramics. One version, scanning laser acoustic microscopy (SLAM) is a through-transmission technique that provides images through the thickness of the sample. Resolution and penetration are determined by the frequency. In a typical advanced ceramic, a 1-MHz beam will penetrate a few millimeters with a resolution of 250 μm, while a 50-MHz beam gives a resolution of about 5 μm. A number of variations of these methods as well as other methods have been summarized by Sheppard.[14] Table 1 lists NDE techniques under development for ceramics, along with advantages and disadvantages of each.

Applications of some of these NDE techniques to reinforced ceramics have been reported. For example, Baaklini and Bhatt [15] described use of X-ray radiography to monitor damage accumulation and failure processes in unidirectional SCS-6 fiber, silicon nitride matrix composites produced by reaction bonding. The fabrication technique used was described in an earlier chapter. Composites that contained from one to five fiber layers were fabricated.

The test system consisted of a 3.2 kW X-ray system in conjunction with a mechanical testing system. The X-ray apparatus was capable of reaching 169

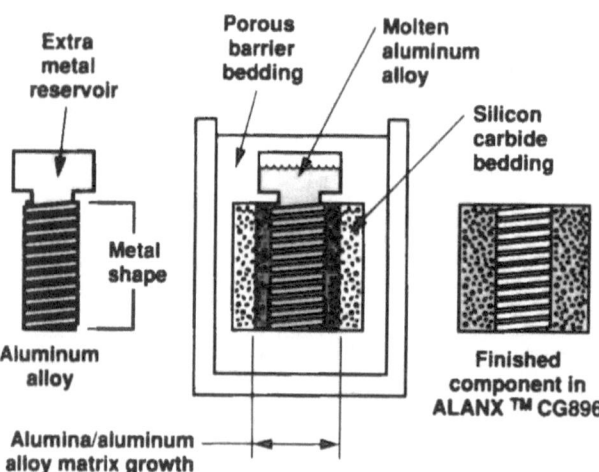

FIG. 13 Schematic illustration of a metal shape replication process that utilizes Lanxide's directed metal oxidation technology. (From Ref. 12.)

TABLE 1. Nondestructive evaluation methods for ceramic heat exchangers.

Technique	Advantages	Limitations	Other comments
Scanning laser acoustic microscopy	Can easily detect cracks in outer surface of tubes	Cannot detect inclusions or voids	Back-reflection method recommended for locating cracks
Ultrasonic C-scan	Can detect both inside and outside surface cracks, inclusions larger than normal, porosity, regions of high porosity, and variations in wall thickness	Low frequency results in low resolution; high frequency causes high scattering	Focused transducers in the range of 20 to 40 MHz are recommended; surface roughness and porosity limit capabilities
Conventional X-ray	Best for detection of high density inclusions and unusual regions of porosity and thickness variation	Limited use for detecting small size of critical defects	Real-time and film radiography should be used together for best results
Time-of-flight acoustic microscopy	Surface connected cracks can be located	Angle between beam and line of intersection of the crack is important	Angled beam—corner reflection method is recommended
Computed tomography	Can measure density and detect irregular surfaces of the interior	Resolution of system limited by resolution of TV image; beam hardening artifacts reduce quality of image	Most features must occupy several pixels to be distinguished from noise
Acoustic emission	Can detect and locate source of sound emitted when grain boundary opens (monitor crack growth)	Multiple sensors are required	Is being investigated for locating potential failure sites for investigation by NDE techniques and characterization of flaw populations which cause failure

*85% SiC and 15% Si (Norton CS10K). Based on research by Babcock and Wilcox, Lynchburg Research Center, Lynchburg, VA.
Source: Ref. 14.

kV or 45 mA with either a 400-μm or a 3-mm focal spot size. Film radiography was used because of its high resolution. A relatively slow rate of either load increase or displacement was used to facilitate radiographic examination during tensile testing. Specimens were also evaluated before loading and after failure by microfocus X-ray radiography.

Panels were machined to remove a carbon layer formed during high temperature vacuum hot pressing of the specimens. The radiographic evaluation of machined panels was used as a guide for preparing tensile specimens having variations in density and impurity particles in the gage section. For the machined specimens, conventional radiography could identify features such as fiber bunching and misalignment, low density areas, and high density impurity particles (chemically identified as iron inclusions).

In a one-fiber-layer specimen that had a cluster of 50-μm diameter inclusions (a region of about 0.5 mm × 1.0 mm), failure apparently initiated at the site of the impurity cluster and propagated across the width of the specimen, as inferred from the size and number of fiber pullouts seen. However, isolated and randomly distributed inclusions less than 150 μm in diameter did not directly affect the fracture of a one-fiber-layer or a three-fiber-layer composite. High density inclusions up to 225 μm in diameter did not influence the fracture behavior of a five-fiber-layer composite specimen. Local density variations did not appear to affect fracture behavior. The radiographic images were also used to estimate matrix crack spacing, which in turn was used to estimate interfacial debond strength (about 5 MPa).

It was concluded that X-ray radiography can provide a basis for interpretation of mechanical test behavior, for validation of analytical models, and for verification of adequate processing procedures.

An X-ray computed tomography imaging system was evaluated for ability to detect voids and characterize fiber orientation of ORNL CVI-produced Nicalon fiber reinforced silicon carbide by Ellingson et al.[16] Specimens reinforced with continuous fibers having various architectures and with chopped fibers were examined. A medical CAT scanner was used for the tests.

Voids of 50–200 μm could not be individually detected. A gradient in density from 2.5 to 2.2 g/cm^3 was detected qualitatively. A clear difference could be seen between plain weave and satin weave architectures.

Ultrasonic acoustic testing has been conducted on lithium aluminosilicate reinforced with about 45 vol% Nicalon fibers.[17] To produce material for NDE with different amounts of damage, specimens of the as-fabricated material were subjected to a different number of 4-point bending fatigue cycles (5 Hz) at 500°C. Several types of measuring arrangements were used, including one where both the transmitting and receiving transducers were mounted on the same side of the specimen, a straight through configuration, and an offset through configuration. Pulsing transducers having frequencies of 2.25, 5.0, and 7.5 MHz were used. Silicone discs 0.25 mm thick were used between the transducers and the specimen. Best results were obtained using the 2.25-MHz transducer.

A concept termed the "stress wave factor" approach, where the stress wave factor is considered to be any useful ultrasonic measurement, was used. Several stress wave factors, including peak amplitude, correlated relatively well with residual strength after fatigue test cycling, indicating that the technique could be useful for quality control. It was also speculated that this type of technique could be useful for evaluation of damage in ceramic composites already in the field.

Acoustic techniques have also been used to measure the modulus of ceramic composites [18] and, in conjunction with fiber/matrix debond tests using an indenter, to identify the instant of debonding.[19]

V. DESIGNING WITH CERAMIC MATRIX COMPOSITES

Designing of structural elements utilizing ceramic matrix composites presents a challenge. Although a continuous fiber reinforced ceramic can have a stress-strain curve resembling the ductile behavior of a structural metal more than the brittle behavior of a monolithic ceramic, metals and reinforced ceramics are quite different in the ductile-appearing strain regime. The metal will, in general, retain most of its original characteristics. On the other hand, after undergoing a strain sufficient to produce the ductile-like behavior, the ceramic matrix composite has undergone matrix microcracking, followed by some fiber/matrix debonding and fiber breakage. This can significantly change various properties, as well as allow atmospheric ingress to the interior, which can have catastrophic results with many systems. Figure 14 [20] reviews the overall elements in failure of a continuous fiber reinforced ceramic.

It is generally agreed that in designing with ceramic matrix composites, the design limit should be the point of first matrix microcracking, with additional stress tolerance before failure providing overload capacity. However, this will utilize a relatively small fraction of the strength values reported throughout this book. Hence, material development should emphasize improvement of stress level to first matrix microcracking rather than overall stress level to failure.

Another problem for the designer at this stage of ceramic composite development is that the extensive databases available for most structural metals and polymeric materials are not yet available. In particular, the most common type of strength measurement has been flexural strength, but this type of measurement is not very useful for design purposes. Hence, designs will generally tend to be conservative initially and will have to involve some engineering judgment. Of course, databases for some of the more developed materials, such as CVI-produced silicon carbide matrix material, have become reasonably extensive.

Even when databases for conventionally measured mechanical properties and other properties are relatively large, the implications of some measured property values in an actual structure are not well established. A case in point is fracture toughness. Although fracture toughness values as high as those for structural metals have been reported for some ceramic matrix composite materials using techniques developed for monolithic ceramics, the validity of the values from such tests for design of actual structures is problematic.

The relatively immature state of development of machining, joining, and nondestructive evaluation of reinforced ceramics also complicates the job of the component designer.

Since whisker reinforced ceramics retain some similarities to monolithic ceramics, design with these materials is perhaps more straightforward than with continuous fiber reinforced materials. However, some of the same considerations outlined above will also apply to whisker reinforced ceramics.

The rather formidable challenges involved in structural designing with ceramic matrix composites have been met in a number of cases and considerable efforts to increase use of these materials are in progress. The remainder of this

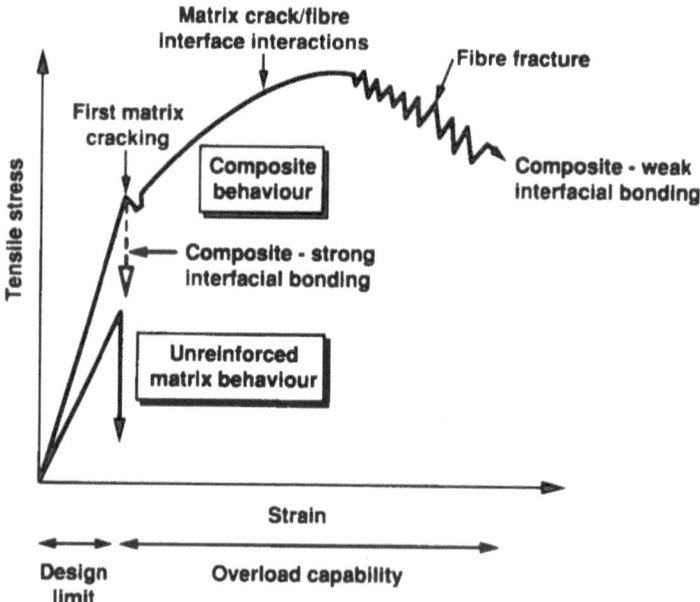

Fig. 14 Generalized stress-strain behavior of a continuous fiber reinforced ceramic. (From Ref. 20.)

section will provide some general strategies that are being developed for designing with these materials.

Some aspects of design of ceramic matrix composite components are not dissimilar to those used for other structural materials. Today's structural designers make extensive use of computer aided design (CAD) programs, thereby minimizing, the "design, build, test, redesign" cycle.[21] Perhaps the most widely used technique is finite element analysis (FEA), which uses a grid of small size structural elements in one-, two-, or three-dimensions to represent a continuum. Mechanical, thermal, and other material properties can be assigned. Appropriate mechanical stresses, temperature conditions, or any number of other factors can be simulated, and the material response calculated. One of the more widely used FEA programs is ANSYS (Swanson Analysis Systems, Houston, PA). ANSYS is used by over 2000 corporations and universities worldwide.

For design of monolithic ceramic components, combination of FEA with a public domain computer program having the acronym CARES (Ceramics Analysis and Reliability Evaluation of Structures) has been widely used. An outgrowth of CARES, having the acronym TCARES (Toughened Ceramics Analysis and Reliability Evaluation of Structures) has been used for whisker reinforced ceramics.[22] In addition to the inherent scatter in strength inherent to monolithic ceramics, the analysis of whisker reinforced ceramics must account for directional properties due to preferred whisker orientation resulting from the particular fabrication process used. Properties can be isotropic, transverse isotropic, or orthotropic.

Even though a whisker reinforced ceramic can have greatly improved reliability relative to a monolithic ceramic, variability in strength is still too high for use of the design philosophy common for structural metals, the "factor of safety approach." Rather, a statistical design methodology is used to take into account the scatter in ultimate strength, as well as to account for decreasing strength with increasing component volume. This approach considers the structural continuum to be a chain composed of links connected in series, so that the overall strength is governed by the strength of its weakest link. It is assumed also that the events leading to failure of a link are independent of events occurring in any other link. Variability in a ceramic material can be accounted for using Weibull statistics.

Since ceramics are sensitive to both surface and volume flaws, the TCARES program has the capability of separately conducting surface and volume reliability analyses. Output of the program includes a summary of input from the finite element code, element statistical properties, element survival probabilities, and an overall component survival probability. An interesting feature of the output is that 15 elements having the highest risk-of-rupture intensity are listed, aiding the designer in identifying the most likely areas for improving the reliability of the proposed structure.

With continuous fiber reinforced ceramics, the structural designer has the possibility of significant alteration of material properties, through fiber architectural design. The material designer has nearly limitless possibilities for placing fibers in multiple directions. Tan et al [23] have reported on the use the Fabric Geometry Model (FGM) for quantifying and modeling fiber architecture, and on the FEA structural analysis procedure. The procedures were exemplified with a hypothetical turbine blade, which needs to survive a centrifugal force loading from the rotation and thermal loading due to the heated gas flow from the combustion chamber. In order to assess fatigue life of the blade, the natural frequencies must be predicted also. Lithium aluminosilicate reinforced with Nicalon fibers was used as the model system.

VI. SUMMARY

Acceptance of ceramic matrix composites for most applications will require suitable methods for machining, joining, nondestructive evaluation, and structural design. Progress is being made in all these areas. Although conventional machining of ceramic matrix composites using superabrasives such as diamond is satisfactory, a newer method, laser machining, has been shown to offer some advantages over conventional machining. Results of work on a number of other newer methods have been reported for monolithic ceramics, and these methods are probably adaptable for reinforced ceramics as well.

Joining of matrix material for a ceramic matrix composite to other matrix material or to a metal is analogous to the joining of monolithic ceramics, and many methods are available, or under development. However, reported work on actual joining of ceramic composite materials is relatively sparse. One constraint on composite joining is that maximum temperature required for the joining process should be below that at which degradation of fibers or the fiber/matrix interfaces occurs. Several methods for which this criterion has been met have been applied to continuous fiber reinforced ceramics.

In some cases, where catastrophic failure cannot be tolerated, it will not be adequate to make simple joints between matrix materials. Lap joints can help this situation, and some of the chemical bonding methods could produce a liquid phase which could result in fibers spanning the joined region. Other methods, such as joining CVI-produced components to each other by putting them in contact part way through the CVI-process to provide a method for providing reinforcement in the joined region, are also under development.

However, in the near future, at least, mechanical joining will probably predominate to ensure joints with a low probability of catastrophic failure. If a metal fastener is used in a conventional manner, differences in thermal expansion coefficients can result in loose joints or overly stressed joints as temperature is varied. However, thermal-stress-free joints have been demonstrated. As an alternative to a metal fastener, ceramic matrix composite fasteners have been produced.

Several nondestructive evaluation techniques are being developed for monolithic ceramics. Although there are a number of variations, the two major types of techniques are the same as those used for metals, X-ray radiography and acoustic techniques. Reported work specific to reinforced ceramics is as yet rather limited.

Some of the modern design tools used for metals or other structural materials, such as finite element analysis, can be used for ceramic composites as well. However, specific design strategies can differ, and there is still relatively little reported in this area. A special consideration with ceramic matrix composite design includes a decision that must be made for each material concerning the portion of the stress-strain curve that is usable for design.

A common problem for the structural designer is the lack of large property databases for many material systems, both because of insufficient measurements and because of a preponderance of measurements not very useful for structural design. The latter include flexural strength and fracture toughness measurements that were developed for monolithic ceramics but which may not be appropriate for composites. For these reasons, designs will undoubtedly be conservative in the near future.

REFERENCES

1. Ridealgh, J. A., Rawlings, R. D., and West, D. R. F., Laser cutting of glass ceramic matrix composite, *Materials Science and Technology*, vol. 6, 395–398 (1990).
2. Guha, J. K., High-pressure waterjet cutting: an introduction, *Ceramic Bulletin*, vol. 69, no. 6, 1027–1029 (1990).
3. Petrofes, N. F., and Gadalla, A. M., Electrical discharge machining of advanced ceramics, *Ceramic Bulletin*, vol. 67, no. 6, 1048–1052 (1988).
4. Moreland, M. A., and Moore, D. O., Versatile performance of ultrasonic machining, *Ceramic Bulletin*, vol. 67, no. 6, 1045–1047 (1988).
5. Cawley, J. D., Joining of ceramic-matrix composites, *Ceramic Bulletin*, vol. 68, no. 9, 1619–1623 (1989).
6. Fernie, J. A., Threadgill, P. L., and Watson, M. N., Progress in joining of advanced materials, *Welding and Metal Fabrication*, vol. 59, no. 4, 179–184 (1991).
7. Rabin, B. H., Joining of fiber-reinforced SiC composites by *in situ* reaction methods, *Materials Science and Engineering A*, vol. A130, L1–L5 (1990).
8. Rabin, B. H., Joining of silicon carbide/silicon carbide composites and dense silicon carbide using combustion reactions in the titanium-carbon-nickel system, *J. Am. Ceram. Soc.*, vol. 75, no. 1, 131–135 (1992).
9. Schjelderup, H. C., and Jones, B. H., Practical influence of fibrous reinforced composites in aircraft structural design. In *Composite Materials: Testing and Design*, ASTM Special Technical Publication 460, American Society for Testing and Materials, 1969, 285–306.

10. Du Pont CVI Ceramic Matrix Composites, Preliminary Engineering Data, Du Pont Composites, Newark, DE.

11. Sawyer, J. W., Blosser, M. L., and McWithey, R. R., Derivation and test of elevated temperature thermal-stress-free fastener concept. In *Welding, Bonding, and Fastening 1984*, NASA Conference proceedings No. 2387, 1984, 101–117.

12. Anon, Alanx: a manufacturing vision, *Ceramic Bulletin*, vol. 68, no. 11, 1917 (1989).

13. Firestone, R. F., NDE: improving reliability of advanced ceramics, *Ceramic Bulletin*, vol. 68, no. 6, 1177–1178, 1186 (1989).

14. Sheppard, L. M., Evolution of NDE continues for ceramics, *Ceramic Bulletin*, vol. 70, no. 8, 1265–1279 (1991).

15. Baaklini, G. Y., and Bhatt, R. T., In-situ X-ray monitoring of damage accumulation in SiC/RBSN tensile specimens, NASA Technical Memorandum 103733, 1991.

16. Ellingson, W. A., Vannier, M. W., and Stinton, D. P., Application of X-ray computed tomography to ceramic/ceramic composites, CONF-8707132-2, Oak Ridge National Laboratory, Oak Ridge, TN, July 1987.

17. dos Reis, H. L. M., Acousto-ultrasonic evaluation of ceramic matrix composite materials, NASA Contractor Report 187073, February 1991.

18. Bridge, B., Round, R., and Green, A., Structural evaluation of phosphate bonded ceramic composite materials from non-destructive ultrasonic velocity and attenuation measurements, *Journal of Materials Science*, vol. 26, 2397–2409 (1991).

19. Rouby, D., and Osmani, H., Characterization of interface debonding in a ceramic-ceramic fibre composite using the indentation method and acoustic emission, *Journal of Materials Science*, vol. 7, 1154–1156 (1988).

20. King, J. E., Failure in composite materials, *Metals and Materials*, 720–726 (December 1989).

21. Couts, J. S., Engineering design with ANSYS software, *Ceramic Bulletin*, vol. 68, no. 12, 2073–2076 (1989).

22. Duffy, S. F., Manderscheid, J. M., and Palko, J. L., Analysis of whisker-toughened ceramic components—a design engineer's viewpoint, *Ceramic Bulletin*, vol. 68, no. 2, 2078–2083 (1989).

23. Tan, T. M., Pastore, C. M., and Ko, F. K., Engineering design of tough ceramic matrix composites for turbine components, *Transactions of the ASME*, vol. 113, 312–317 (April 1991).

17

Applications of Ceramic Matrix Composites

I. BACKGROUND

Only one fiber or whisker reinforced ceramic material has been routinely used for some time in a major application, silicon carbide whisker reinforced alumina as a cutting tool. However, a number of other applications are actively being pursued for whisker reinforced ceramics, and many components are under development for continuous fiber reinforced ceramics. As experience is gained in these applications and as composite properties are further improved and cost reduced, many future applications are anticipated. In this chapter, projections for future markets for ceramic matrix composites will be summarized and major sources of funding for ceramic composite development described. More details on whisker reinforced alumina for cutting tools will be given and examples of development programs for continuous fiber reinforced ceramic applications discussed.

II. MARKET SIZE FOR CERAMIC MATRIX COMPOSITES

A number of projections are available for ceramic matrix composite markets. These differ somewhat and, of course, only time will tell the accuracy of the forecasts. Nonetheless, a few of these projections will be cited here. In a summary [1] of reports released by Business Communications, Inc., total market for ceramic matrix composites in 1988 was given as about $77,000,000, with about two-thirds commercial and one-third defense-related. Presumably, this includes particulate reinforced ceramics, which have not been covered in this book. Predicted market for 1993 is about $200,000,000, with over 80% for commercial applications, and predicted market for the year 2000 is over $625,000,000, with over 90% for commercial applications. These forecasts correspond to annual growth rates over the period covered of over 5% for defense-related markets and over 20% for commercial markets.

    To put these numbers in perspective, polymer matrix composites are projected to grow at a much slower overall rate, but the total market (commercial and defense-related) in 1988 dwarfed the ceramic matrix composite market, with a value of over $3,000,000,000. Total market for the year 2000 is forecasted to be over $9,000,000,000. Metal matrix composites had a market of about $20,000,000 in 1988, about 85% defense-related, and this is projected to increase to over $225,000,000 by the year 2000, with roughly equal increases in commercial and defense-related markets.

    A summary [2] in 1991 of a report by Frost & Sullivan Inc. gave the market at that time as nearly $135,000,000 and predicted a ceramic matrix composite market of over $280,000,000 by 1995. A summary [3] of yet another report, by the Freedonia Group Inc., projected a market of about $245,000,000 by 1995. Both of these reports apparently include particulate reinforced ceramics in their data and predictions.

III. FUNDING SOURCES FOR APPLICATIONS RELATED R&D

The total level of industrial funding for ceramic matrix composite R&D is not accurately known, but a clearer picture is available for government-sponsored R&D. For example, Persh [4] has reviewed U.S. Department of Defense (DOD) funding for ceramic matrix, metal matrix, and carbon/carbon composites from 1979 to 1989. This included funding from the Army, Navy, Air Force, and the Defense Advanced Research Projects Agency (DARPA). For ceramic matrix composites, funding increased from only $1,400,000 in 1979 to nearly $28,000,000 in 1989. In comparison, metal matrix composites were funded at a much greater level in 1979, about $13,000,000, but funding increased to only about $25,000,000 in 1989. Carbon/carbon composite funding was about $6,000,000 in 1979 and increased to only about $14,000,000 in 1989.

    The growth in ceramic matrix composite R & D funding was the result of a number of factors, including the mounting evidence that monolithic ceramics were not going to be able to adequately function for many DOD applications. Another factor was that for many important high temperature applications, such as advanced gas turbine engines, metal matrix composites and carbon/carbon composites have serious shortcomings.

A DOD-sponsored effort has been started on development of the National Aerospace Plane (NASP), which will travel at hypersonic (>Mach 5) speeds. This will be an "air breathing" (as opposed to rocket powered) vehicle capable of attaining orbital velocity. The few metals that are capable of retaining sufficient mechanical properties at the highest temperatures that will be encountered are subject to oxidative degradation, are very dense, or have other shortcomings. Carbon fiber reinforced carbon, which is used to a considerable extent in the Space Shuttle program, is the most well developed material for this type of application, but coatings to satisfactorily overcome the oxidation problem at very high temperatures have not yet been developed. Continuous fiber reinforced ceramics have the potential for use at the higher temperatures required for NASP materials. A NASP demonstration vehicle is to be constructed by 1993-94.

A military aircraft engine oriented program is the DOD-sponsored Integrated High Performance Turbine Engine Technology (IHPTET) program. This program has a goal of doubling the thrust-to-weight ratio of today's most advanced engines by the year 2000. To meet this goal, it is anticipated that materials with the capability of being used at up to about 2200°C will be needed. Initial R&D contract work on providing ceramic matrix composite systems to meet this objective has emphasized high temperature chemical compatibilities of potential reinforcement and matrix materials.

Complementary to the NASP and IHPTET programs is the National Aeronautics and Space Administration (NASA) High Temperature Engine Materials Technology Program (HITEMP), which is aimed at assuring the competitiveness of civil transport aircraft into the 21st century. Goals are to improve fuel economy, improve reliability, extend life, and reduce operating costs. Coated carbon/carbon is specifically excluded from this program because of the low probability for success for a gas turbine engine application. Ceramic matrix composites will be emphasized for structures needed to operate in the 1100° to 1650°C temperature range. The HITEMP program is intended to provide materials for the High-Speed Civil Transport, projected to have a speed of Mach 2-5, a range of 6500 nautical miles, and a passenger load of 300. Goals are a 40% reduction in fuel consumption and a 20% reduction in direct operating costs, relative to today's technology.

In addition to the aerospace programs outlined above, there are other defense-related rocket and missile programs in which ceramic matrix composites are expected to play an important role. Besides the attributes of continuous fiber reinforced ceramics previously described, many fiber-matrix combinations result in composites with low radar observability (LO), in common with organic matrix composites, but in contrast to metals or carbon/carbon.

In late 1990, a ceramic matrix composite strategic plan (part of a broader plan for all types of composites) was presented by the Aerospace Industries Association (AIA) and the National Center for Advanced Technologies

(NCAT).[5] The plan proposes an integrated program to develop a technology base in the United States that will be required for effective utilization of ceramic matrix composites in the aerospace industry. A ten-year, $674,000,000 plan, involving coordinated efforts by universities, national laboratories, and industry, was proposed to overcome present technology gaps. The report distinguished among low to moderate temperature (<1200°C), high temperature (1200°–1650°C), and ultrahigh temperature (>1650°C) ceramic matrix composite systems and proposed demonstrations for each type of material. Basic research, database development, materials and processing technologies, and a number of support technologies were listed as requirements for successful demonstrations of each category of ceramic matrix composite material.

Additional R&D funding for ceramic matrix composites that is not defense and/or aerospace related is provided by the Continuous Fiber Ceramic Composite (CFCC) initiative of the Department of Energy (DOE). The National Research Council examined the role of materials in eight major U.S. industries employing seven million people and having sales of $1,400,000,000 in 1987, and concluded that there is a need for lighter, stronger, more corrosion-resistant materials capable of performing at higher temperatures.[6] Ceramic matrix composites, particularly those having continuous fiber reinforcement, are leading candidates for many of these applications.

IV. SILICON CARBIDE WHISKER REINFORCED ALUMINA CUTTING TOOLS

Large sums of money are spent in machining ferrous and nonferrous alloys. Annual cost has been estimated to be $100 billion, about 1% of which is tool cost.[7] Alumina tools were used as early as 1905, but the lack of toughness and reliability relative to other types of tools, such as high speed steel and cemented carbides, was a severe limitation. However, the improvement in toughness and reliability by whisker reinforcement has led to considerable usage since the mid-1980s. At that time, Greenleaf Corporation (Saegertown, PA) in cooperation with the silicon carbide whisker facility of Arco Chemical Corp. (now Advanced Composite Materials Company, Greer, NC) developed a cutting tool material having the trade name WG-300.[8]

For most metals, the highest practical cutting speed is limited by the availability of cutting tool materials that can survive over sustained periods under the conditions of high speed machining. Wear mechanisms are complex and no single wear mechanism can explain observed tool wear at high cutting speeds.[9] Although early ceramic tools had high hardness at elevated temperatures and chemical stability with respect to the most common engineering materials, these tools were poor in terms of tensile strength, toughness, and transverse rupture strength. Therefore, use was restricted to continuous cutting at moderate speeds and feeds.

The incorporation of silicon carbide whiskers into an alumina matrix results in a ceramic composite tool with significantly improved properties. This material has the abrasion resistance of a ceramic but the strength and thermal shock resistance of cemented carbides. Thermal shock resistance is an important consideration because of the heat generated in high speed machining. A large amount of shear stress is required to cause plastic deformation and shear to occur in the workpiece, and this results in the generation of high temperatures. As much as 80% of the heat generated in cutting is produced in this way. The other 20% comes from sliding of the chip over the tool faces. Thermal shock resistance is especially important when lubrication is used, since large temperature differentials can result. The increased thermal shock resistance of silicon carbide whisker reinforced alumina can derive both from increased thermal conductivity and toughening due to the whisker reinforcement. Toughness increases attainable with whisker reinforcement have been described in an earlier chapter.

Heat generated in the shear zone can be used to advantage. There is an optimum speed where the heat generated lessens the cutting forces by aiding in grain boundary dislocation. This is often outside the range of stability for a tool of cemented carbide, which begins to soften at about 850°C. On the other hand, silicon carbide whisker reinforced alumina can be operated at a temperature of at least 1400°C. The resulting increase in cutting speed can be dramatic, from a few hundred feet per minute with a cemented carbide tool to thousands of feet per minute with the composite tool. Speed can be increased to such an extent using the composite material that in light finishing cuts a chip may not even be formed; the removed metal can be so hot that it oxidizes to a powder.

There must be a carefully balanced feed/speed relationship, since a reduction in speed without a corresponding decrease in feed results in a thicker, cooler chip and can result in reduced tool life or failure. To minimize heat related problems when the chip is too thin to carry away enough heat, coolant must be used. The excellent thermal shock resistance of silicon carbide whisker reinforced alumina means that the tool can be used wet, dry, or even intermittently cooled. Figure 1 [10] indicates feed and cutting speed ranges for several cutting tools used for machining nickel base superalloys. Silicon carbide whisker reinforced alumina can operate at higher speeds than other ceramic materials such as monolithic sialon and titanium carbide containing alumina.

Some examples of benefits from substitution of silicon carbide whisker reinforced alumina for other tool materials in machining nickel base superalloys have been given.[8] As one example involving WG-300, a major gas turbine manufacturer had been using a monolithic sialon material to rough cut Inconel 718 and a tungsten carbide tool to finish machine. Using WG-300 for both operations, machining time was reduced from five hours to twenty minutes, which amounted to a savings of $250,000 annually. Another manufacturer experienced dramatic improvements in machining a blade lock groove on a turbine hub. By using WG-300, machine time was cut from three hours to eighteen seconds. Production rate increased from three parts per shift to the equivalent of two weeks of parts in one shift.

In addition to increasing cutting speed, tool life has been increased in some applications. Figure 2 [11] gives examples of tool lives and metal removal rates for silicon carbide whisker reinforced alumina, titanium carbide particulate containing alumina, and cemented carbide in machining Inconel 718. A desirable feature of silicon carbide whisker/alumina composites is that tool degradation mode is gradual wear, compared to catastrophic failure for some other ceramic materials, such as alumina reinforced with particulate titanium carbide.

Details of the mechanisms for tool wear under various conditions for Inconel 718 have been reported in considerable detail by Wayne and Buljan[12] for a number of ceramic materials including silicon carbide whisker reinforced alumina and silicon carbide whisker reinforced silicon nitride. The sensitivity of both alumina and silicon nitride to depth-of-cut notching is significantly reduced by the whisker additions, which is primarily attributed to the increase in fracture toughness. The ceramic composites also exhibited a higher resistance to nose and flank wear.

Improvements have been observed with other superalloys also. For example, stock removal with Inconel 901 was five times faster than with conventional cemented carbides.[13] However, it has been reported elsewhere [10] that

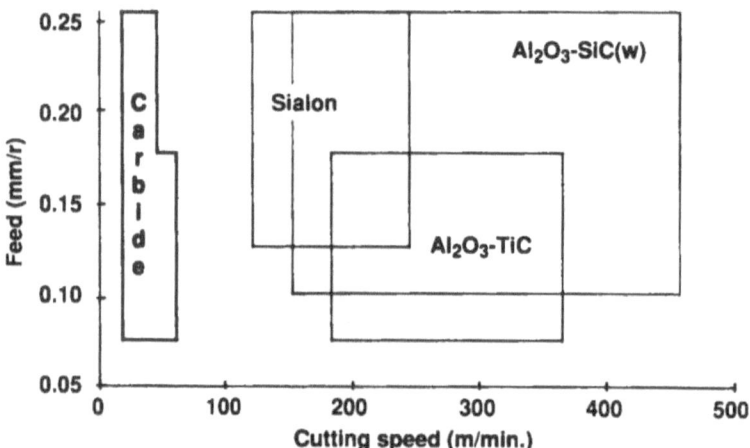

Fɪɢ. 1   Feed and cutting speed ranges for several cutting tool materials used in machining nickel base superalloys. (From Ref. 10.)

with Inconel 901, as well as with Waspaloy, sialon tools outperformed whisker reinforced alumina, and that whisker reinforced alumina is not particularly good with ferrous alloys due to chemical reaction with the alumina.

Other producers of silicon carbide whisker reinforced alumina cutting tools include GTE Valenite Corp. (Troy, MI) [14] and Kennametal (Latrobe, PA).[15] Kennametal's Kyon 2600 was reported to be undergoing commercial testing. Kyon 4000 contains an additional reinforcing phase of zirconia. It is reported to be the first ceramic that can operate effectively on low carbon steels and ductile cast iron. Tool life on ductile cast iron was found to be over 50% higher than for ceramic coated carbides and over three times that of unreinforced silicon nitride. Kyon 4000 is aimed at the automobile and truck manufacturing industry, which has a goal of reducing weight by switching from gray cast iron castings to ductile cast iron. An effort by Advanced Composite Materials Corporation and SCT Incorporated to develop ceramic matrix composite tooling for the can-making industry has also been reported.[16]

There are many other applications where the properties of whisker rein-forced ceramics can be used to advantage, but there is relatively little docu-mentation of commercial applications. Material manufactured by Advanced Composite Materials Corporation (ARtuff™) has been reported to be used com-mercially in nozzle inserts and durable wear resistant pump components, but no

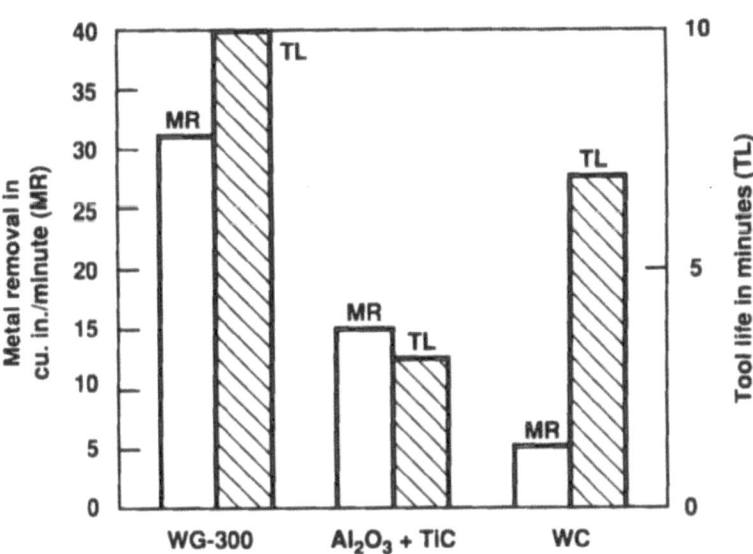

FIG. 2 Relative tool life and metal removal in machining Inconel 718. (From Ref. 11.)

details were given.[16] Since the requirements for parts of this type are generally not as stringent as those for cutting tools, particulate reinforced ceramics, which are currently less expensive, have been emphasized for wear part applications. However, as improvements are made in processing of whisker reinforced ceramics, and cost is reduced, it is likely that there will be a variety of commercial applications.

V. APPLICATIONS FOR CONTINUOUS FIBER REINFORCED CERAMICS

A. Aerospace Applications

Because of the relatively high cost of continuous fiber reinforced ceramics, emphasis has been on aerospace programs, where increased material temperature capabilities and decreased weight can result in tremendous increases in performance. Since military and space programs tend to be less cost sensitive than commercial plane development programs, initial emphasis has been further narrowed to military and space applications. As material performance becomes better documented and manufacturing costs are reduced, it is likely that commercial aerospace applications will be developed.

Structural and engine parts on proposed planes and missiles traveling at much higher speeds than today's versions will be required to withstand much higher temperatures. In the case of exterior structural parts, this derives from increased frictional heating from the atmosphere with increased speed. With engine parts, the higher temperature will derive from operation at close to adiabatic combustion temperatures to increase efficiencies. An indication of increased temperature requirements over the years is given in Figure 3.[17] An example of temperatures that will be encountered in future high speed aircraft is given in Figure 4.

Although continuous fiber reinforced polymers have very high specific strengths, use is limited to about 400°C. Metal alloys have a maximum application temperature of about 1000°C, although this range is extended somewhat by incorporation of reinforcing fibers. Monolithic ceramics such as sialon and silicon carbide can be used at higher temperatures, but their catastrophic failure mode makes them unattractive. Glass matrix composites and glass-ceramic matrix composites (which have been considered as a class of ceramic matrix composites in this book) have higher specific strengths than metals and have about the same maximum application temperatures. State-of-the-art crystalline ceramic matrix composites extend the temperature range to about 1400°C. Carbon/carbon has the highest maximum use temperature and nonbrittle fracture, but oxidation protection at elevated temperatures is a formidable problem.

Because of the increasing temperature requirements, ceramic matrix composites, as well as metal matrix composites, are expected to be used to increas-

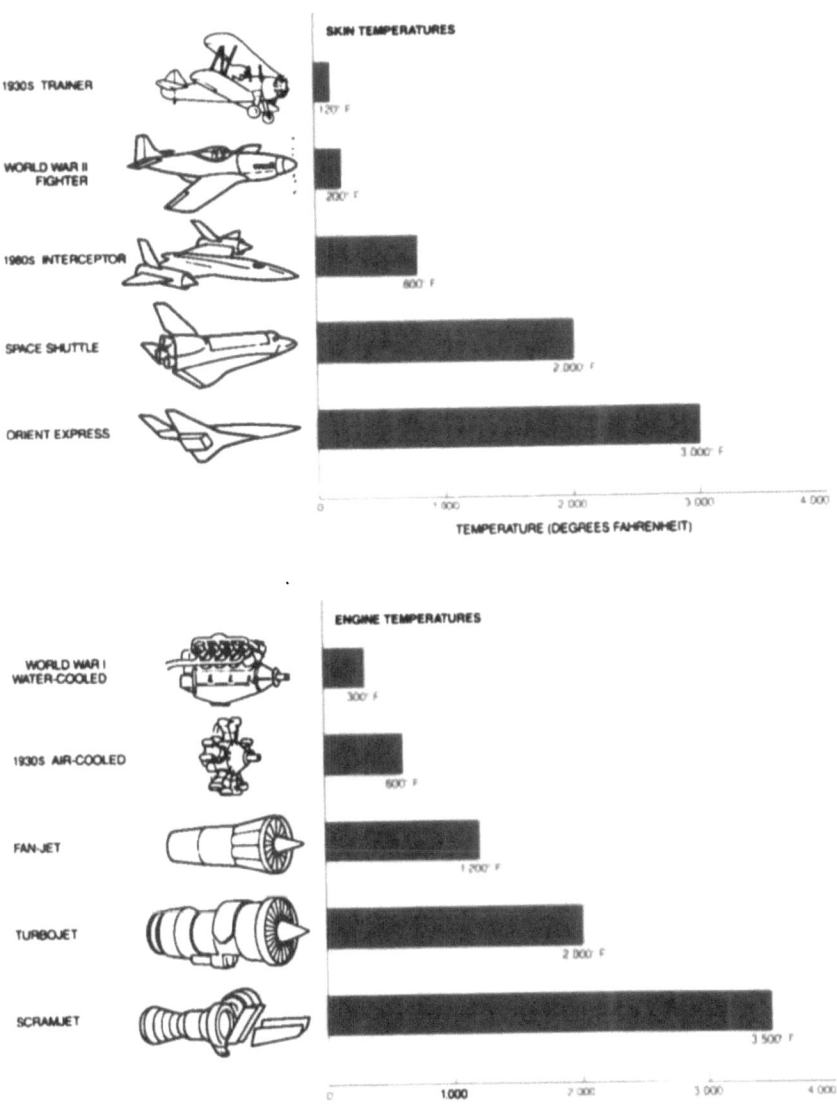

FIG. 3 Increases in temperature requirements for aircraft skin and engine components over the years. (The Orient Express was a transatmospheric craft proposed by former president Reagan to take off from a standard runway, rise through the atmosphere, and cruise at great speed in space. The name is dated but the general concept remains, as the National Aerospace Plane.) (From Ref. 17.)

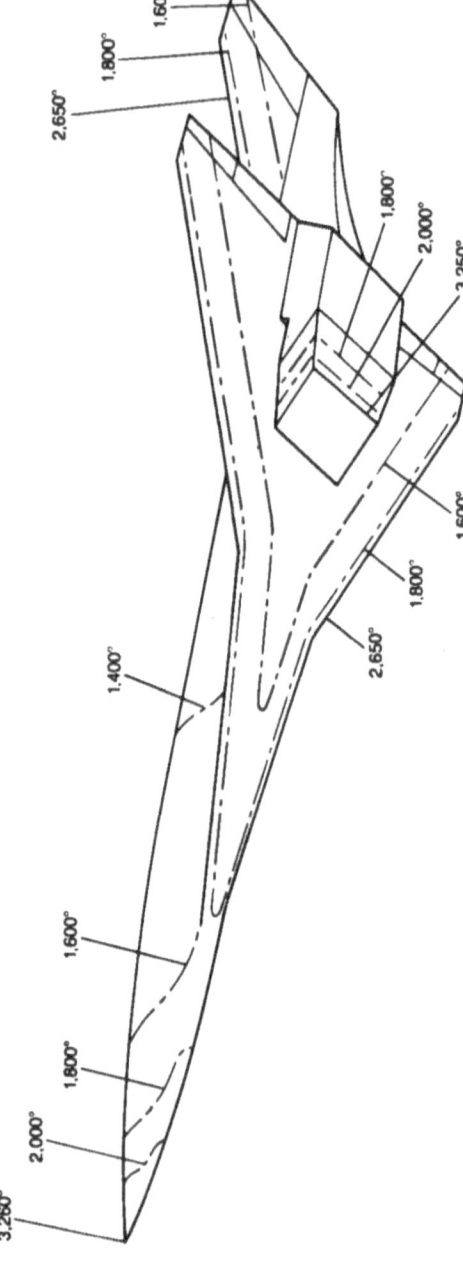

Fig. 4 Temperatures, in °F, predicted to be encountered at various locations on a hypothetical transatmospheric aircraft due to frictional heating and heat generated from the propulsion system. (From Ref. 17.)

ingly greater extents in aerospace engines.[18] In jet engines, for example, nickel content is expected to peak at about 45wt% in the mid-1990s and titanium content is expected to peak at about 30wt% in the same period. By the year 2010, quantity of each of these metals in jet engines is forecast to be about 10wt%. On the other hand, percentages of ceramic and metal matrix composites are projected to increase from near zero levels today to over 25wt% each by the year 2010.

Undoubtedly, the most extensive work on utilization of ceramic matrix composites for aerospace applications has been by Société Européenne de Propulsion (SEP). Both CVI-produced silicon carbide reinforced with carbon fibers, Sepcarbinox™, and CVI-produced silicon carbide reinforced with silicon carbide base fibers, Cerasep™, have been tested extensively for a variety of components.

In the late 1970s, SEP began to develop ceramic matrix composites for solid propulsion applications.[19] In 1984, it began using these materials on bipropellent engines. In 1988, the work served as the basis for an attitude and orbit control thruster equipped with a combustion chamber and nozzle made from reinforced ceramics. A second version of this thruster is under development for the Hermes European Spaceplane.

For the Hermes Spaceplane, SEP and Marcel Dassault Brequet Aviation Company produced and tested ceramic composites for nose cones, leading edges, fins, rudders, and body shingles.[20] A leading edge prototype was tested to a maximum temperature of 1400°C for several hours total time. Advantages relative to the U.S. Orbitor tile thermal protection system have been listed as lower weight, improved impact strength, higher temperature capability, and better maintainability due to mechanical attachment.[21] Components were delivered from SEP to Dassault in mid-1990 for final assembly and testing. An outer skin part made from Sepcarbinox had a mass of 12 kg, and was 1.6-m long by 0.6-m wide, with thicknesses varying from 1.9 mm to 9.9 mm. Figure 5 illustrates a Hermes leading edge ceramic matrix composite component.

The parts were installed with high temperature resistant fasteners, although the fasteners presented a temperature limitation (1300°C) lower than the ceramic composite material. Efforts are underway to produce integrated skin-stiffened unified parts, which will reduce the number of required parts, increase thermal life performance by avoiding fastener restrictions, and decrease weight by reducing local thicknesses needed for mechanical fasteners.

Saucereau et al [22] described work on a ceramic matrix composite nozzle extension for an HM7 engine for the third stage of the Ariane space vehicle. This engine uses liquid oxygen, with liquid hydrogen as the fuel. Compared with a metal nozzle extension, the ceramic composite material was reported to offer manufacturing simplicity with a single wall, instead of a spirally welded tube structure. This resulted in lower weight and greater specific impulse.

Overall component dimensions were a total length of 987 mm, an entrance inner diameter of 309 mm, and an exit inner diameter of 930 mm. Inner profile was an optimized parabolic shape. Mass was about 25 kg, but the composite component could probably be lighter, since design was conservative. The component was subjected to both mechanical and thermomechanical loads.

Sepcarbinox, with Novoltex™ three-dimensional fiber architecture, was selected for this application. Although more susceptible to oxidation, Sepcarbinox has a higher temperature capability than Cerasep and is less expensive. Since the Novoltex has a fine three-dimensional texture, fabrication of a component having a large variation in thickness is facilitated. The thickness of this part ranged from 44 mm at the forward interface, to 10 mm just beyond the attachment location, to 2.4 mm near the exit area.

Successful test firing of this component was carried out. Thermomechanical analysis indicated that the general steady state temperature was reached after 150 s in the thickest part of the nozzle. A maximum temperature of over 1550°C on the inner wall was calculated at about 50 mm from the forward end. Maximum thermal gradient was 60°C/mm after 10 s.

Examination of the component after testing showed no distortion. Mass loss during testing was only about 20 g, or less than 0.1%. No flaws were

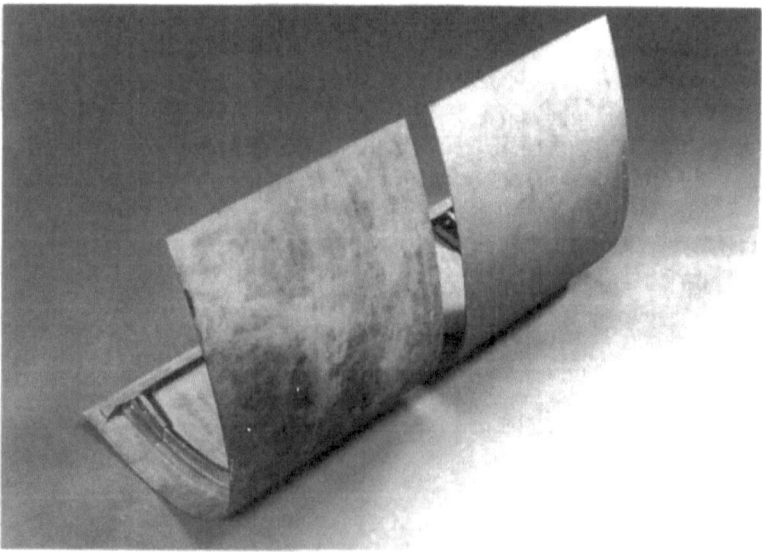

FIG. 5 Ceramic matrix composite leading edge component for the wings of the Hermes spaceplane. (Photo courtesy of SEP.)

detected by X-ray analysis. There was some discoloration of a portion of the inner wall, which was found to be due to a thin silicon dioxide oxidation product. A carbon gasket and metal studs that were used in joining the component were intact after the tests. Some examples of ceramic matrix composite combustion chambers and nozzles for different thrust levels are shown in Figure 6.[23]

Melchior et al [20] also reported on other SEP-produced nozzles, exit cones, pipes, and valves for various aerospace applications. For jet engines, SEP in cooperation with SNECMA produced nozzle parts and afterburner flameholders from both Sepcarbinox and Cerasep materials. The initial reported test flight of a ceramic matrix composite component was at the June 1989 Paris Air Show. A Mirage 2000 fighter plane was equipped with a SNECMA M-53 engine with ceramic matrix composite hot and cool engine nozzle flaps.[24] Figure 7 shows a plane of this type and Figure 8 [23] shows ceramic matrix composite nozzle flaps. Nozzle flaps are also being produced for the M-88 engine for the Rafale fighter.

Since 1984, SEP has had a program to demonstrate feasibility of storable propellant rocket engines with Sepcarbinox and Cerasep combustion chambers

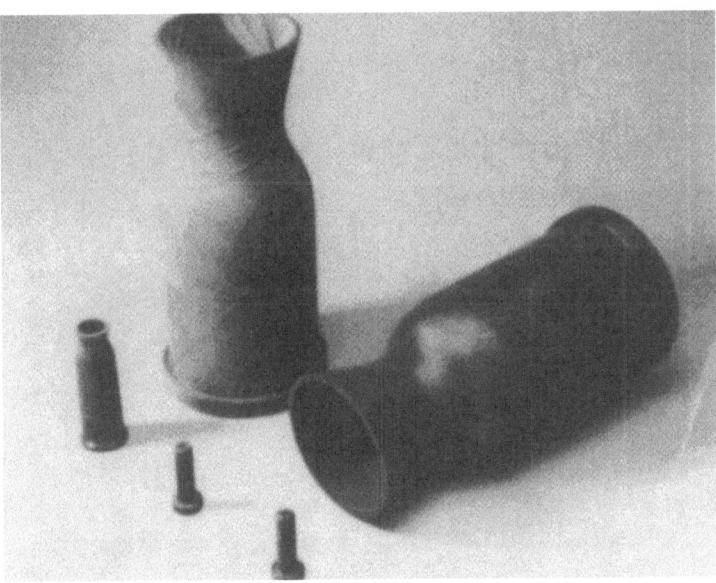

FIG. 6 Examples of CVI-produced silicon carbide matrix composite rocket engine combustion chambers. (From Ref. 23.)

and nozzles. First tests were on a 5 Newton (5 N) chamber. Thirty one chambers were tested for a total time of over 33 h. The tests led to use of Cerasep for long firing durations (about 3 h to 30 h) and the less expensive Sepcarbinox for medium firing durations (about 15 min).

Two Cerasep chambers accumulated over 10 h firing time at a combustion pressure of 1 MPa (10 atm). The chamber walls had no visible attack and there was little erosion in the throat area. Calculated inner wall temperature was as high as 1700°C. A third Cerasep chamber accumulated about 7 h firing duration in 400 thermal cycles (cold starting and 1 min firing). There was no wall erosion and a throat erosion depth of only 20 μm. Other chambers of up to 6000 N have been fabricated and tested also.

In the United States, Du Pont has installed, in Newark, DE, a dedicated ceramic matrix composite facility based on the SEP technology. Initially, production was limited to components about 1 m in diameter by about 1.3 m long. In 1992, size capability increased to components about 1.3 m in diameter by about 2.1 m long. X-ray and C-scan equipment are used for NDE.[23] Du Pont has produced Nicalon/silicon carbide composite structures such as a combustor sub-element (funded by the Garrett Engine Division of the Allied-Signal Aer-

FIG. 7   A Mirage 2000 fighter plane of the type equipped with ceramic matrix composite engine flaps and flown at the 1989 Paris Air Show. (Photo courtesy of the French Air Force and Dassault Aviation.)

ospace Company). The component includes laser drilled holes for cooling. Turbine engine and turbopump rotor components (Figure 9) have also been fabricated.

In addition to aerospace components fabricated by CVI, production of prototype components by other methods has also been pursued. Du Pont Lanxide has manufactured a test combustor made from a ceramic matrix composite.[25] The combustor was used in a Williams International Co. expendable unmanned turbine engine. The work was performed under a DARPA-funded program administered by the Office of Naval Research. In addition, afterburner flameholders for a man-rated engine were fabricated. The parts withstood repeated thermal cycling from 300° to 1000°C on engine test rigs. A capability of up to 1300°C is required, which may require materials other than those used.

The Compglas material of United Technologies has also been tested for aerospace applications.[26] Pratt & Whitney has tested this material in gas tur-

FIG. 8 Close-up of CVI-produced ceramic matrix composite engine flaps on a Mirage 2000 fighter plane. (From Ref. 23.)

bines, and test components are being incorporated in both current engines and in advanced experimental engines.

A prototype turbine blade produced by sol-gel processing has also been demonstrated (Figure 10).[27]

B. Other Applications

The characteristics of high temperature capability, low density, and non-catastrophic failure that make continuous fiber reinforced ceramics ideal candidates for advanced aerospace applications can be utilized to advantage in other applications as well. Since these attributes tend to be more valuable for aerospace applications, routine use of continuous fiber reinforced ceramics for other applications will probably lag behind aerospace uses. However, as material cost decreases, it is anticipated that continuous fiber reinforced ceramics will find many applications.

Fɪɢ. 9   Turbine engine and turbopump motor components fabricated from CVI-produced ceramic matrix composites. (From Ref. 23.)

An application that has received considerable attention is radiant burner tubes for indirect heating with natural gas.[28] Silicon carbide is a good candidate for tube material, and monolithic silicon carbide tubes are in operation for applications up to about 1150°C. However, monolithic silicon carbide tubes must have relatively thick walls and can also fail due to thermal variations and to vibration. Hence, fiber reinforced silicon carbide has been investigated as a tube material.

A joint program was started in 1984 by Amercom, Inc., Columbia Gas System, and Industrial Furnace Services on continuous Nextel 312 reinforced silicon carbide produced by CVI. In 1986, 3M Company joined the effort. Sponsors were Columbia Gas Systems, Consolidated Natural Gas Service Company, and Southern California Gas Company. Amercom developed the first full size tubes and supplied other test materials. Columbia Gas evaluated tube performance and coordinated furnace and component evaluations. Industrial Furnace Services builds, sells, and services furnaces using radiant burner tubes. The Nextel 312 fibers were furnished by 3M Company, which also characterized materials and established a process development plan. In Phase I of the program, four tubes were produced using braided Nextel 312. Tubes were about 20 cm

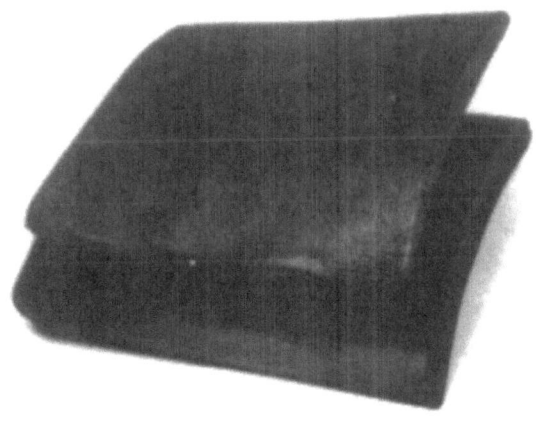

FIG. 10 Ceramic matrix composite turbine engine blades fabricated by sol-gel vacuum impregnation of fiber preforms. (From Ref. 27.)

in diameter and 200 cm long. These were laboratory tested as well as field tested in melting furnaces at Arrow Aluminum Company. Productivity increased, even though only two of the four tubes in the system were composite tubes. However, the tubes were more permeable than desirable.

In Phase II of the program, a new preform procedure using wound cloth instead of braiding was tried. In laboratory tests, the new tubes had about one-tenth the leakage of the original tubes. Preform cost was also lower. One of these tubes was used for 18 months at Arrow Aluminum, although a crack had appeared after 15 months. A second tube also developed a crack, and was re-placed after 11 months. Residual stresses in the seam regions were a problem with this material. Industrial Furnace Services developed a procedure to change these relatively light (about 4.5 kg) tubes while the furnace was in operation.

In Phase III of the program, improvements in tubes using cloth wrapped preforms were sought and a new triaxial braiding process was investigated. The triaxial braiding process was found to be more reproducible than the cloth wrap-ping procedure, but the material produced using the braided preform was still not gas tight.

Expected burner requirements for commercial applications are a length up to about 250 cm, a diameter up to about 20 cm, and a wall thickness up to about 1 cm. Working temperatures near 1400°C, with excursions to about 1550°C, are desirable. Life should be up to at least two years. High thermal conductivity and diffusivity, and low permeability are necessary. The material resulting from this program has been commercialized under the trade name Si-conex™ Fiber Reinforced Ceramic. Figure 11 shows the excellent thermal shock resistance of this material.[29]

Continuous fiber reinforced ceramic heat exchangers have also been in-vestigated.[30] In 1985, DOE became involved with two heat exchanger pro-grams conducted by Babcock & Wilcox Company and by Thermal Electron Corporation.

In 1988, work on a program called HiPHES (high pressure heat exchanger systems) was initiated. Two contracts, to Babcock & Wilcox and to Solar Tur-bines Incorporated, were awarded. These were for air heaters operating at greater than 800°C and 700 kPa, feeding an indirectly fired gas turbine. The heating source in the air heater was to be the exhaust gases from a municipal or haz-ardous waste incinerator. This would allow high temperature operation of the gas turbine without introducing corrosive gases into the turbine. A HiPHES contract was also awarded to Stone & Webster for a convective steam reformer integrated with a directly fired gas turbine. A catalyst filled tubular reactor pro-ducing hydrogen at elevated pressures would be used for subsequent processing to methanol or ammonia. A ceramic composite tube would allow for higher reaction temperature than the 870°C temperature in existing systems. Improve-ments in performance due to use of continuous fiber reinforced ceramic tubes are based on improvement in heat transfer or chemical reactions that occur with a combination of increased temperature and pressure. Reaction rates for the selected applications increase dramatically with an increase in temperature. Increasing the pressure can decrease rate slightly, but this is more than compensated for by the increased throughput with increasing pressure.

The projected benefit of using continuous fiber reinforced ceramics for heat exchangers for power generation include a 33% increase in efficiency. The projected benefits by increasing pressure in reforming systems are an 11% re-

FIG. 11   Illustration of the thermal shock resistance of a Siconex™ ceramic matrix composite tube. (From Ref. 29.)

TABLE 1. Recommended Industrial Applications for the CFCC Initiative.

Product area	Examples	Likely industrial markets
Advanced heat engines	Combustors, liners, wear parts, etc.	Primarily high-temperature gas turbines; possible adiabatic diesels, S.I. engines (promising market is gas turbine combustor retrofits)
Heat recovery equipment internals	Air preheaters, recuperators	Any indirect heating uses; energy-intensive industrial processes (e.g., aluminum remelters, steel reheaters, glass melters)
Burners and combustors	Radiant tube burners	Potentially any indirect-fired, high temperature and/or controlled-atmosphere heating/melting/heat-treating industrial application.
Burners and combustors	Catathermal combustors	Low-NO$_x$ clean fuel heating applications—including gas turbine combustors, industrial process heat
Burners and combustors	Low-temperature radiant combustors	Low NO$_x$ clean fuel heating applications—including small scale (space heating) and large
Process equipment	Reformers, reactors, HIP equipment	scale (industrial process) applications
Waste incineration systems	Handling equipment, internals, cleanup	Chemical process industry, petroleum refining
		Conventional MSW/RDF facilities, advanced toxic/hazardous waste facilities, with or without energy recovery
Separation/filtration systems	Filters, substrates, centrifuges	Gas turbine, combined cycle, and IGCC configurations; particulate traps for diesel exhausts; molten metal filters; sewage treatment
Refractories and related	Furnace linings, crucibles, flasks, etc.	High-temperature industrial heating/melting/heat treating processes
Structural Components	Beams, panels, decking, containers	Possible niche applications for EMI shielding, corrosive/abrasive environments, fire-protection, missile protection [e.g., gas turbine shrouds]; and major infrastructure applications

CFCC = continuous fiber reinforced ceramic composites. (Source: Department of Energy.)
Source: Ref. 31.

duction in fuel use, an increase in hydrocarbon feed conversion from 80% to 99.7%, and a 26% reduction in required throughput.

Many other potential applications for continuous fiber reinforced ceramics have been identified as part of the CFCC initiative mentioned earlier. Table 1 [31] summarizes these potential applications.

VI. SUMMARY

Although applications for fiber and whisker reinforced ceramics are as yet limited, well funded efforts for applications R&D are underway and appreciable market growth is projected.

As a cutting tool material, silicon carbide reinforced alumina offers dramatic increases in cutting speeds and tool life for machining some nickel base superalloys and other materials. New developments in whisker reinforced ceramics promise to extend the range of cutting tool applications as well as to make inroads into other applications such as wear parts.

Impetus to develop applications for continuous fiber reinforced ceramics is greatest for the aerospace industry, where weight savings and increases in material temperature capabilities can enhance performance tremendously, and where the brittle failure mode of monolithic ceramics or even whisker reinforced ceramics can ordinarily not be tolerated. The major developmental efforts have involved CVI-produced silicon carbide matrix composites for a variety of components for aerospace applications.

Large energy savings are projected with the use of ceramic matrix composites for burner tubes, heat exchangers, and other non-aerospace applications. Although brittle failure is tolerable for some of these applications, so that monolithic whisker reinforced ceramics could be used, adequate resistance to failure due to thermal shock or vibration and the ability to produce thin walled components dictates the use of continuous fiber reinforcement for many of these applications.

REFERENCES

1. Anon., Advanced matrix composites cover the spectrum: polymers, metals, ceramics, *Ceramic Bulletin*, vol. 68, no. 2, 1120–1121 (1989).
2. Anon., US markets for ceramic and metal matrix composites, *Advanced Ceramic Reports*, 13–14 (November 1991).
3. Anon., No major success for ceramic matrix composites this decade, *Advanced Ceramic Reports*, 15 (July 1991).
4. Persh, J., Department of Defense directions for engineering ceramics, *Ceramic Bulletin*, vol. 68, no. 6, 1174–1176 (1989).

5. Geiger, G., Progress continues in composite technology, *Ceramic Bulletin*, vol. 70, no. 2, 212–218 (1991).

6. Richlen, S., Opportunities for the industrial application of continuous fiber ceramic composites, *Ceram. Eng. Sci. Proc.*, vol. 11, no. 7–8, 576–577 (1990).

7. Baldoni, J. G., and Buljan, S. T., Ceramics for machining, *Ceramic Bulletin*, vol. 67, no. 2, 381–387 (1988).

8. Confer, H., Greenleaf Corporation, *Cutting Tool Engineering*, vol. 39, no. 1, 54–56 (1987).

9. Whitney, E. D., and Vaidyanathan, P. N., Microstructural engineering of ceramic cutting tools, *Ceramic Bulletin*, vol. 67, no. 6, 1010–1014 (1988).

10. Billman, E. R., Mehrota, P. K., Shuster, A. F., and Beeghly, C. W., Machining with Al_2O_3-SiC-whisker cutting tools, *Ceramic Bulletin*, vol. 67, no. 6, 1016–1019 (1988).

11. Rhodes, J. F., Whisker reinforced ceramic composites. In *Proceedings of the Fifth Annual Conference on Materials Technology*, Materials Technology Center, Southern Illinois University, Carbondale, IL, 1988, 205–219.

12. Wayne, S. F., and Buljan, S.-T., Wear of ceramic cutting tools in Ni-based superalloy machining, *Tribology Transactions*, vol. 33, no. 4, 618–626 (1990).

13. Gruss, W. W., Ceramic tools improve cutting performance, *Ceramic Bulletin*, vol. 67, no. 6, 993–996 (1988).

14. Lewis, C. E., Ceramic matrix composites: the ultimate materials dream, *Materials Engineering*, vol. 105, no. 9, 41–45 (1988).

15. Anon., Ceramics cut harder into tool market, *Materials Edge*, no. 25, 4 (June 1991).

16. Anon., Ceramic composite being developed in tooling of packaging systems, *Industrial Heating*, vol. 55, no. 12, 10 (1988).

17. Steinberg, M. A., Materials for aerospace, *Scientific American*, vol. 225, no. 4, 67–72 (1986).

18. Highton, D. R., and Crispin, W. J., Future advanced aero-engines—the materials challenge. In *Application of Advanced Material for Turbomachinery and Rocket Propulsion*, NATO Advisory Group for Aerospace R&D, AGARD Conference Proceedings No. 449, March 1989, 4-1 to 4-4.

19. Mathieu, A., Monteuuis, B., and Gounot, V., Ceramic matrix composite materials for a low thrust bipropellant rocket engine. Paper AIAA 90-2054 presented at the AIAA/SAE/ASME/ASEE 26th Joint Propulsion Conference, July 16–18, 1990, Orlando, FL, American Institute of Aeronautics and Astronautics, Washington, DC.

20. Melchior, A., Pouliquen, M. F., and Soler, E., Thermostructural composite materials for liquid propellant rocket engines. Paper AIAA-87-2119 presented at the AIAA/SAE/ASME/ASEE 23rd Joint Propulsion Conference, June 29–July 2, 1987, San Diego, CA, American Institute of Aeronautics and Astronautics, Washington, DC.

21. Lacombe, A., and Bonnet, C., Ceramic matrix composites, key materials for future space plane technologies. Paper AIAA-90-5208 presented at the AIAA Second International Aerospace Planes Conference, October 29–31, 1990, Orlando FL, American Institute of Aeronautics and Astronautics, Washington, DC.

22. Saucereau, D., Beaurain, A., Demonstration of carbon/silicon carbide Novoltex reinforced composite nozzle on a LH_2-LO_x engine. Paper AIAA 90-2180 presented at the AIAA/SAE/ASME/ASEE 26th Joint Propulsion Conference, July 16–18,

1990, Orlando, FL, American Institute of Aeronautics and Astronautics, Washington, DC.

23. Du Pont CVI Ceramic Matrix Composites, Preliminary Engineering Data, Du Pont Composites, Newark, DE.

24. Bacon, M., Gas turbine engines: taking the heat, *Materials Edge*, vol. 17, 21–24 (June 1990).

25. Ashley, S., Tailor-made ceramic-matrix composites, *Mechanical Engineering*, vol. 113, no. 7, 44–49 (1991).

26. Drewer, C., Ceramics fire the future, *Flight International*, vol. 134, 26–31 (1988).

27. Hyde, A. R., Ceramic matrix composites—a new generation of materials for mechanical and electrical applications, *GEC Journal of Research*, vol. 6, no. 1, 65–71 (1988).

28. Richards, R. E., Bodkins, D. W., and Copes, J. S., Progress towards a cost effective, thin walled RBT. In *Energy Technology: Proceedings of the Energy Technology Conference*, vol. 15, Energy Technology Conference, February 17–19, 1988, Washington, DC, Government Institute Inc., Washington, DC, 1988, 749–756.

29. Nextel™ Textiles, *Ceramic Fiber Products for High Temperature Applications*, Ceramic Materials Department, 3M (1991).

30. Richlen, S. L., Overview of DOE's Office of Industrial Programs development program in continuous fiber composites. In *International SAMPE Symposium and Exhibition—Book 2*, 34th International SAMPE Symposium and Exhibition, May 8–11, 1989, Reno, NV, SAMPE, Covina, CA, 1989, 2326–2332.

31. Sheppard, L. M., Progress in composites processing, *Ceramic Bulletin*, vol. 69, no. 4, 666–673 (1990).

Index